Cyber Insecurity

Cyber Insecurity: Examining the Past, Defining the Future deals with the multifaceted world of cybersecurity, starting with the premise that while perfection in cybersecurity may be unattainable, significant improvements can be made through understanding history and fostering innovation. Vladas Leonas shares his journey from Moscow to Australia, highlighting his academic and professional milestones.

This book covers the evolution of cybersecurity from the late 1960s to the present, detailing significant events and technological advancements. The author emphasises the importance of simplicity in technology projects, citing complexity as a major hindrance to success. The book also discusses the impact of the digital revolution, using the example of a global IT outage caused by a faulty software update.

Project management methodologies are explored, tracing their origins from ancient civilisations to modern techniques such as CPM and PERT. The concept of cloud computing is examined, highlighting its benefits and potential security issues. The evolution and advantages of SaaS solutions are also discussed, noting their increased adoption during the COVID-19 pandemic.

The author then addresses supply chain challenges, using real-world examples to illustrate vulnerabilities. He traces the history of communication methods leading up to TCP/IP and discusses the development and importance of DNS. The differences between compliance and conformance in cybersecurity are clarified, emphasising that compliance does not equate to security.

Key cybersecurity standards such as the NIST CSF and ISO/IEC 27000 series are examined. The book also covers the Essential 8, a set of cybersecurity controls developed by the Australian Signals Directorate. The convergence of OT and IoT is discussed, highlighting the cybersecurity risks associated with this integration.

Emerging threats from AI and quantum computing are explored, noting their potential to both advance and threaten cybersecurity. The evolving legal landscape of cybersecurity is also covered, emphasising the need for international cooperation and innovative legal solutions.

In conclusion, the book stresses the importance of critical thinking and a holistic approach to cybersecurity, advocating for simplicity and foundational practices to enhance security.

Security, Audit and Leadership Series

Series Editor: Dan Swanson, Dan Swanson and Associates, Ltd., Winnipeg, Manitoba, Canada

The Security, Audit and Leadership Series publishes leading-edge books on critical subjects facing security and audit executives as well as business leaders. Key topics addressed include leadership, cybersecurity, security leadership, privacy, strategic risk management, auditing it, audit management and leadership.

Team Intelligence: A New Method Using Swarm Intelligence for Building Successful Teams
Mohammad Nozari

The Gardener of Governance: A Call to Action for Effective Internal Auditing
Rainer Lenz and Barrie Enslin

Navigating the Cyber Maze: Insights and Humor on the Digital Frontier
Matthias Muhlert

Safeguarding the Future: Security and Privacy by Design for AI, Metaverse, Blockchain, and Beyond
Alan Tang

Security Relationship Management: Leveraging Marketing Concepts to Advance a Cybersecurity Program
Lee Parrish

A Cybersecurity Leader's Journey: Speaking the Language of the Board
Edward Marchewka

Cyber Risk Management in Practice: A Guide to Real-World Solutions
Carlos Morales

Cyber Insecurity: Examining the Past, Defining the Future
Vladas Leonas

For more information about this series, please visit: https://www.routledge.com/Security-Audit-and-Leadership-Series/book-series/CRCINTAUDITA

Cyber Insecurity
Examining the Past, Defining the Future

Vladas Leonas

With Contribution from Sorin Toma

CRC Press
Taylor & Francis Group
Boca Raton London New York

CRC Press is an imprint of the
Taylor & Francis Group, an **informa** business

Designed cover image: Shutterstock

First edition published 2025
by CRC Press
2385 NW Executive Center Drive, Suite 320, Boca Raton FL 33431

and by CRC Press
4 Park Square, Milton Park, Abingdon, Oxon, OX14 4RN

CRC Press is an imprint of Taylor & Francis Group, LLC

ISBN: 978-1-032-67256-4 (hbk)
ISBN: 978-1-032-67257-1 (pbk)
ISBN: 978-1-032-67260-1 (ebk)

DOI: 10.1201/9781032672601

Typeset in Sabon
by SPi Technologies India Pvt Ltd (Straive)

To My Parents

Contents

Foreword

It is my privilege to have met and known Professor Vladas Leonas now for over five years. During that time, I have come to regard him as not only an exceptionally knowledgeable and technical expert, but also a pragmatic thinker who sees and cuts through the fog of cyberwar and the often self-imposed blind spots that cloud the cyber realm.

His deep understanding of the complexities within digital security is matched by his clarity of vision and ability to address these issues head-on. It was a pleasure to have Vladas critique my books, and I am honoured to contribute, in my own small way, to this exceptional work and to write its foreword.

In this book, Vladas delves into the intricate evolution of the digital world – charting its growth from a tool for academic collaboration to a colossal industry processing and transacting trillions of dollars. Throughout this transformation, the infrastructure has often been built on questionable and insecure foundations as the rush for commercialism and functionality was, and is, the main driver by a country mile. Security played, and continues to play the poor cousin, if played at all.

Today, technologies like IoT, AI, Blockchain and even Quantum Computing, which promise to revolutionize computing, are frequently deployed without basic security principles or capabilities in place. The result? A landscape where cybercrime has surged to the scale of a global economic superpower, generating over $10 trillion annually, effectively positioning itself as the third-largest "economy" in the world.

Organised cybercrime continues to outpace a disjointed cyber security defence and response – a situation compounded by the very own goals of the digital world.

Vladas does not shy away from discussing how unchecked digital dominance, surveillance tactics and the convenient labels of "conspiracy theorist" were introduced by intelligence agencies to obscure and perpetuate these issues.

I deeply thank Vladas for his dedication, knowledge, conviction and courage in authoring this magnificent book. His unwavering commitment to

"Fight the Good Fight" serves as an inspiration, even as governments too often collude in the very activities they claim to combat.

This excellent publication by Professor Leonas serves as a chronology and deep insight into the often misunderstood stages and intricacies of computing and their connectiveness to the Internet. It is a book you will want to read twice, or at least parts to truly appreciate and understand its depth (I always mark mine with a marker pen for reference). That is testimony to the book's incredible granularity and comprehensiveness. This book should be part of all Cyber Security Academia Curriculum and for anyone and everyone interested in history and computing.

It is my privilege to commend this work and to applaud my friend, Vladas, for his continued efforts for his unwavering efforts to make the digital world a safer place.

Andy Jenkinson
CEO of Cybersec Innovation Partners

Preface

After the Bay of Pigs fiasco, John F. Kennedy famously noted, "Success has many fathers, but failure is an orphan." It seems this quote was originally from Tacitus, which in Latin has a slightly different tone: "This is an unfair thing about war: victory is claimed by all, failure to one alone."

This book is clear evidence of the fact that this saying, though it sounds great, is often not true.

Recent chain of disastrous events in cybersecurity is impossible to attribute to a single particular failure. It is (as it often happens with disasters) a result of a chain of seemingly unrelated events, each of which separately may look relatively innocent, but their cumulative effect is fascinating and is a clear illustration of the adage attributed to Aristotle: "The whole is greater than the sum of the parts."

Step by step this book paints a picture of multiple, unrelated from the first glance, events that led us to the current appalling state of where cybersecurity is today. We go into reasonable detail describing historical perspectives that have contributed to the current state. For each and every stream discussed in this book, decision(s) made at the time have been made with the best intentions, but as we look back, we can see where combined effect of these decisions brought us to.

Those who may start wondering why do we need to go back in time, should read this short, but very educational essay: "4 feet 8.5 inches, The Space Shuttle and a Horse's Ass" (https://www.linkedin.com/pulse/4-feet-85-inches-space-shuttle-horses-ass-william-batch-batchelder#:~:text=Space%20Shuttle%20Solid%20Rocket%20Boosters,4%20feet%208.5%20inches%20wide).

The author expects that a number of rocks will be thrown into him by zealots of every stream discussed. And that's okay. Debate is useful and expected. Let's have these debates.

"The road to hell is paved with good intentions," St. Bernard of Clairvaux allegedly wrote c. 1150. And it looks like this is still true.

Acknowledgements

Special thanks go to Sorin Toma who has contributed Chapters 15 and 16 and kindly written Chapter 20.

And a special "Thank You" goes to Andy Jenkinson, CEO of Cybersec Innovation Partners, for his ongoing support, insightful comments and for kindly writing a foreword to this book.

This book would not have happened without advice, comments, support and help from:

Alex Ezrakhovich, MD of AEConformity

Mike Gruntman, Professor of Astronautics at USC

Steve Novis, former colleague and friend

Jim Hegarty, former colleague and friend

Sergei Karelov, former Interquadro colleague

Author

Vladas Leonas has over 45 years of experience in ICT and cybersecurity, which includes a variety of public and private sector industries as well as tertiary education organisations. He is also a member of the Commonwealth and NSW Governments ICT Assurance Panels and a formally trained auditor (ISO 9001 and ISO27001). Over the last seven years, Vladas consulted a number of organisations, including UNSW, icare (NSW State Workers Compensation Insurance Company) and a variety of NSW Government Departments. He is also the author of 50+ publications and an internationally recognised speaker. He is an Adjunct Professor at the Australian Graduate School of Leadership and a Principal Supervisor of four doctoral students. He is a subject matter expert and specialises in ICT strategies, their implementation and ICT operations, gateway reviews and internal audits, enterprise risk management, cybersecurity, governance, procurement and compliance. Over the last 25 years, Dr Leonas has held eight CIO and CTO positions. He earned a doctoral degree at the Moscow Aviation Institute and a GAICD from AICD via AGSM in Sydney. He is a Fellow of the Australian Computer Society and a Fellow of the Institute of Engineers Australia.

Chapter 1

The beginnings

I was born in Moscow in 1956, the year when Nikita Khrushchev denounced Stalin's personality cult in his famous report at the XX Congress of the Communist Party (CPSU) of the USSR. Five years later, Nikita Khrushchev declared from the podium of the XXII Congress of the CPSU: "The Party solemnly proclaims: the present generation of Soviet people will live under the communism."

When on September 1, 1963, I started school, Nikita Khrushchev was still at the helm of the USSR. The very first thing I noticed in the foyer was the poster stating: "The Party solemnly proclaims: the current generation of Soviet people will live under communism!" Image of this poster can be seen at: https://mizugadro.mydns.jp/t/index.php/File:PriKommunizme.jpg.

Like many Soviet kids, I was sent to a kindergarten that I did not mind, apart from the food – they used to put a spoon of fish oil in the soup and it was absolutely unpalatable! They also demanded that kids should eat meals in full, otherwise kids were threatened that whatever was left on a plate would be deposed under their collars…. Those days kids – boys and girls – were wearing stockings on a suspender belt attached around the chest…

I have lived with my Parents and my Maternal Grandmother in a one-bedroom (in the USSR it was called two-room) apartment in an apartment block on Kutuzovskiy Prospekt, Moscow. The 9-story building was known as "House of Toys" as the ground floor was fully occupied by a toy store. Image of this can be seen at: https://pastvu.com/p/2155755.

School No. 5 was a privileged one with extensive English studies (1 h a day each day of the week, starting from year two and later on with lessons in English Literature). I was lucky that I was in this school's catchment as it was literally next door to the block of units we lived in at the time. There were two classes in each year – class A and class B. I landed in class B. Just to give the reader a feeling of who else went to this school – in class A of my year was a granddaughter of Mikhail Suslov (https://en.wikipedia.org/wiki/Mikhail_Suslov), the Second Secretary of the Communist Party of the Soviet Union

DOI: 10.1201/9781032672601-1

from 1965 and the unofficial chief ideologue of the party – and the younger son, Arthur, of the famous Soviet illusionist Arutyun Akopyan (https://ru.wikipedia.org/wiki/%D0%90%D0%BA%D0%BE%D0%BF%D1%8F%D0%BD,_%D0%90%D1%80%D1%83%D1%82%D1%8E%D0%BD_%D0%90%D0%BC%D0%B0%D1%8F%D0%BA%D0%BE%D0%B2%D0%B8%D1%87); Brezhnev's granddaughter Viktoria was also studying at this school – she was 4 years older than me and was selling (cheaply ☺) at school Philip Morris cigarettes, probably pinched from her Parent's or Grandparents' drawer.

My mother was an architect and my father was a physicist, and like many boys, I wanted to be like my father – a physicist. At that time in the USSR one had to study for 10 years to be eligible to start tertiary education and to enter into a university course (8 years to continue with secondary vocational education).

So, after 8 years at this privileged school, I sat an exam resulting in me being admitted to a special Physics and Mathematics class at another school, No. 710 (also known as " school-laboratory No. 1 at the Academy of Pedagogical Sciences of the USSR"). Director of the school was Vadim Konstantinovich Zhudov and now this school is named after him (https://smapse.com/gymnasium-no710-named-after-zhudova-moscow/). Image of the school can be seen at: https://pastvu.com/_p/d/2/f/t/2ft6r1srkj8k0xskxt.jpg.

After successfully graduating from this school (in 1973) and attending "The Evening Mathematical School" and having had an additional math tutor, I was eager to become a student of the Physical Faculty of the Moscow State University. The way it worked in the USSR, one had to get sufficient marks in four exams (two written – Mathematics and Composition and two oral – Physics and Mathematics) plus an average mark for the school program. Competition was pretty stiff – 20 candidates for each spot and I have ended with 0.5 mark short to get in. Moscow State University (as well as Moscow Institute of Physics and Technology) had entry exams in July, while all other universities in August.

I had to get into a university that same year, otherwise I would be potentially (and most likely!) conscripted for 2 years' service in the Soviet Army. Together with my parents we looked at the options and decided that I should try Applied Mathematics Faculty at the Moscow Aviation Institute. Image of one of the entrances to the Moscow Aviation Institute and the main administration building can be seen at: https://yandex.com/maps/org/moscow_aviation_institute/1110161245/gallery/?ll=37.502489%2C55.811713&photos%5Bbusiness%5D=1110161245&photos%5Bid%5D=urn%3Ayandex%3Asprav%3Aphoto%3A167436098&z=5.

Faculty of Applied Mathematics there had the closest curriculum to the Physics Faculty of the Moscow State University, which allowed me to attempt transfer after year one.

I could use my marks from the earlier exams at the Moscow State University (there was very high probability that this will be enough to get in as competition was much lower – around seven candidates for each spot), but I decided to sit the exams again and subsequently was admitted as a full-time student.

I also started working part time at one of the Experimental Physics labs within the Aviation Engines Faculty. Everything was new and unusual, I have learned a lot of new terms, which was exciting. But the work itself was boring – I had to measure the length of various experimental curves and graphs using curvimeter (https://en.wikipedia.org/wiki/Opisometer). Having said this, this job gave me an opportunity to make a presentation at a student conference, which was my first exposure to attending and presenting at a conference.

After finishing year one, I realised that my ideas about Physics were too idealistic and that I probably didn't want to proceed in this direction. What I realised was that I was interested in Programming. We had a course that required us to write a program for M20 computer – a 45-bit word computer with only 4KWords of memory (https://www.computer-museum.ru/english/m220.htm) and this was a fascinating experience for me that determined my future career. Image of M20 computer can be seen at: https://it.wikireading.ru/61638.

I married early at the tender age of 19, the same year my daughter Katia was born. This put some financial pressure on me – I had to support the family. My scholarship was 40 roubles per month, while average salary was 120 roubles per month (1 rouble = 100 kopeks). So, my part-time work was giving me another 40 roubles per month, bringing the total income to 80 roubles per month. Just some stats on the cost of living: communal payments for the apartment – about 14–15 roubles per month; electricity, 4 kopeks for 1 KW-h; bus or metro ride – 5 kopeks; tram ride – 3 kopeks; a box of matches – 1 kopek; loaf of rye bread – 14 kopeks; the most expensive loaf of white bread – 28 kopeks; butter – 3.6 roubles per kilo; the most expensive cheese – 3.9 roubles per kilo; ten eggs – between 0.9 and 1.3 roubles. You've got a feeling, I hope.

So, when I heard in 1975 that one of the guys from my year (there were 125 students in my year divided into 5 groups) started working as an operator on one of the institutes' modern computers (M20 was not considered modern), I decided that I want to do the same. As a result, I started as a night shift operator on an M4030 mainframe computer (https://www.atariarchives.org/bcc2/showpage.php?page=12). Image of M4030 mainframe computer (1973) can be seen at: https://www.computer-museum.ru/articles/universalnie_evm/987/.

This mainframe computer had an interesting history. M4030 was a Soviet copy of Siemens System 4004, which was essentially a rebadged RCA Spectra 70, designed to be partially compatible with the successful IBM System/360 (https://de.wikipedia.org/wiki/Siemens_System_4004), but used a different operating system (OS); it was featured in a couple of movies, including Willy Wonka and the Chocolate Factory in 1971). (https://www.starringthecomputer.com/computer.html?c=160).

The one I worked on had 128 KB of ferrite RAM (4 × massive "fridges"), five hard drives (7.25 MB each), five tape drives, a drum printer, a card reader and a couple of other devices. Normally, operators used to run one job at a time. I developed a technique to load multiple decks on a tape and then run two to three jobs in parallel – this increased productivity by a factor of 2 to 3.... I also learned how to manually cut out holes in punched cards and how to "close" them with small punched-out rectangular pieces of cards....

After I mastered controlling and running M4030, I was lucky to become an IT person (still being a full-time student) at the small Laboratory of Memory Systems Bionics at one of the research institutes of the Academy of Sciences of the USSR where I had to run and write software for a rare for the USSR beast – HP-21MX (https://www.hpmuseum.net/exhibit.php?class=3&cat=33). Image of HP21MX computer can be seen at: https://www.hpmuseum.net/exhibit.php?class=3&cat=33 (Figure 1.1).

During this time, I also got exposure to my first time-sharing computer HP-3000 (https://en.wikipedia.org/wiki/HP_3000) and this was an eye-opener after playing with batch mode and single-user mini-computer! Image of HP-3000 computer can be seen at: https://en.wikipedia.org/wiki/HP_3000#/media/File:HP_3000_Series_III.jpg.

I also diversified my experience starting as a fireman at the Moscow Drama Theatre named after K. S. Stanislavskiy, where my best schoolmate's

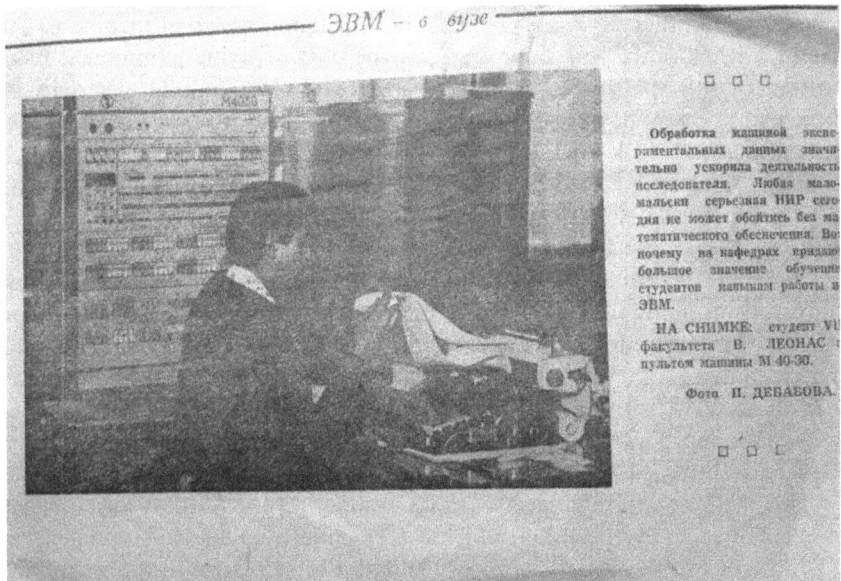

Figure 1.1 Author behind the console of M4030 computer, 1975.

Source: author's archive.

father was a director. Image of the theatre can be seen at: https://izi.travel/ru/browse/eb8ac480-3269-468c-ba16-c225aa964284.

Each fireman shift consisted of two people who were supposed to stay on the shift for 24 h, then 3 days of rest. Duties included: opening the theatre in the morning, hourly patrols, locking the theatre up for the night. Of course, this rule has never been followed – one person stayed there during the day, and the other stayed during the night. I always worked night shifts and this paid me 90 roubles per month! I loved to work on public holidays as this was double pay!

There were no Bachelor's or Master's degrees in the USSR – there was a single degree and it used to take 5.5–6 years for STEM degrees and 4–5 years for arts degrees. For STEM degrees, depending upon their nature, one either had to do a project or had to write a thesis. My thesis was devoted to a developed by me virtual memory system for HP-21MX.

Now, a little bit about security ☺. Moscow Aviation Institute was a so-called "closed facility," like all military–industrial facilities. One could not get in without a pass. The Institute was located next to metro station Sokol through which the vast majority of its 30,000+ full-time students were coming. The pass to the university was usually shrink-wrapped in plastic to preserve it and students usually put their monthly travel ticket (6 roubles for unlimited travel on all transportation modes) inside the plastic. Guards were usually retired people who did not pay a lot of attention to passes – it was almost always possible to get in showing your travel ticket (as well as to get into the metro showing your university pass).

Security was a bit more stringent on the second territory, where M4030 data centre and Military Kathedra were located. We were all trained to do something with ballistic missiles. This required the lowest level of security clearance. I was trained to aim a so-called "article 8K84" or UR100 or SS-11 Sego in NATO nomenclature (https://en.wikipedia.org/wiki/UR-100). Security was quite tight – nothing (apart from pens) was allowed in the class; notebooks had numbered pages and were sawn; they were stored in a security room in a sealed suitcase that at the beginning of the class was brought in and taken back to security room at the end of the day. One day a week of military training was followed by a month in Military Camps at Ostrov-3 near Pskov (https://rvsn.info/training/uc_047_35600_36700.html), where we were stationed in a "tiny" bedroom that accommodated 500 students on two-level banks, after which I was commissioned as a second lieutenant. Image of the article 8K84 (or UR-100) ICBM, known to the west as the SS-11 Sego, can be seen at: https://pioneer-club.at.ua/publ/strategicheskij_raketnyj_kompleks_ur_100_s_raketoj_8k84/10-1-0-61.

Applied Mathematics Faculty's program was designed for 5 .5 years – 2 years of exactly the same curriculum for everyone, then for the next 3 years students could choose one of five specialisations, then 6 months to write a thesis. I chose "Systems Programming" and focused on OS. For whatever reason compilers and databases did not attract my interest, though I had several subjects on these topics too.

I was extremely fortunate to be able to attend a 1-year "Operating Systems Design" subject delivered by Victor Petrovich Ivannikov (https://en.wiki pedia.org/wiki/Viktor_Ivannikov) – the man who created OS for the mightiest Soviet Supercomputer BESM-6 (https://en.wikipedia.org/wiki/BESM-6) that was heavily used by military.

On February 29, 1980, I was given my degree, finishing the fifth in my year, and was recommended to continue my studies as a postgraduate student doing a doctorate.

Based on this recommendation, I was able to find a spot for a full-time PhD course at the Institute of Mathematics and Cybernetics of the Lithuanian Academy of Sciences in Vilnius, Lithuania. They had a spot for my speciality (Software for Computers and Systems), but they did not have a Principal Supervisor to supervise me. With the help of one of my university lecturers I found a Principal Supervisor, who happened to be at the time be her husband, his name was Evgeny Andreevich Zhogolev (https://www.computer-museum.ru/english/galglory_en/Zhogolev.htm). This was convenient as this allowed me to stay in Moscow, subject to finding a Principal Supervisor myself, and visit Vilnius 2–3 times a year. As a full-time PhD student, I had to study three subjects (Computers and Software, Philosophy, and English) and subsequently sit three exams; I studied all these courses at the Academy of Sciences of the USSR.

My relationship with my first Principal Supervisor ended up abruptly around 24 months after it started. Those days access to a computer was very important, but he was not able to provide me with this. Fortunately, I got access to several computers at the Space Research Institute of the Academy of Sciences of the USSR (https://en.wikipedia.org/wiki/Russian_Space_Research_Institute) in exchange for developing a distributed experiment automation system. I developed the first version of such a system using Data General MicroNova microcomputers (https://retrocomputingforum.com/t/the-data-general-micronova/1289), CAMAC (https://en.wikipedia.org/wiki/Computer_Automated_Measurement_and_Control) and a 4 Mb coaxial comms module during 1980. This included a star architecture, new bootstrap (on EPROM) to boot from the remote disk, fast real-time protocol and enhancements to CATY programming language (https://www.sciencedirect.com/science/article/abs/pii/0167508782900333) allowing upload/download of programs and data. In 1981, I developed the next version of this system making it heterogeneous allowing use of both Data General MicroNova and LSI-11 microcomputers (https://gunkies.org/wiki/LSI-11). Image of Space Research Institute of the Academy of Sciences of the USSR can be seen at: https://indicator.ru/label/iki-ran.

It was suggested to me by the owner of the system that I should present my work at a conference which I agreed to do. After this presentation at a conference in St. Petersburg (which was Leningrad those days), my Principal

Supervisor created a scene around why he was not a co-author of this presentation. Considering that he did not contribute to this work in any form or shape, I decided that I couldn't use him as a Principal Supervisor any longer.

Parting ways with my Principal Supervisor turned out to be a big problem for me. It was not an easy situation for a PhD student with an almost finished PhD thesis to find a new Principal Supervisor – Principal Supervisors always want their students to contribute to their work and research. Eventually, when my 3-year full-time PhD course was almost over, I managed to find a new Principal Supervisor – for his next promotion he needed to have a certain number of PhD students completing their courses under his supervision and as such it was a win–win situation.

It took me almost another year to finalise my thesis (including adding an additional chapter with some mathematics) and eventually early in 1984 I submitted my completed thesis. By this time (since September 1983) I had been working as a Senior Research Fellow in a research lab with focus on developing a Unix-like OS for a special BESM-6 compatible CPU within Elbrus system and teaching Unix kernel at the Advanced Training Institute. After the tragic death of the head of our laboratory (he was hit by a car while exiting a tram on his way back home from work), I was invited by its inaugural director Alfred Karlovich Ailamazyan to join the recently founded Program Systems Institute of the Academy of Sciences of the USSR (https://www.psi-ras.ru/) to head its Portable Operating Systems Research Laboratory. I was also in charge of the Institute's data centre where I put in the very first in USSR non-military fibreoptics link between two Institute buildings (Figure 1.2).

But back to my PhD story. Getting a PhD in the USSR was a much longer and more complicated process compared to the Western countries. After one's Principal Supervisor was comfortable with the thesis there was a need to find the relevant Scientific Council for the so-called "defence" of the thesis. This council had to be approved by the Higher Attestation Commission of the USSR (https://en.wikipedia.org/wiki/Higher_Attestation_Commission#:~:text= Higher%20Attestation%20Commission%20(Russian%3A%20%D0% 92%D1%8B%D1%81%D1%88%D0%B0%D1%8F,awarding%20 of%20advanced%20academic%20degrees) and had to be able to hear the "defence" within the relevant discipline. The Scientific Council typically consisted of 12–16 experts in their relevant disciplines. After such a Scientific Council is located, relevant paperwork needed to be submitted and waiting time (typically 6–12 months) then started. Relevant documentation included two copies of the thesis, including a hundred copies of the extended abstract to be sent to a hundred main libraries in the country. This also included names and agreement of two so-called principal opponents and one leading organisation. By the time of the "defence," principal opponents and leading organisation were supposed to provide the Scientific Council with written reviews of the thesis. These could also be augmented by any number of unsolicited reviews of the extended abstract that was sent earlier to the hundred main

Figure 1.2 Author in front of the wooden institute building (former orphanage), 1985.

Source: author's archive.

libraries. At the "defence," the author of the thesis had to make a short 10–20 min presentation, followed by reviews of the principal opponents and leading organisation and questions from the members of the Scientific Council. The author of the thesis was also required to answer all questions asked earlier. In addition to this, there was a secret ballot (I got one "negative" or "black ball") and completed paperwork was sent for scrutiny to the Higher Attestation Commission of the USSR. It could take anywhere between 3 and 15 months to get approval (or not) from the Higher Attestation Commission of the USSR. I was lucky, my approval came through just 4 months after the "defence."

After spending 4 years at the Program Systems Institute of the Academy of Sciences of the USSR, I moved to the very first Soviet–French–Italian joint venture Interquadro to head its UNIX department. The company proudly called itself systems integrator, but in fact 90% of the activities included flogging PCs that were in high demand at the time. Companies' UNIX-related activities were a result of the French owner's (Alexandre Kaplan) passion for UNIX and Motorola 68000/68010 based minicomputers that his French company Aniral U.T.E.C. built.

During this time, I also became the founding Chair of the Soviet UNIX Users Group (SUUG) and in 1988 organised the very first UNIX conference in the USSR with presentation by a dozen of invited Western computer scientists. In 1989, I attended USENIX Conference in San Diego, CA.

Despite my multiyear involvement with computers and networks, the word "cybersecurity" was not heard by me or used within the industry. Security was centred around physical security of data centres and anti-wiretapping.

As my mother was an architect, I was exposed to architecture from early childhood and knew the names of all famous architects. So, as I was starting this book, I remembered what was once said by the famous Swiss–French architect Charles-Édouard Jeanneret, known as Le Corbusier (October 6, 1887 to August 27, 1965). By the late 1920s, Le Corbusier was done with Western Europe and got attracted by the new country and new opportunities offered by the USSR and visited Moscow three times – in 1928, 1929 and 1930. In 1932, he participated in the competition for the master plan for the reconstruction of Moscow. Le Corbusier responded with a 59-page report "Response to Moscow," in which he sketched his ideas for the city. What Le Corbusier said at the time was:

> There is no way to dream about combining the city of the past with the present or with the future; and in the USSR more than anywhere else... In Moscow, apart from a few precious monuments of former architecture, there are still no solid foundations; it is all piled up in disarray and without a definite purpose... In Moscow, everything needs to be remade, having first destroyed everything.

He did not win this competition and this ended his relationship with the USSR. Interesting parallels....

In 1991 I moved to Australia, which is home since then.

Chapter 2

Cybersecurity or cyber security?

The answer to this question is - it depends ☺. It depends. The Oxford and Merriam Webster dictionaries spell cybersecurity as one word. NIST spells it as one word; however, many organisations (like, e.g., UK's NCSC) spell it as two separate words. Spell checkers generally don't flag either method of spelling as incorrect. The only conclusion is that both spellings, cybersecurity and cyber security, are correct.

UK's NCSC uses a very short and simple definition of cybersecurity: "Cybersecurity is how individuals and organisations reduce the risk of cyberattack." It then elaborates:

> Cybersecurity's core function is to protect the devices we all use (smartphones, laptops, tablets and computers), and services we access - both online and at work - from theft or damage. It's also about preventing unauthorised access to the vast amounts of personal information we store on these devices, and online.

Cybersecurity is the application of technologies, practices, policies, processes and controls to protect systems, networks, programs, devices and data from cyberattacks, or mitigating their impact. It aims to protect computer systems, applications, devices, data, financial assets and people against ransomware and other malware, phishing scams, data theft and other cyberthreats, or, in other words, to reduce the risk of cyberattacks and protect against the unauthorised exploitation of systems, networks and technologies. Cybersecurity refers to every aspect of protecting an organisation, its assets and its employees against cyberthreats. Cyberattacks are usually aimed at accessing, changing or destroying sensitive information, extorting money via ransomware or interrupting normal business processes.

Today everything relies on computers and the Internet – communication (e.g., email, smartphones, tablets), entertainment (e.g., interactive video games, social media, apps), transportation (e.g., navigation systems), shopping (e.g., online shopping, credit cards), medicine (e.g., medical equipment, medical records), banking and the list goes on. Rapid rise in number, variety and complexity of cyberattacks during the last decade made cybersecurity a

DOI: 10.1201/9781032672601-2

priority for organisations and individuals. For organisations, cybercrimes can lead to financial loss, operational disruption, data breaches and a loss of trust, while individuals face identity theft, financial fraud and privacy invasion. This elevates importance of cybersecurity in the most significant way. Cybersecurity is a very wide field that includes:

- **Network security:** Practice of securing a computer network from intruders, whether targeted attackers or opportunistic malware.
- **Application security:** Practice of keeping software and devices free of threats, as compromised application(s) could provide access to data that it is supposed to protect. Successful security begins in the design stage, well before a program or device is deployed.
- **Information security:** Practice of protecting integrity and privacy of data, both in storage and in transit.
- **Operational security:** Practices, processes and decisions for handling and protecting data assets. The permissions users have when accessing a network and the procedures that determine how and where data may be stored or shared all fall under this umbrella.
- **Disaster recovery and business continuity:** Practices, policies and procedures that define how organisations respond to cybersecurity incidents or any other event that causes the loss of operations or data. Disaster recovery policies dictate how organisations restore their operations and information to return to the same operating capacity as before the event. Business continuity is the plan organisations fall back on while trying to operate without certain resources.
- **End-user education:** Practice that addresses the most unpredictable cybersecurity factor: people. Anyone can accidentally introduce a virus to an otherwise secure system by failing to follow good security practices. Teaching users to delete suspicious email attachments, not plug in unidentified USB drives and various other important lessons is vital for the security of any organisation.

Global cyberthreats continues to evolve at a rapid pace, with a rising number of data breaches each year. As cyberattacks become more common and sophisticated and corporate networks grow more complex, a variety of cybersecurity solutions are required for each of the above-mentioned areas to mitigate corporate cyber risks. With the scale of the cyberthreats set to continue to rise, global spending on cybersecurity solutions is naturally increasing. Gartner predicts cybersecurity spending reached in excess of $188 billion in 2023 and will surpass $260 billion globally by 2026.

Beginnings of cybersecurity can be traced back to the 1940s. To be more precise, 1945, when the first general-purpose electronic digital computer Electronic Numerical Integrator and Computer ((ENIAC) was released. Although networks were not available in the 1940s, some people hypothesised what they could look like in the future. So, while there were no connections between various pieces of equipment yet, John von Neumann was already thinking about what a virus could look like. His paper, called "Theory of Self Reproducing Automata" would not be published until the 1960s, but the seeds of virus theory were there 20 years earlier. The idea was that there could be some sort of mechanical organisms able to copy itself and spread to new hosts.

The initial intent of hacking did not encompass computer information collection. It might be argued that the origins of computer hacking can be traced back to the early utilisation of telephones. The prominence of "phone phreaking" (ability to hijack telephone protocols with the goal of allowing people to make cheaper or free phone calls) emerged during the 1950s. The term "phone phreaking" covers a range of methodologies employed by individuals known as "phreaks," who used a specific knowledge of the inner workings of telephone systems, in order to manipulate the protocols enabling telecommunications professionals to remotely operate on the network, and as a result to make cheaper or free phone calls and circumvent long-distance call fees. The practice of "phone phreaking" gained popularity throughout the latter part of the 1950s. Despite a gradual decline in the practice during the 1980s, telephone service providers were unable to effectively suppress the activities of individuals known as "phreaks." There have been speculations on the involvement of Apple's co-founders Steve Jobs and Steve Wozniak, who basically started the company following a successful run of building and selling phone phreaking equipment.

In the 1960s, the cyberthreat landscape was in its infancy, and the notion of cyberattacks as we understand them today was not prevalent. The primary concerns were still focused on the physical security of the hardware and preventing unauthorised access to the limited computing resources. However, there were no standardised security practices, and the idea of cybersecurity as a distinct discipline had not yet emerged. During this time earlier computer hacking attempts were mostly focused on gaining access to certain systems. For example, in 1967, IBM asked students to test drive their new computer. Through this process (something we typically refer to as "user testing" today), IBM learned about possible vulnerabilities. This may have been the first example of what is called "ethical hacking" today. So, there was already a concern about security measures. The result was the development of a defensive mindset, that computers required security measures to keep hackers out. This was an important step in the development of cybersecurity strategies.

The true birth of cybersecurity occurred in the 1970s with creation of ARPANET (Chapter 11 and 12). In 1971, Robert Thomas, a researcher at BBN technologies (who is widely regarded as the father of cybersecurity), realised the possibility of creating a program capable of moving through a network and leaving behind a trail. This discovery led to the invention of the first computer worm. The worm was called Creeper. It printed the message "I'M THE CREEPER: CATCH ME IF YOU CAN." Creeper was actually designed as a test to see if a self-replicating program was possible. Creeper is often regarded as the very first computer virus. In response to Creeper, Ray Tomlinson (who gained the fame for his development of email) developed a program he called Reaper – to chase and delete Creeper. Reaper was the very first example of antivirus software.

In 1979, a 16-year-old Kevin Mitnick managed to hack into the Ark, the computer system operated by Digital Equipment Corporation (DEC) for the development of its RSTS/E operating systems, and made copies of the software. Mitnick actually carried out one of the first, if not the very first, social engineering attack by tricking the administrators of the Ark to give him employee credentials. He was later the first cybercriminal to be arrested.

For most of the 1970s and 1980s, when the Internet was still under development, computer security threats was still easily identifiable. Majority of the threats were from malicious insiders who gained access to documents that they weren't supposed to view. Therefore, computer security in software programs and the security involving risk and compliance governance evolved separately.

The 1980s definitely became the decade of high-profile attacks, both against private companies and government systems. It was also the time when computers became personal. A number of high-profile attacks took place in this decade. These included attacks on AT&T, the Los Alamos National Laboratory and National CSS. In 1983, new terms were developed to describe these attacks. Among them were "computer virus" and "Trojan Horse."

Use of Bulletin Board Systems (BBS) was very popular in the 1980s. BBS allowed users to connect their personal computers to a host system via a modem. So, as sharing became easier, security challenges exploded. This is when explosion of viruses and malware (including the Elk Cloner virus (1982), which targeted Apple II computers, and the infamous Brain virus (1986), which affected IBM PC-compatible systems) occurred. In 1983, computer hack of ARPANET systems by the 414s hacker group highlighted vulnerability of early computer networks. It also prompted heightened security concerns, spurring organisations to reassess their cybersecurity strategies.

A big fear at this time was the threat from other governments. It was the middle of the Cold War and fear of cyber espionage reached very high level. The threat was real and this pushed the US government to create new guidelines and resources for managing such events and threats. The Trusted Computer System Evaluation Criteria was developed in 1985 by the US Department of Defense. It was later called The Orange Book. Network breaches and malware already existed at this time. But they were used for purposes other than financial gain. For instance, the Soviets used them to deploy cyber power as a form of weapon. Similarly, in 1986 German computer hacker Marcus Hoss hacked into an Internet gateway. Marcus Hoss used the gateway located in Berkeley to connect to the Arpanet. He then proceeded to access 400 military computers, including the Pentagon's mainframes. Marcus Hoss' primary intent was to acquire information and sell it to KGB. However, astronomer Clifford Stoll used honeypot systems to detect the intrusion and foil the plot. Notably, these attacks marked the start of severe computer crimes utilising virus intrusion. Viruses were no longer used just for academic purposes.

Then in 1988, the Morris Worm, one of the earliest instances of widespread malware, was unleashed by a graduate student from Cornwell University Robert Tappan Morris. Robert Morris was curious about the Internet size and created a worm to gauge it. The worm was designed to infect UNIX systems such that it would count the total connections present on the web. Morris thus wrote a worm program that would propagate across a set of networks, use a known vulnerability to infiltrate UNIX terminals and then replicate itself. Robert Morris became the first person to be charged successfully under the Computer Fraud and Abuse Act. He was fined $10,000, sentenced to probation of three years and dismissed from Cornwell (although he went on to become an MIT tenured professor). The act further led to the development of a Computer Emergency Response Team, the predecessor of US-CERT.

The Morris worm triggered the start of an entirely new field in computer security. It led to more people researching on how they can create deadlier and more effective worms and viruses. The more worms evolved, the greater their effect on networks and computer systems. Morris worm paved the way for newer types of malicious programs. Viruses were more aggressive programs that came into light in the 1990s. Majority of the virus attacks were primarily concerned with financial gains or strategic objectives. However, inadequate security solutions at the time caused a huge number of unintended victims to be affected. Worms and viruses, in turn, led to the rise of antivirus solutions as a means of countering the worm and virus attacks. In a very short span of time cyberthreats and cyberattacks became a huge concern necessitating creation of an immediate solution. This problem gave birth to antivirus software solutions. These programs were designed to detect the presence of viruses and to prevent them from accomplishing their intended tasks. At the time, the primary delivery method for viruses was the use of malicious email attachments. The virus attacks, most importantly, caused increased awareness, especially with regard to opening email messages originating from unknown people.

Commercial antivirus products were first developed and released in 1987, just a year after the Pentagon attack. What's confusing is determining who actually developed the very first commercial product, as many claims exist. Some of the most notable to consider include the development by John McAfee (British–American computer programmer who in 1987 founded commercial antivirus software company McAfee Associates) VirusScan. An antivirus product was released for the Atari ST by German inventors (who in 1985 founded G Data Software) Kai Figge and Andreas Luning. Also, in 1987, the first version of NOD32 (MS DOS-based program called NOD-ICE) antivirus solution was released in Czechoslovakia by Miroslav Trnka and Peter Paško. Significant prominence was later gained by Norton AntiVirus (based on the use of signatures and heuristics) developed and released in January 1990 by Peter Norton and distributed by Symantec (now Gen Digital).

Early 1990s brought a sharp growth of companies creating and retailing antivirus products. The products were scanning computer systems for the presence of viruses or worms. At the time, the available antivirus solutions scanned systems and tested them with signatures written in a database. Although the signatures were initially file-computed hashes, they later incorporated strings similar to those present in malware. However, two significant problems had high impacts on the effectiveness of these early antivirus solutions. The issues persist today in some of the current cybersecurity solutions. The problems included the intensive use of resources and a large number of false positives. The former caused the most problems since antivirus solutions scanning systems used a lot of the available resources such that they interrupted user activities and productivity.

In 1989, PC Today magazine through a disk offered to subscribers released DiskKiller virus. It infected a boot sector of the computer, gradually destroying the hard disks onto which it was released. It infected thousands of computers. The magazine later stated that it was an accident, and they did not know the risk was present. As the Internet gained widespread adoption in the 1990s, the history of cybersecurity entered a new era. The interconnectivity of global networks brought unprecedented opportunities but also introduced new cyberthreats. During this period, cybercriminals became increasingly sophisticated, exploiting vulnerabilities in software and systems to gain unauthorised access, steal data and disrupt operations. This decade saw incredible growth and development of the Internet. The cybersecurity industry grew with it. During this period malware samples produced every day increased in size and scope. Whereas only a few thousands of malware samples existed in the 1990s, the number had grown to at least 5 million by 2007. As a result, legacy antivirus solutions could not handle such a capacity, as cybersecurity professionals were unable to write signatures that would keep up with the problems as they emerged. The challenge called for a newer approach that would offer adequate protection to all systems. The problem was further exacerbated by emergence of polymorphic viruses. In 1990, the first code that mutates as it infects systems, but that also keeps the original algorithm in place, was developed. The polymorphic virus was designed to avoid detection and posed significant challenge for traditional antivirus defences. That made it harder to determine its presence. Growing popularity of text-based chat system for instant messaging, like Internet Relay Chat (IRC), and online communities, like America Online (AOL), gave rise to new forms of cyberthreats, including unauthorised access, social engineering and distributed denial-of-service (DDoS) attacks.

In 1995, the first version of the Data Encryption Standard (DES) was adopted for securing electronic communications, marking a milestone in cryptography and laying the groundwork for modern encryption standards.

By 1996, new stealth capability was developed. This year also saw macro viruses being released. Both created more challenges and required new developments of antivirus software. From the first antivirus on, the goal was to increase ways to protect against risks. As hacker groups were mushrooming, organisations faced a lot of challenges to improve cybersecurity to minimise disruption and losses. More types of malicious programs were on the way. The ILOVEYOU virus and Melissa virus infected millions of computers in the 1990s targeting Microsoft Outlook. Created in 1999 by David Lee Smith, Melissa virus spread across Internet and earned notoriety as the fastest-spreading infection of its time. It inflicted damages estimated at $80 million, compelling organisations to invest heavily in clean-up and repair efforts to mitigate its widespread impact on affected systems. These viruses caused significant slowdowns and failures of email systems. New strategies were developed to help with growing problems. One of those was Secure Socket Layer (SSL). It was developed as a way to make use of Internet more safe and secure. SSL was put in place in 1995. It helped to protect activities like online purchases. Netscape developed the protocol for it. It would later become the foundation for the development of HyperText Transfer Protocol Secure (HTTPS).

In the 2000s, cybersecurity faced a surge in cyberattacks, prompting organisations to prioritise compliance with regulations and standards. The first hacker groups also developed at this time. These groups typically included people with specific hacking skills. They could launch a cyberattack campaign for various goals. One of the first to become recognised when it hacked the Church of Scientology using DDoS attack. The group was called Anonymous and it continued to attack various high-profile targets. The era saw a rise in high-profile breaches, highlighting the importance of robust security measures to protect sensitive data. In the early 2000s, crime organisations started to heavily fund professional cyberattacks and governments began to clamp down on the criminality of hacking, giving much more serious sentences to those culpable. New type of infection emerged where there was no longer a need to download files. Just going to a website infected with the virus was enough. This type of hidden malware was damaging. It also infiltrated instant messaging services.

In the 2000s, cyberattacks started being more targeted. One of the most memorable attacks during this period includes the first reported case of serial data breaches targeting credit cards. These were perpetrated between 2005 and 2007 when Albert Gonzales created a cybercriminal ring for compromising credit card systems. The Albert Gonzales group managed to steal confidential information for 45.7 million credit cards that belonged to customers of TJX retailers and caused TJX a loss amounting to a staggering amount of $256 million. They gained access through retailer's database. This created a broader need to focus on cybersecurity by various sectors, including retailers.

Technology progress in different industries propelled cyber laws to emerge. These laws intend to protect systems and confidential data. Some of the notable regulations in cybersecurity history include the Health Insurance Portability and Account Act (HIPAA). HIPAA became law on August 21, 1996. The bill was amended over the years to focus more on protecting employee personally identifiable information (PII). In 1999, Gramm–Leach–Bliley Act (GLBA), also called the Financial Modernisation Act, was enacted to protect the personal data of customers of financial institutions. The law requires financial institutions to provide detailed information on the strategies they intend to use when securing a customer's private data. To comply with the law, financial institutions must always alert customers on how they will share their personal information. More so, the law stipulates that customers have the right to deny financial institutions the rights to share sensitive data. Also, financial institutions must maintain a documented information security program for protecting customers' sensitive data. In 2003, the Federal Information Security Management Act (FISMA) was legislated to provide organisations with guidance for securing information systems. The law defines a complex framework to be applied in securing government IT assets, data and operations from natural or manmade disasters. The act followed the enactment of the E-Government Act (Public Law 107-347), which outlined the main threats that affect information systems. The E-government Act also outlined the need for adopting effective security measures for securing against the threats. FISMA falls under the E-Government Act. According to the FISMA act, all federal agencies must develop and document agency-wide programs for protecting information systems. Organisations can be fined at least 4% of their annual profits for failing to properly secure PII information or using customer data without their permission. They can also be fined 4% when a breach occurs due to inadequate security measures. For an agency to be FISMA compliant, it must observe the following guidelines:

- Conduct frequent inventories of current security measures
- Analyse existing or anticipated threats
- Design working security plans
- Designate security professionals for observing the implementation of the security plans and continuously monitor its effectiveness
- Document plans for reviewing the security plans and periodically assess its security operations

The first standard, modelled after the Visa Cardholder Information Security Program (CISP), was released on December 15, 2004, called the Payment Card Industry Data Security Standard (PCI-DSS) v1. PCI DSS is an information security standard used to handle credit cards from major card brands. The standard is administered by the Payment Card Industry Security Standards Council, and its use is mandated by the card brands. It was created to better control cardholder data and reduce credit card fraud. Validation of compliance is performed annually or quarterly with a method suited to the volume of transactions:

- Self-assessment questionnaire (SAQ)
- Firm-specific Internal Security Assessor (ISA)
- External Qualified Security Assessor (QSA)

Another example is the General Data Protection Regulation (GDPR) that entered into force in 2016 after passing European Parliament, and as of May 25, 2018, all organisations are required to be compliant. This regulation provides mandatory guidelines for institutions handling PII data and imposes hefty fines on any incidence of non-compliance. The GDPR protects data specifically belonging to members of the European Union. The fundamental part of the regulation is ensuring that organisations implement adequate data protection controls, which include encryption for both data in transit and data at rest.

These regulations resulted in widespread adoption of cybersecurity measures in the healthcare and finance sectors. Despite all these measures the number of credit card hacks kept increasing and there have been massive credit card data leaks. In 2012, Union Savings Bank (Danbury, Connecticut) saw an odd pattern of fraud on about a dozen of the debit cards it had issued at the beginning of March 2012. It also noticed that many of the cards had recently been used at a cafe at a neighbouring private school. This breach was limited to a small number of people, and it was made clear to the card holders that they wouldn't be responsible for any fraudulent card use.

And everyone heard about Stuxnet – a malicious computer worm first uncovered in 2010 and thought to have been in development since at least 2005. Stuxnet targets supervisory control and data acquisition systems (SCADA) and is believed to be responsible for causing substantial damage to the nuclear program of Iran.

In 2011, hackers broke into Sony's PlayStation network and stole the personal information of millions of PlayStation users, taking the network offline for several weeks. Driving force behind this attack was anger over Sony suing an American hacker who tried to reverse-engineer the PlayStation 3 to enable customers to play unofficial third-party games. During this time there were also attacks on Yahoo. They were found in 2013 and 2014 and in one incident, hackers gained access to the Yahoo accounts of over 3 billion users. In 2013, hacktivist group Anonymous launched Singapore cyberattacks, which were a series of assaults in retaliation for Singapore's web censorship laws. An Anonymous member going by the online alias "The Messiah" claimed leadership of the attacks.

State-sponsored attacks became another area of concern. They are monitored by CIA and NSA. These attacks put governments, businesses and individuals (and their sensitive data) at risk. One example of this occurred in 2014, when Lazarus Group (sponsored by North Korea) took aim at consumers. It hacked into Sony at the time. It resulted in the release of videos for new films, including actor's images. There have been many cases of state-sponsored attacks. One hundred and forty-four universities within the United States were attacked in 2018 using different types of attacks. The attacks were executed over three years and led to the loss of intellectual properties amounting to $3 billion and at least 31 terabytes of data. Investigations revealed that Iran was behind the attack. The United States identified and prosecuted nine hackers of Iranian descent.

Lateral movement attack techniques allow cybercriminals to run codes, issue commands and to spread across a network. Such methods have been in play for several years. Lateral movement vulnerabilities have been present for many years, enabling cybercriminals to execute lateral stealth attacks. EternalBlue constitutes of a notable example of lateral movement vulnerability. For example, EternalBlue vulnerability allows an attacker to exploit SMB protocols used to share files across a network. Shadow Brokers leaked the protocol on April 14, 2017, and the notorious Lazarus group used it as an exploit for the infamous WannaCry attack on May 12, 2017. The WannaCry attack was a global ransomware attack targeting health institutions mostly in Europe. The attack was quite devastating as it caused health services to halt for almost a week. EternalBlue exploit has also been used to execute other high-profile cyberattacks. On June 27, 2017, this vulnerability was exploited in the NotPeyta attacks which targeted banks, ministries, electricity and newspaper firms across Ukraine. The attack spread in other countries, including France, the United States, Russia, Poland, Italy, Australia, and the United Kingdom. It was also used to execute Retefe banking trojans.

The threats from cyberattacks are numerous. They continue to be present. Phishing, personal data loss online and ransomware attack events take place around the world often. Yet, finding a way to minimise security breaches became more important than ever. The threats countered by cyber-security today are three-fold:

- Cybercrime includes single actors or groups targeting systems for financial gain or to cause disruption.
- Cyberattack often involves politically motivated information gathering.
- Cyberterrorism is intended to undermine electronic systems to cause panic or fear.

The history of cybersecurity is a reminder of the ongoing need for vigilance, innovation and collaboration in the face of constantly evolving cyberthreats. By understanding the lessons of the past and embracing the latest advancements in cybersecurity, one can better prepare for the challenges that lie ahead, as new threats continue to emerge due fast technological progress.

Examples of these new threats are Artificial Intelligence (see Chapter 18) and Quantum Computing (see Chapter 19).

The cybersecurity industry is continuing to grow at the speed of light. According to Statista, global cybersecurity market size is forecast to grow to $345.4 billion by 2026.

Today, according to Netskope, two-thirds of attributable malware comes from state-funded attack groups (https://itwire.com/business-it-news/security/cyber-threat-data-66-of-attributable-malware-comes-from-state-funded-attack-groups,-reveals-netskope.html). The largest share of malware attacks come from North Korean threat groups, with Chinese and Russian groups as second and third most prevalent – and a growing number of attacks use cloud applications as a point of entry and exfiltration. As Sanjay Beri, CEO and co-founder of Netskope said: "There is no doubt that we are witnessing a global escalation of cyber-attacks carried out by nation state actors as a form of 'quiet war' on nation states that are currently officially at peace."

Chapter 3

History of computing
(von Neumann architecture)

EARLY HISTORY

Let's make a brief excursion back in time.

We will start with the abacus (https://alohamindmath.com/different-abacus/#:~:text=11%20Apr%20Abacus%2C%20the%20oldest%20calculator&text=Using%20a%20tool%20to%20do,the%20oldest%20calculator%20in%20existence), allegedly the oldest calculator used for addition, subtraction, multiplication and division. Sumerian abacus, the oldest known form of the counting device, was created approximately 5000 years ago sometime in the period between 2700 and 2300 BC in Mesopotamia area and was followed by Roman abacus, Chinese Suanpan and Russian abacus.

We will then go as far back as 2000 or so years to Antikythera Mechanism discovered in 1901 (https://www.nature.com/articles/s41598-021-84310-w#:~:text=Abstract,since%20its%20discovery%20in%201901) and pictured below.

The mechanism has been described as an astronomical calculator as well as the world's first analogue computer. It is made of bronze and includes dozens of gears. It is understood that the hand-powered Ancient Greek device is thought to have been used to predict eclipses and other astronomical events. Unfortunately, only a third of the device survived, leaving researchers guessing how it worked and how exactly it looked like. The workings of the back of the mechanism were solved by earlier studies, but the nature of its complex gearing system at the front has remained a mystery (Figure 3.1).

DOI: 10.1201/9781032672601-3

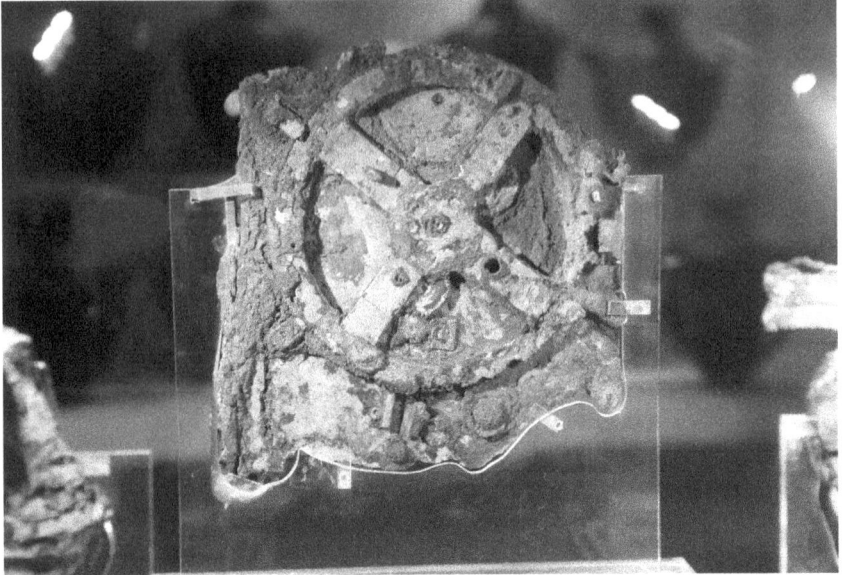

Figure 3.1 Antikythera Mechanism discovered in 1901.

Source: https://www.flickr.com/photos/101561334@N08/28453473976, Bronze Gallery, National Archaeological Museum of Greece, Athens, Greece. Complete indexed photo collection at WorldHistoryPics.com.

We will then mention cylindrical calculator invented by Philip Matthaus Hahn in 1773 (only five or six devices have been manufactured and only two of them survived to the present). Image of cylindrical calculator can be seen at: https://www.google.com/imgres?q=cylindrical%20calculator%20 invented%20by%20Philip%20Matthaus%20Hahn&imgurl=https%3A% 2F%2Fwww.kopykitab.com%2Fblog%2Fwp-content%2Fuploads% 2F2023%2F07%2Fimage-4008.png&imgrefurl=https%3A%2F%2Fwww. kopykitab.com%2Fblog%2Fweb-stories%2Foldest-calculators-ever-invented%2F&docid=BjI7Bjl5Gr8f8M&tbnid=gQ7DhYh1Mo75cM& vet= 12ahUKEwiIi9WF4bmJAxWQzjgGHdBUH08 QM3oECHEQAA..i&w=924&h=1200&hcb=2&ved=2ahUKEwiIi9WF4b mJAxWQzjgGHdBUH08QM3oECHEQAA.

In 1775, Lord Stanhope designed a pinwheel calculating machine (its image can be seen at: https://collection.sciencemuseumgroup.org.uk/objects/ co8412027/collection-of-ferranti-calculalating-instruments) and further improved it in 1777 (its image can be seen at: https://collection.science museumgroup.org.uk/objects/co8412027/collection-of-ferranti-calculalating-instruments).

Also see the arithmometer invented by Charles Xavier Thomas De Colmar in 1820 while he was serving in the French army, pictured below (Figure 3.2).

Figure 3.2 Arithmometer invented by Charles Xavier Thomas De Colmar in 1820, exhibit in the Science Museum, London, UK.

Source: https://commons.wikimedia.org/w/index.php?curid=153080243.

This first mechanical calculator gained widespread use and became a commercial success and was still being used up to World War I.

These devices had amazingly long life as one of their descendants was produced in the USSR from 1929 till 1978 (https://elektronika.su/en/calculators/felix-m/) (Figure 3.3).

All these machines technically were not programmable devices.

The era of programmable computers starts with analytical engine (https://www.britannica.com/technology/Analytical-Engine) conceived in 1833 and designed by English mathematician and computer pioneer (known as "the father of computers") Charles Babbage (Figure 3.4).

The analytical engine was first described in 1837, it incorporated an arithmetic logic unit, control flow in the form of conditional branching and loops, and integrated memory, making it the first design for a general-purpose computer that could be described in modern terms as Turing-complete, or in simple terms, one that can approximately simulate the computational aspects of any other real-world general-purpose computer or computer language. In other words, the structure of the analytical engine was essentially the same as that which has dominated computer design in the electronic era. Unfortunately, Charles Babbage, who had worked on the analytical engine until his death in 1871, was never able to complete its construction due to conflicts with his chief engineer and inadequate funding.

As we are diving into the history and talking about analytical engine, it would be a big mistake not to mention Hon. Augusta Ada Lovelace

Figure 3.3 Arithmometer Felix produced in the USSR.
Source: https://www.flickr.com/photos/31679151@N00/29221019204.

Figure 3.4 Part of Charles Babbage's Analytical Engine.
Source: https://commons.wikimedia.org/w/index.php?curid=28024313.

(daughter of the famous English poet Lord Byron), who worked on Charles Babbage's proposed mechanical general-purpose computer, the Analytical Engine. She was the first to recognise that this machine had applications beyond pure calculation.

Between 1842 and 1843, Ada Lovelace translated an article by the military engineer Luigi Menabrea (later Prime Minister of Italy) about the Analytical Engine, supplementing it with an elaborate set of seven notes, simply called "Notes." Her notes are important in the early history of computers, especially since the seventh one contained what many consider to be the first computer program – that is, an algorithm designed to be carried out by a machine.

Moving to the 20th century, it is important to mention cryptography achievements in Poland. Precursor of the Polish Cipher Bureau was created in May 1919, during the Polish–Soviet War (1919–1921), and played a vital role in securing Poland's survival and victory in that war. During the Polish–Soviet War, approximately a hundred Russian ciphers were broken by a sizable group of Polish cryptologists who included army Lieutenant Jan Kowalewski and three world-famous professors of mathematics – Stefan Mazurkiewicz, Wacław Sierpiński and Stanisław Leśniewski. Soviet army was still following the same disastrously ill-disciplined signals-security procedures as Russian Tsarist army during World War I. As a result, during the Polish–Soviet War, the Polish military was regularly kept informed by Russian signals stations about the movements of Russian armies and their intentions and operational orders. In mid-1931, the Cipher Bureau was formed by the merger of pre-existing agencies. In December 1932, the Bureau began breaking Germany's Enigma ciphers. Over the next 7 years, Polish cryptologists overcame the growing structural and operating complexities of the plugboard-equipped Enigma. The Bureau also broke Soviet cryptography. Five weeks before the outbreak of World War II, on July 25, 1939, in Warsaw, the Cipher Bureau revealed its Enigma-decryption techniques and equipment to representatives of French and British military intelligence, which had been unable to make any headway against Enigma. This Polish intelligence-and-technology transfer would have given the Allies an unprecedented advantage (Ultra) in their ultimately victorious prosecution of World War II.

In 1938, Polish Cipher Bureau cryptologist Marian Rejewski designed bomba or bomba kryptologiczna (Polish for "bomb" or "cryptologic bomb") – a special-purpose machine to break German Enigma ciphers. How the machine came to be called a "bomb" has been an object of fascination and speculation. The most credible explanation of why it was called this way was given by a Cipher Bureau technician, Czesław Betlewski: worker at B.S.-4, the Cipher Bureau's German section, christened the machine a "bomb" (also, alternatively, a "washing machine" or a "mangle") because of the characteristic muffled noise that it produced when operating. One other theory, most likely apocryphal, originated with Polish engineer and army officer Tadeusz Lisicki (who knew Rejewski and his colleague Henryk Zygalski in wartime Britain but was never associated with the Cipher Bureau). He claimed that Jerzy Różycki (the youngest of the three Enigma cryptologists, and who had died in a Mediterranean passenger-ship sinking in January 1942) named the "bomb" after an ice-cream dessert of that name. This story seems implausible, since Lisicki had not known Różycki. Rejewski himself stated that the device had been dubbed a "bomb" "for lack of a better idea"

The next important device to be mentioned is the Bombe (https://www. britannica.com/topic/Bombe and https://bletchleypark.org.uk/our-story/6-facts-about-the-bombe/), electromechanical code-breaking machine created by cryptologists in Britain during World War II to decode German messages that were encrypted using the Enigma machine. The Bombe was derived from a device called the bomba – Polish for "bomb" – that was invented in Poland during the 1930s (there are several theories about the origin of the name, among them being the units' bomb-like clicking as they did their work). Alan Turing (https://www.britannica.com/biography/Alan-Turing) developed the Bombe in 1939 at Bletchley Park, and the first Bombe machine was installed there in March 1940. Eventually there were 211 Bombes built (https://www.tnmoc.org/bombe#:~:text=The%20engineering%20and%20 construction%20of,keys%20on%20a%20daily%20basis). The Bombes were an important intelligence tool for the Allies, who used the machines to turn the tide of World War II against Nazi Germany.

The Bombes were followed by Colossus (https://en.wikipedia.org/wiki/Colossus_computer) developed in 1943–1945 and used help with the cryptanalysis of the Lorenz cipher. There were only two Colossus machines built by William (Bill) Tutte at Bletchley Park. Some sources erroneously stated that Turing designed Colossus to aid the cryptanalysis of the Enigma, while in fact Turing's machine that helped to decode Enigma was the electromechanical Bombe, not Colossus (https://www.gchq.gov.uk/news/colossus-80#:~:text=The%20Colossus%20computer%20was%20created,after%20 six%20decades%20of%20secrecy.).

The complete Bombe unit was 2 m tall, 2 m long, and 1 m wide (about 6.5 feet by 6.5 feet by 3.2 feet). The first Bombe was delivered to Bletchley Park on March 14, 1940. Built by the British Tabulating Machine Company, it was named Victory, but it proved to be too slow. A new design was created, and the new Bombe machine, called Agnus Dei, arrived in August. The Germans had, as expected, already stopped repeating the message key, but Agnus Dei proved to be the machine capable of facing the new German challenge, and many more were built during the following months. This version of the Bombe consisted of about 100 rotating drums, 16 km (10 miles) of wire, and about one million soldered connections. Hundreds of Wrens (members of the Women's Royal Naval Service, or WRNS) operated the Bombe machines, working long shifts in dark, stuffy rooms. The Bombes began finding keys in less than an hour.

Developing and using the Bombes, Turing and others working at Bletchley Park pioneered machine-driven cryptanalysis and the industrialisation of the code-breaking process. These accomplishments had a lasting impact on modern cryptography and computer science. The Bombe machines played a key role in deciphering the Enigma codes and were ultimately crucial to the Allies' victory over Nazi Germany in World War II.

Although the Bombe was a single-purpose (rather than universal) device, considering Alan Turing's contribution to computer science (his ideas led to early versions of modern computing), both deserve at least brief mentioning. Alan Turing was highly influential in the development of theoretical computer science, providing a formalisation of the concepts of algorithm and computation with the Turing machine, which can be considered a model of a general-purpose computer. He is widely considered to be the father of theoretical computer science.

Before we dive into computer architectures, it is important to mention not very well-known German computer Z3 built by German engineer Konrad Zuse, who worked in complete isolation from developments elsewhere. Z3 used 2300 relays and used 22-bit word length. It was able to perform floating point binary arithmetic and was used for aerodynamic calculations, but was destroyed in a bombing raid on Berlin in 1943. Zuse later supervised a reconstruction of the Z3 in the 1960s, which is currently on display at the Deutsche Museum in Munich.

COMPUTER ARCHITECTURES

Now, after we have schemed through the history from 5000 years back to 1940s, the really interesting stuff begins.

Historically there have been two types of computers:

- Fixed Program Computers – Their function is very specific and they couldn't be reprogrammed, for example, Calculators (apart from Charles Babbage's analytical machine, all devices in the previous section fall into this category)
- Stored Program Computers – These can be programmed to carry out many different tasks, applications are stored on them, hence the name (and this is what we will be focusing on now)

One of the most important names in modern computers is John von Neumann born in Hungary as Neumann János Lajos (https://www.britannica.com/biography/John-von-Neumann and https://www.ias.edu/von-neumann). Not everyone knows about the important role he has also played in the Manhattan Project – he is pictured below with Robert Oppenheimer (Figure 3.5).

John von Neumann is perhaps best known for his work in the early development of computers: as director of the Electronic Computer Project at

Figure 3.5 John von Neumann (right) and Robert Oppenheimer next to him during Manhattan Project days.

Source: https://openverse.org/search/?q=+oppenheimer+and+von+neumann& license_type=commercial,modification.

Princeton's Institute for Advanced Study (1945–1955), he developed MANIAC (mathematical analyser, numerical integrator and computer), which was at the time the fastest computer of its kind. He also made important contributions in the fields of mathematical logic, the foundations of quantum mechanics, economics and game theory.

Von Neumann was a founding figure in computing with significant contributions to computing hardware design, to theoretical computer science to scientific computing, and to the philosophy of computer science; he consulted for the Army's Ballistic Research Laboratory, most notably on the ENIAC project. Although the single-memory, stored-program architecture is commonly called von Neumann architecture, the architecture was based on the work of J. Presper Eckert and John Mauchly, inventors of ENIAC and its successor, EDVAC.

As always, history is full of patent disputes, so let's look at one of them. After successfully demonstrating a proof-of-concept prototype in 1939, Professor John Vincent Atanasoff received funding to build a full-scale machine at Iowa State College (now University). The machine was designed and built by Atanasoff and graduate student Clifford Berry between 1939 and 1942. The Atanasoff–Berry Computer (ABC) became the centre of a patent dispute related to the invention of the computer, which was resolved in 1973 when it was shown that ENIAC co-designer John Mauchly had seen the ABC shortly after it became functional. The legal result was a landmark: Atanasoff was declared the originator of several basic computer ideas, but the computer as a concept was declared un-patentable and thus freely open to all. A full-scale working replica of the ABC was completed in 1997, proving that the ABC machine functioned as Atanasoff had claimed. The replica is currently on display at the Computer History Museum.

While we are talking about early computers, it would be wrong not to mention Whirlwind computer was developed at the Massachusetts Institute of Technology (1944–1959). It became operational in 1951 and was the first real-time high-speed digital computer using random-access magnetic-core memory. Whirlwind featured outputs displayed on a CRT, and a light pen to write data on the screen. Whirlwind's success led to the US Air Force Semi-Automatic Ground Environment (SAGE) system and to many business computers and minicomputers.

Let's have a look at von Neumann (or Princeton) and Harvard computer architectures (Figure 3.6).

The von Neumann architecture – also known as the Princeton architecture – is a computer architecture based on a 1945 description by John von Neumann, and by others, in the First Draft of a Report on the EDVAC. The document describes a design architecture for an electronic digital computer with these components:

- A processing unit with both an arithmetic logic unit and processor registers
- A control unit that includes an instruction register and a program counter
- Memory that stores data and instructions
- External mass storage
- Input and output mechanisms

Von Neumann architecture is defined as a standard design of a computer system (released 1945–1951) in which there is a control unit, arithmetic

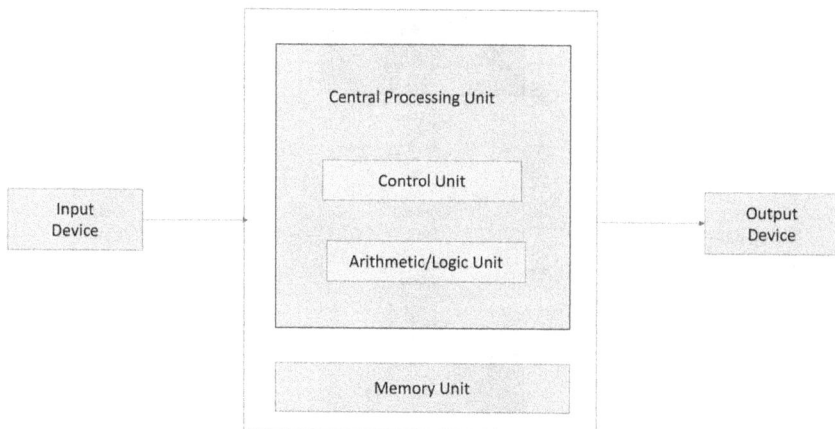

Figure 3.6 Von Neumann computer architecture.

logic unit (ALU), a memory unit (all within CPU) and input/output devices. These entities are connected over a series of busses.

- There is only one data bus which is used for both instruction fetches and data transfer from the memory which also is used for storage of both instructions and data
- Data/instructions can pass in half duplex (scheduled/one at a time) mode to and from CPU
- Also called stored program concept:
  - The memory is addressed linearly; that is, there is a single sequential numerical address for each memory location
  - Memory is split into small cells of equal sizes each with address numbers (i.e., same word size used for all memory)
  - Program instructions are executed in the order in which they appear in the memory, the sequence of instructions can only be changed by unconditional/conditional jump instructions
  - All instructions/data are in binary form

The term "von Neumann architecture" has evolved to refer to any stored-program computer in which an instruction fetch and a data operation cannot occur at the same time (since they share a common bus). This is referred to as the von Neumann bottleneck, which often limits the performance of the corresponding system.

Alternative to the von Neumann (Princeton) architecture is the Harvard architecture that dates back to 1943. The Harvard architecture is a computer architecture with separate storage and buses for instructions and data. It is often contrasted with the von Neumann (Princeton) architecture, where program instructions and data share the same memory and buses (Figure 3.7).

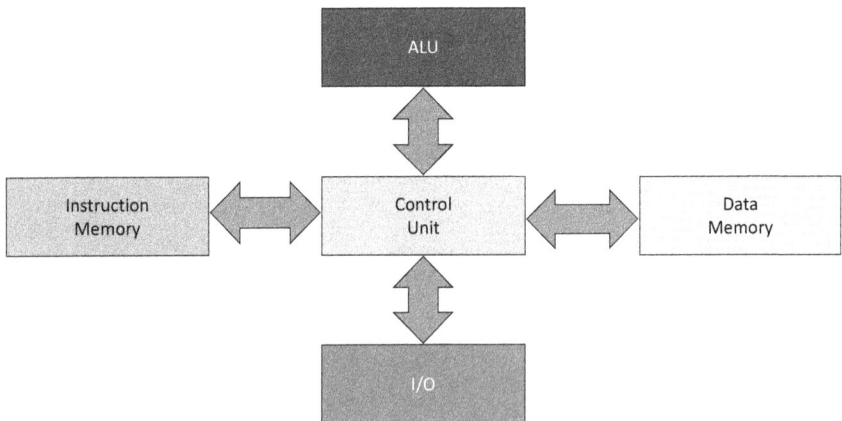

Figure 3.7 Harvard computer architecture.

In computer that follows von Neumann (Princeton) architecture, instructions and data both are stored in the same memory. So, the same bus is used to fetch instructions and data. This means the CPU cannot do both things together (read the instruction and read/write data). Harvard architecture is the computer architecture that contains separate buses for instruction and data. It was basically developed to overcome the bottleneck of von Neumann's Architecture. The main advantage of having separate buses for instruction and data is that the CPU can access instructions and read/write data at the same time.

The term Harvard architecture is often stated as having originated from the Harvard Mark I relay-based computer, which stored instructions on punched tape (24 bits wide) and data in electro-mechanical counters. Mark 1 was used in the Manhattan Project. These early machines had data storage entirely contained within the central processing unit, and provided no access to the instruction storage as data. Programs needed to be loaded by an operator and the processor could not initialise itself. The first computer with Harvard architecture was introduced in 1947. It separated memory for instructions and data and had separate buses for instruction fetches and data transfer. The Memory cell sizes for instructions and data are different. A more complex Control unit is required to handle two buses. Both instruction fetches and data transfer can take place simultaneously.

However, in the only peer-reviewed published paper on the topic – The Myth of the Harvard Architecture published in the *IEEE Annals of the History of Computing* (https://metalup.org/harvardarchitecture/The%20Myth%20of%20the%20Harvard%20Architecture.pdf) – the author demonstrates that:

- "The term 'Harvard architecture' was coined decades later, in the context of microcontroller design and only retrospectively applied to the Harvard machines and subsequently applied to RISC microprocessors with separated caches";
- "The so-called 'Harvard' and 'von Neumann' architectures are often portrayed as a dichotomy, but the various devices labeled as the former have far more in common with the latter than they do with each other";
- "In short [the Harvard architecture] isn't an architecture and didn't derive from work at Harvard."

Modern processors appear to the user to be systems with von Neumann (Princeton) architectures, with the program code stored in the same main memory as the data. For performance reasons, internally and largely invisible to the user, most designs have separate processor caches for the instructions and data, with separate buses into the processor for each. This is one form of what is known as the modified Harvard architecture (https://en.wikipedia.org/wiki/Modified_Harvard_architecture). So, although strictly speaking the vast majority of modern computers follow modified Harvard architecture, they all have one key feature of von Neumann (Princeton) architecture: use of the same memory for both data and instructions.

Essentially, historical debates about pros and cons of these architectures were focused primarily on:

- Design simplicity (1 bus vs 2 buses)
- Cost (1 bus vs 2 buses)
- Performance – sequential or parallel instructions and data access (1 bus vs 2 buses)
- Unified memory structure vs two separate memories
- Bus width (maybe the same or different for instructions and data)
- Flexibility and effectiveness of memory utilisation
- Architecture regularity and instruction sets

However, some people started to talk about cybersecurity implications associated with the use of each of these architectures. Let's have a look at what this actually means.

Yes, though use of von Neumann (Princeton) architecture allows storing both instructions and data in the same memory which in turn allows lower implementation costs, more flexible memory utilisation, higher design simplicity and regularity, but also brings some unpleasant side effects, like, for example: "Devised by John von Neumann around 1945, von Neumann architecture is insecure by design. Lack of separation of instructions memory from data memory (unlike in Harvard architecture) allows the transfer of control into data space that may contain malicious code". No one at this stage could predict the sheer number of computers today nor the creation of the Internet. In the famous words of Thomas George Watson, then president of IBM: "I think there is a world market for maybe five computers." (https://cybertheory.io/why-are-we-here-and-what-to-do-about-it/).

One may ask: "What is the danger of storing instructions and data in the same memory?." The answer to this question is very simple: this architecture offers very easy way to create malware by hiding malicious code within data and then under certain conditions transferring control to this malicious code. These certain conditions can be multiple ranging from time of the day to the N-th time the program is being executed to the name of the user to multiple other ones. Detection of such malicious code is difficult without explicit and detailed code reviews. Moreover, one can use another level of malicious code hiding by moving it from one area in data to another before control is transferred to this malicious code.

We have used the term malware (malicious software) without introducing it first. So, for the purposes of this discussion, malware is a generic term that encapsulates all threats – viruses, worms, botnets, ransomware, spyware, etc. – anything malicious that is software-related. Good and more formal definition of malware was offered in 2007 by Robin Sharp (https://backend.orbit.dtu.dk/ws/portalfiles/portal/4918204/malware.pdf):

Software which is used with the aim of attempting to breach a computer system's security policy with respect to Confidentiality, Integrity or Availability.

And again, it all goes back to John von Neumann. As the very first modern computers were being built, John von Neumann developed the concept of a program that could reproduce and spread itself throughout a system. Published posthumously in 1966, his work, *Theory of Self-Reproducing Automata* (https://cba.mit.edu/events/03.11.ASE/docs/VonNeumann.pdf), serves as the theoretical foundation for computer viruses.

Five years after John von Neumann's theoretical work was published, a programmer by the name of Bob Thomas created the first proof of John von Neumann's concept – an experimental program called Creeper, designed to move between different computers on the ARPANET, a precursor to the modern Internet. His colleague Ray Tomlinson, considered to be the inventor of email, modified the Creeper program to not only move between computers, but to also copy itself from one to another. Thus, the first computer worm was born, it was built at Bolt Beranek and Newman (BBN) – an American research and development company later acquired by Raytheon.

The history of actual malware starts in 1971 with a worm called "Creeper." Its original version was designed to move between DEC PDP-10 mainframe computers running the TENEX operating system. It is the first known example of a worm. Creeper was a test created to demonstrate the possibility of a self-replicating computer program that could spread to other computers. As a proof of concept, Creeper wasn't made with malicious intent and didn't damage or disrupt the systems it infected, instead only displaying the message: "I'M THE CREEPER: CATCH ME IF YOU CAN." Taking up his own challenge, in 1972 Ray Tomlinson created Reaper, the first antivirus software designed to delete Creeper by similarly moving across the ARPANET.

Contrary to what every non-technical person says, "Macs are not susceptible to viruses," the first computer virus found in the wild, dubbed "Elk Cloner," was designed to target Apple II computers. It was written in 1982 by a then-15-year-old, who wrote such programs to play pranks on his friends.

The very first PC virus, dubbed "Brain" was born in 1986. It changed the information security world as we know it today. It originated in Pakistan but quickly spread worldwide to Europe and North America. Ironically, the virus had replicated from machine to machine because of an anti-piracy countermeasure. The Brain virus was developed by two brothers from Pakistan - Amjad Farooq Alvi and Basit Farooq Alvi.

And then Internet has taken over the world and brought with it new faster and wider-reaching malware distribution methods. However, independently of malware distribution methods, a significant number of types of cyber attacks are conceptually based on the fundamental feature of von Neumann (Princeton) architecture – lack of any discrimination between instructions and data which allows easy substitution/insertion of maliciously behaving components.

Now, back to computer architecture. In April 2021, in a very interesting publication (https://www.thebroadcastbridge.com/content/entry/16767/computer-security-part-5-dual-bus-architecture), John Watkinson elaborated on the benefits of Harvard architecture for cybersecurity. After discussing von Neumann (Princeton) architecture and its evolution into modified Harvard architecture he clearly states that "the true Harvard machine has two" buses.

Then he continues:

> This leads to a more recent and useful definition of the Harvard machine, which is that it always has two address spaces. From a computer security standpoint, that definition is especially meaningful, because it is the separation of address spaces that gives the Harvard architecture an edge in security measures where it allows better executable space protection. Unfortunately, that definition is widely ignored.

The improved executable space protection is very beneficial as it also protects against such things as cache overflows and partially against memory leaks. In the extreme case, like high-security applications, one can load operating system's code in read only memory which prevents on the fly changes to it.

With such executable space protection, the operating system cannot be modified and code cannot be written to it or changed by malware. The opportunity is there to build machines that are practically unhackable and no new technology is needed to create them. All that is needed is the commitment to proceed in that direction. Logic suggests that any IT problem should define what the software needs to do and then hardware needs to be found that will provide an environment for that software.

COMPILERS AND INTERPRETERS

Today most programs are written in a high-level programming language like, for example, C, Perl, or Java. High-level programming languages can be broadly categorised into two types based on how they are processed: compiled and interpreted. Every program is a set of instructions, whether it's to add two numbers or send a request over the internet. Compilers and interpreters take human-readable code and convert it to computer-readable machine code.

Compiled programming languages are those in which the source code is compiled into machine code before it is executed. Machine code is a low-level language that can be executed directly by the computer's CPU. When a program written in a compiled language is compiled, the source code is converted into an executable file that can be run on the target machine. The compiled code is optimised for the specific hardware and operating system of the machine on which it is intended to run.

Compiled languages are converted directly into machine code that the processor can execute. As a result, they tend to be faster and more efficient to execute than interpreted languages. They also give the developer more control over hardware aspects, like memory management and CPU usage. Compiled languages need a "build" step – they need to be manually compiled first. One needs to "rebuild" the program every time one needs to make a change.

Interpreted programming languages, on the other hand, are those in which the source code is executed directly by an interpreter, without being compiled into machine code first. The interpreter reads the source code line by line and executes each line as it is read. The interpreter is responsible for translating the source code into machine code at runtime.

The main difference between compiled and interpreted programming languages is in how they are being processed. Compiled languages are translated into machine code before they are executed, while interpreted languages are translated into machine code at runtime by an interpreter. Both types of languages have their advantages and disadvantages, and the choice between them depends on the specific needs of the project. If one needs maximum speed and performance, one may want to choose a compiled language, while if one needs portability, faster development and ease of use, an interpreted language may be a better choice.

Statistics show that in 2024 (https://www.orientsoftware.com/blog/most-popular-programming-languages/) more than 50% of software is being written using interpreted programming languages, which was not the case 20–30 years ago.

Without going into detailed discussion of compiled and interpreted programming languages, it is obvious that interpreted programming languages offer even more opportunities to create malware.

CONCLUSION AND TAKEAWAYS

In essence ability to create malware is a "design feature" of von Neumann architecture and generations of various computer languages, from assembler to high-level ones. This ability, as you will learn later in the book, has been further enhanced by various networking protocols within TCP/IP family including DNS and BGP (see Chapters 11 and 12).

As Ken Thompson said in his Turing Award Lecture (https://www.cs.cmu.edu/~rdriley/487/papers/Thompson_1984_ReflectionsonTrustingTrust.pdf): "You can't trust code that you did not totally create yourself."

Chapter 4

Complexity is your enemy

It is probably right to start this chapter quoting Richard Branson: "Complexity is your enemy. Any fool can make something complicated. It is hard to make something simple." Humans have a natural tendency to over-complicate things, and it has ever been so, even Confucius knew this as he said "Life is really simple, but men insist on making it complicated."

Steve Jobs once said: "Simple can be harder than complex: You have to work hard to get your thinking clean to make it simple. But it's worth it in the end because once you get there, you can move mountains." It is also worth mentioning here another quote, this time, from Tony Robbins, "Complexity is the enemy of execution." In fact, one can go further and say that complexity is the enemy of success, because if we cannot execute, then we definitely cannot be successful.

Why all these quotes? Because:

- Complexity steals our focus, it's impossible for us to focus on complex things.
- Complexity steals our confidence, and without confidence we seriously impact our ability to be successful.
- Complexity steals our understanding, and without understanding the quality of our solutions and actions is reduced.
- One of the main reasons technology projects fail so often is their (underestimated) complexity.
- The rate of complexity is rapidly increasing with uncontrolled proliferation of new tools, our interconnectedness and globalisation.

One may ask what this has got to do with cybersecurity? The answer is very simple, according to Bruce Schneier (who is famous for his saying: "If something is free, you're not the customer; you're the product") as he has written in his book "Data and Goliath: The Hidden Battles to Collect Your Data and Control Your World": "Complexity is the worst enemy of security, and our systems are getting more complex all the time."

Another quick exercise in time travel and one can easily see how original concept of centralised (mainframe) monolithic computing applications

DOI: 10.1201/9781032672601-4

evolved to partially or almost de-centralised (mini-computers' era) to a mixture of centralised monolithic solutions and desktop applications to a total mess of today's microservices world, when nobody within any organisation can confidently describe (not to say understand end-to-end) organisation's IT ecosystem with numerous Software-as-a-Service (SaaS), Cloud-based, data centre and desktop solutions.

The ever-growing complexity of organisations' IT ecosystems brought us to the current situation where in any reasonably sized organisation nobody has a full and detailed understanding of the ecosystem and its interdependencies. This results in an enormous amount of effort going into maintaining these ecosystems and often results in unpatched software required to maintain interoperability as replacement/upgrade is costly and takes a long time. This also exponentially increases the risk of potential supply chain attacks and makes PKI management even more challenging.

There are two ways one can look at complexity. One is a micro-level, or application-level complexity, the other one is a macro-level, or IT ecosystem level of complexity. Micro-level has been described by Larry Tesler using the law of conservation of complexity as described in the next paragraph. Our focus, however, will be on IT ecosystem level of complexity.

The law of conservation of complexity, also known as Tesler's Law, or Waterbed Theory, is an adage in human–computer interaction stating that every application has an inherent amount of complexity that cannot be removed or hidden. Instead, it must be dealt with, either in product development or in user interaction. This poses the question of who should be exposed to the complexity. For example, should a software developer add complexity to the software code to make the interaction simpler for the user or should the user deal with a complex interface so that the software code can be simple? Larry Tesler argued that, in most cases, an engineer should spend an extra week reducing the complexity of an application vis-à-vis making millions of users spend an extra minute using the program because of the extra complexity. However, Bruce Tognazzini proposed that people resist reductions to the amount of complexity in their lives. Thus, when an application is simplified, users begin attempting more complex tasks.

It is a well-known fact, for example, that software complexity in aerospace systems is increasing exponentially. Source lines of code in aerospace systems are doubling about every four years. That trend has been in place for at least five decades and applies to both commercial and to military aircraft (Source: https://savi.avsi.aero/about-savi/savi-motivation/exponential-system-complexity/#:~:text=Software%20complexity%20in%20aerospace%20systems,commercial%20and%20to%20military%20aircraft). A good example of this is growth of the code size for space missions (https://www.gregorybufithis.com/2024/04/25/a-most-incredible-story-about-software-and-technology/) that shows growth from about 50 lines of code Mariner mission in early 1960s to about 5,000,000 million lines of code for Mars science lab in early 2010s. More proofs of this can be found at https://www.isasi.org/Documents/library/technical-papers/2018/Wed/The%20Growing%20Level%20of%20Aircraft%20Systems%20Complexity%20and%20Software%20Investigation%20-%20Paulo%20Soares%20Oliveria%20Filho.pdf. It is also worth to understand relationship between complexity, usefulness and bugs in the final software product, as after initial growth of all of them, after a certain point usefulness of software starts to decrease with its complexity (https://blog.pdark.de/2012/07/14/software-development-costs/).

This is the right point to remember Frederic P. Brooks, Jr and his famous book "The Mythical Man-Month. Essays on Software Engineering." This book on software engineering and project management was first published in 1975, with subsequent editions in 1982 and 1995. Its central theme is that adding manpower to a software project that is behind schedule delays it even longer. This idea is known as Brooks' law. Brooks's observations are based on his experiences at IBM while managing the development of OS/360. He had added more programmers to a project falling behind schedule, a decision that he would later conclude had, counter-intuitively, delayed the project even further. He also made the mistake of asserting that one project – involved in writing an ALGOL compiler – would require six months, regardless of the number of workers involved (it required longer). The tendency for managers to repeat such errors in project development led Brooks to quip that his book is called "The Bible of Software Engineering, " because "everybody quotes it, some people read it, and a few people go by it."

Brooks' law has been also formulated by Robert Metcalf in 1980. In software development, the importance of effective communication is often overlooked. It's easy to think that software development is a purely technical matter, however, when working in a team, communication becomes a critical component. A linear increase in the number of team members can lead to an exponential increase in communication complexity. This correlation is known as Metcalfe's law and can have a significant impact on team productivity:

Solution to this challenge was proposed by Amazon CEO Jeff Bezos and is known as The Two Pizza Rule: this concept suggests that teams should be small enough to be fed with two pizzas. This keeps the team size manageable and communication channels to a minimum.

We are all used to hearing the phrase "It's not black and white." This phrase is typically used to describe an issue or situation as being complex or having more than one perspective or answer. And, indeed, everything in this world has pros and cons. It is well known that solutions to various problems usually come with various side effects and medication is one of the best examples of this situation.

So, when multiple organisations got sick and tired of complexity, inflexibility ("one size fits all"), high cost and long "time to market" (associated with changes) of monolithic systems, solution called microservices has been invented. Microservices are an architectural and organisational approach to software development where software is composed of small independent services that communicate over well-defined Application Programming Interfaces (APIs). This solution enabled organisations to make changes to individual parts of microservices-based solution faster, cheaper and easier. The explosion of APIs driven by digital expansion has 41% of organisations managing at least as many APIs as applications.

However, as everything is not black and white, microservices-based solutions come with numerous side effects. One of them is potential for uncoordinated development of various subcomponents that in time may (and likely will) result in a messy architecture. But there are two immediate side effects that are worth a discussion.

The first of these two side effects is linked to exponential growth of complexity that is best illustrated in Figure 4.1. When instead of a monolithic solution organisation moves to a microservices-based one, the number of connections explodes. And this has an impact both on supportability (including, but not limited to, patching and testing) and on cybersecurity. In fact, one may argue that similar to Brooks' law moving to microservices architecture does not improve "time to market" in case if organisation wants to perform adequate rigorous testing and maintain its cybersecurity posture.

The other side effect is associated with introduction of APIs that present a new attack surface. At this stage there is insufficient understanding of risks associated with APIs and ability to secure APIs, as even properly configured APIs can be exploited. API security is important because businesses use APIs to connect services and to transfer data, so a hacked API can lead to a data breach. Malicious actors love APIs because they often hold the keys to a lot of valuable information. If not properly secured, APIs can potentially expose

Robert Metcalfe's Law

The value of a network is proportional
to the square of the number of nodes.

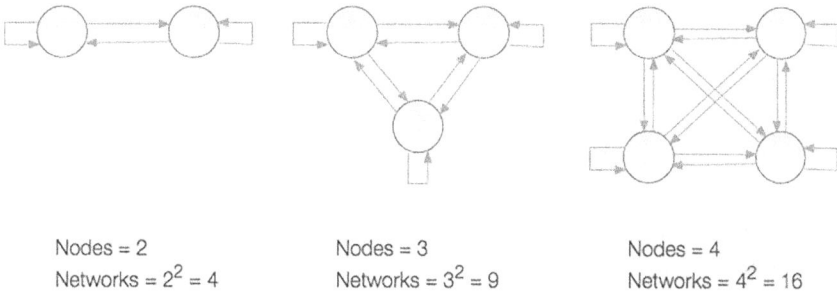

Nodes = 2
Networks = $2^2 = 4$

Nodes = 3
Networks = $3^2 = 9$

Nodes = 4
Networks = $4^2 = 16$

Figure 4.1 Illustration of Metcalf's law which states that the value of the network is proportional to the square of the number of the nodes.

sensitive data. By targeting API endpoints in a distributed denial-of-service (DDoS) attack, malicious actors could significantly disrupt organisation's operations. More importantly, as various components of microservices-based solution evolve, the same happens to APIs and they start living their own life.

The rise of cyberattacks is becoming increasingly concerning with the rise of API breaches. APIs provide digital access between applications, meaning that any breach of security can have a huge impact on organisations and their customers. With this in mind, the potential for API breaches to cause catastrophic damage to the economy is a growing concern that needs to be addressed.

It has been predicted by Gartner that API attacks would be the most common type of cyberattack by 2022 and beyond. This is because applications are getting more APIs, and there are more cloud-based services that can be used to deploy them. For these reasons, over the next few years, API breaches are likely to become much more common and sophisticated. This will affect the security of both organisations and their customers.

The top 10 most common API security risks exploited by malicious actors include:

- **Injection attacks:** Injection attacks occur when malicious code or data is injected into an API request. This can include SQL injection, where an attacker injects SQL code into an API request to gain unauthorised access to a database, and cross-site scripting (XSS), where an attacker injects malicious code into a web page that is accessed through an API.
- **Broken authentication and session management:** APIs that lack proper authentication and session management can be vulnerable to attacks where an attacker can gain unauthorised access to the API. This can include guessing or cracking passwords, stealing session cookies and other forms of identity theft.
- **Insecure communication:** APIs that transmit data over unencrypted connections can be vulnerable to attacks where an attacker intercepts the data and reads or alters it. This can include Man in the Middle (MITM) attacks, where an attacker intercepts the data and reads or alters it, and eavesdropping attacks, where an attacker listens in on the communication between the API and the client.
- **DDoS attacks:** APIs can be vulnerable to DDoS attacks, where an attacker floods the API with a large number of requests in order to overwhelm the server and make the API unavailable.
- **Misuse of API keys:** API keys are unique, secret strings that are provided to authorised users and systems. If these keys are compromised, they can be used to gain unauthorised access to the API.
- **Lack of input validation:** APIs that do not properly validate the data that is sent in requests can be vulnerable to attacks where an attacker sends malicious data in the request.

- **Unvalidated redirects and forwards**: APIs that allow unvalidated redirects and forwards can be vulnerable to attacks where an attacker redirects the user to a malicious website or API.
- **Unvalidated forward input**: APIs that do not properly validate the data that is sent in requests can be vulnerable to attacks where an attacker sends malicious data in the request.
- **Lack of access control**: APIs that do not properly control access to their resources can be vulnerable to attacks where an attacker gains unauthorised access to the API.
- **Lack of monitoring and logging:** APIs that do not monitor and log API requests and responses can be vulnerable to attacks where an attacker uses the API for malicious purposes without being detected.

As one can easily see from this list, APIs are often used as a starting point or a staging area for almost all known types of attacks.

Let's have a brief look at some of the API-related recent major cybersecurity breaches in Australia (a list of 13 biggest data breaches in Australia is available here: https://www.upguard.com/blog/biggest-data-breaches-australia):

- Vinomofo Data Breach – on October 10, 2022, Australian online wine retailer Vinomofo experienced a major data breach, affecting millions of customers. The company reported that the breach potentially exposed user names, emails, phone numbers, addresses and encrypted passwords. There is strong evidence suggesting that it was an API attack.
- Medibank Data Breach – Medibank, one of Australia's largest health insurers, suffered a massive data breach in October 2022. The breach exposed customer data, including full names, addresses, phone numbers, email addresses, dates of birth and bank account details. It seems like malicious actors gained access to the company's customer data through a weak endpoint. The event shows once again how important it is to use strong security protocols and keep an eye on APIs all the time.
- MyDeal Data Breach – Australian online retailer, MyDeal (a Woolworths Group subsidiary), identified on October 14, 2022 that it suffered a data breach that exposed the private information of millions of customers. The breach occurred when malicious actor gained access to an endpoint and stole personal data including names, addresses, payment card numbers and email addresses.

Complexity comes with high cost too. According to Hackett Group's benchmark study on the cost of complexity documented that:

- Organisations with higher-than-average technology complexity spend 25% more than average companies and 58% more than organisations with low complexity.

- The most significant cost factor is the number of applications per end user:
  - Organisations with high numbers of applications employ 27% more FTE's than average companies.
  - World class IT functions support 44% fewer applications per end user than typical organisations.

To illustrate the number of applications per user getting out of control in today's world of Digital Transformations (see Chapter 5) and Cloud computing/SaaS (see Chapters 7 and 8), the author of this chapter can bring to the readers' attention just a couple of recent first-hand experiences:

- Organisation that employed 15,000 staff and contractors with 4,000+ known applications.
- Organisation that employed 300–400 staff and contractors with more applications than employees.

According to Hackett Group, there are many root causes of complexity, including, but not limited to:

- Lack of standards or adherence to standards.
- Outdated, inadequate technology/data architecture.
- Business cases that fail to identify compatibility issues, redundancies or other conflicts.
- Mergers and acquisitions without systems consolidation or integration.
- Rapid growth.
- Lack of system sunsetting or asset management program.
- Deferred maintenance, updates, upgrades.
- Rampant customisations to applications.
- Poor data governance.
- Shadow IT.

But, now back to the complexity of IT ecosystems. What drives growth complexity of IT ecosystems? As many organisations embark on digital transformations (see Chapter 5) and race to adopt new Cloud computing technologies and systems (see Chapter 7 and 8), they quickly discover that with new systems, applications and processes comes monumental new complexity. Current trend of replacing monolithic systems with microservices-based solutions contributes to complexity growth too. And, as we know, complexity increases exponentially with scale.

An organisation's IT ecosystem is the network of services, systems, providers and other organisations connected to the organisation that creates and delivers information technology products and services. This ecosystem includes entities that are connected to but not always controlled directly by the organisation, such as, for example, a third party. Cloud computing resources used by the organisation are also part of its IT ecosystem. All of the assets associated with all of the IT ecosystem entities define the organisation's attack surface and as such size and complexity of organisation's IT ecosystem matters.

Now, let's have a quick look at human's ability to remember. Memory is the ongoing process of information retention over time. The idea of separate memories for short-term and long-term storage originated in the 19th century. A model of memory developed in the 1960s assumed that all memories are formed in one store and transfer to others store after a small period of time. Short-term memory does not have a large capacity and holds information for seconds or minutes. The final storage is long-term memory, which has a very large capacity and is capable of holding information possibly for a lifetime.

Human memory, as considered by cognitive neuroscience, which is a blend of cognitive psychology and neuroscience, is composed of three parts: sensory, short-term and long-term. Sensory memory relays messages to the brain from the senses; iconic memory is relayed by sight, echoic memory by sound and haptic memory by touch. Stimuli received by sensory memory is passed on to short-term memory by attention, meaning that the only things that get saved to sensory memory are the things that we pay attention to. Short-term memory acts as a "scratch pad" for the brain and deals with temporary stimuli. It works in limited frames of time and space, and can be easily interrupted, which is why we become easily distracted and lose our train of thought if we try to do too many things at once. Information that is constantly repeated in short-term memory gets converted to long-term memory, which then stores information for long periods of time and is composed of two types, episodic and semantic. Episodic memory deals with memory of serial experiences and events and is used to construct the way we remember events. Semantic memory is full of "facts, concepts, and skills" that we have learned and gathered over periods of time, and is derived from episodic memory.

Short-term memory (or "primary" or "active" memory) is the capacity for holding a small amount of information in an active, readily available state for a short period of time. For example, short-term memory holds a phone number that has just been recited. The duration of short-term memory (absent rehearsal or active maintenance) is estimated to be on the order of seconds. The commonly cited short-term memory capacity of 7 ± 2 items was discovered by a cognitive psychologist George A. Miller of Harvard University's Department of Psychology and published in 1956 in Psychological Review. It is often interpreted that the number of objects an average human can hold in short-term memory is 7 ± 2. This has occasionally been referred to as Miller's law.

Working memory is one of the most widely used terms in psychology. The term "working memory" was coined in the 1960s by Miller, Galanter and Pribram. Working memory is a cognitive system with a limited capacity that can hold information temporarily. Working memory is the retention of a small amount of information in a readily accessible form. It facilitates planning, comprehension, reasoning and problem-solving. It is important for reasoning and the guidance of decision-making and behaviour. Working memory is often used synonymously with short-term memory, but some theorists consider the two forms of memory distinct, assuming that working memory allows for the manipulation of stored information, whereas short-term memory only refers to the short-term storage of information. Working memory is a theoretical concept central to cognitive psychology, neuropsychology and neuroscience.

Whether we accept that working memory and short-term memory are the same, the well-accepted fact is that working memory is the small amount of information that can be held in mind and used in the execution of cognitive tasks. It is often connected or related to intelligence, information processing, executive function, comprehension, problem-solving and learning in people. Some researchers argue that is not really about storage or memory per se, but about the capacity for controlled, sustained attention. Without going too deeply into cognitive neuroscience we must acknowledge that our attention is limited and that our memory is imperfect.

Having this in mind, let's have a look at what does this mean for our ability to grasp, understand and manage complex IT ecosystems. Let's have a look at one of the real examples that have been witnessed first-hand by the author of this chapter. Consider organisation that employs approximately 15,000 staff and contractors with IT department of a bit over 160 staff and contractors, some of whom have been with the organisation for up to 30 years. This organisation uses more than 4,000 applications. How many connections (not counting for network infrastructure components, domains, subdomains, websites) are there? If we use on average 4–5 connections per application, this amounts at least to 15,000–16,000 connections. Is there a human being who can understand this complexity? Not surprisingly, nobody in the IT department of this organisation had full end-to-end visibility and understanding of IT ecosystem. It is important to note that this organisation went through a series of M&As and divestments. Partially as a result of this, or partially as a result of complexity of IT ecosystem, majority of documentation is not up to date and significant portions of documentation were simply non-existent. As such, the next question is – can this IT ecosystem be secured and if yes, what is the level of confidence that it is secured at any given point in time? Can any level of Essential 8 maturity (see Chapter 15) be achieved in such environment? Can ISO/IEC 27001 (see Chapter 14) be meaningfully implemented in such environment? The answer to these three questions is no and the reason for such an answer is very simple – it's complexity.

According to Zachary Ginsburg, Senior Director, Research of Gartner Risk & Audit Practice, "The two new emerging risks relate to complexities of the IT and political environment made highly visible to executives and boards by current events."

When we are talking about complexity, we should not forget complexity of DNS (see Chapter 12).

One more thing to remember – the bigger one's organisation's IT ecosystem, the higher is the supply chain risk (see Chapter 9).

To increase our success, we need to increase simplicity, reduce complexity and the clutter that surrounds our daily life. It is possible, but to do so simplicity should be our starting point not our goal. We cannot engineer simplicity into complex situations so easily. It's much easier to start simple and try to and avoid complexity as much as possible.

We need to have a simplicity first mindset, that is the only way we will end up with simplicity.

One of the challenges is that simplicity is more difficult to achieve than complexity, but if we deal with the difficulty at the start we will profit from simplicity at the end. **If we give in to complexity at the start then we will be paying for it forever!**

Chapter 5

Digital revolution and its consequences

As these lines are being typed in the world keeps struggling with the consequences of a massive world-wide IT outage caused by issues with CrowdStrike software that has caused havoc with computer systems around the world. Computer systems across Australia and overseas have failed in the afternoon after an update was pushed out by global security software provider CrowdStrike.

Reports of the outage in Australia began flooding in about 3 pm AEST on Friday July 19, 2024. Airport check-in systems have been disrupted and businesses have reported the "blue screen of death" and IT outages. Outages hit banks and payment systems, forcing some supermarkets and petrol stations to close.

The problems have emerged across the world, but were first noticed in Australia, and possibly felt most severely in the air travel industry, with more than 3,300 flights cancelled globally. These are just some of the examples showing the impact of this event:

- UK airports saw delays, with long queues at London's Stansted and Gatwick.
- Ryanair said it had been "forced to cancel a small number of flights today (19 July)" and advised passengers to log-on to their Ryanair account, once it was back online, to see what their options are.
- British Airways also cancelled several flights.
- Several US airlines, notably United, Delta and American Airlines, grounded their flights around the globe for much of Friday; Delta has lost track of some crew members and is asking flight crews to report to the airline their location.
- Australian carriers Virgin Australia and Jetstar also had to delay or cancel flights.
- Airports in Tokyo, Amsterdam and Delhi were also impacted.
- Railway companies, including Britain's biggest which runs Southern, Thameslink, Gatwick Express and Great Northern, warned passengers to expect delays.

DOI: 10.1201/9781032672601-5

- In Alaska, the 911 emergency service was affected, while Sky News was off air for several hours on Friday morning, unable to broadcast.
- Payment and payroll systems disruptions: there were reports that cloud accounting software Xero was caught up in the crisis (Xero services more than four million businesses in Australia, New Zealand and the United Kingdom).

Airport check-in systems across the globe have been disrupted, while banks, supermarkets and media companies, healthcare providers are among the other businesses reporting the "blue screen of death" and network outages. The big four Australian banks, Telstra and major media organisations including the ABC and Foxtel have had services go offline. Customers are not able to use EFTPOS to pay for goods and services in many businesses. The number of businesses that have ceased operation is staggering. Major airlines, banks, shops and many other businesses have been forced to suspend trading or providing services. Thousands of people were stranded at airports around the world on a Friday evening.

Many experts call it one of (if not) the largest IT failure(s) in history. Conservatively the cost of this outage is estimated as at least A$1.5 billion in losses, A$200m in damages in NSW alone (https://www.news.com.au/finance/money/costs/crowdstrike-global-it-outage-led-to-at-least-15-billion-in-losses-200m-in-damages-in-nsw-alone/news-story/04f58103463d97d20ecd1e8009bf6181). According to (https://www.itnews.com.au/news/insured-losses-from-crowdstrike-outage-could-reach-us15-billion-610122?eid=1&edate=20240726&utm_source=20240726_AM&utm_medium=newsletter&utm_campaign=daily_newsletter) global insured losses from last week's massive IT outage are likely to range from US$400 million to US$1.5 billion.

The Australian government said the outage is not the result of a cybersecurity incident but have been caused by a CrowdStrike update (an update to one of CrowdStrike's pieces of software, Falcon Sensor). The prime minister says there has been no impact to critical infrastructure in Australia, such as 000 services and core emergency services, whilst outages continue to impact health services and air travel around the world.

Having said that this massive world-wide IT outage is not the result of a cybersecurity incident, it is important to understand what consequences a major cybersecurity incident can have worldwide. Humanity has just got the first-hand experience what consequences worldwide a major cybersecurity incident may have.

The problem appears to have been caused by a software update gone wrong. A newly released version of CrowdStrike's cybersecurity software (which is anecdotally designed to protect Microsoft Windows devices from malicious attacks) reportedly caused Windows computers to crash and display a "blue screen of death" – a standard error screen that happens when the operating system cannot load correctly.

CrowdStrike's CEO says a defect in a recent update for Windows hosts has been identified and a fix has been deployed but some systems could be down for "some time." Microsoft has said apps and services are still experiencing residual impacts and that it is taking "mitigation action" to deal with "the lingering impact" of the outage.

How was Microsoft involved? When Windows computers everywhere started to crash with a "blue screen of death" message, early reports stated the IT outage was caused by Microsoft. In fact, Microsoft confirmed it experienced a cloud services outage in the Central United States region, which began around 6 pm Eastern Time on Thursday, July 18, 2024. This outage affected a subset of customers using various Azure (Microsoft's proprietary cloud services platform) services.

The Azure outage had far-reaching consequences, disrupting services across multiple sectors, including airlines, retail, banking and media. Not only in the United States but also internationally in countries like Australia and New Zealand. It also impacted various Microsoft 365 services, including PowerBI, Microsoft Fabric and Teams.

As it has now turned out, the entire Azure outage could also be traced back to the CrowdStrike update. In this case it was affecting Microsoft's virtual machines running Windows with Falcon installed.

The world was lucky that only MS devices and services were impacted and none of Unix/Linux derivatives were. Otherwise, the impact would have been much worse.

However, the process of fixing affected computers might be very time-consuming. CrowdStrike advised customers that an affected machine needs to be booted into "safe mode," and then a specific file will need to be deleted. This process is likely to need to be done manually, so there is no easy fix that can be applied to many machines at once.

For those who haven't heard about CrowdStrike, it is a US firm based in Austin, Texas. It is listed on NASDAQ (NASDAQ: CRWD), featuring in both the S&P 500 and the high-tech NASDAQ indexes. Since its inception in 2011, the company has grown rapidly as it began to offer a range of security services using cloud-based software. Like a lot of modern technology companies, it hasn't been around that long. Today it employs nearly 8,500 people around the globe serving about 29,000 customers (https://www.cbsnews.com/news/what-is-crowdstrike-global-microsoft-outage/) in over than 170 countries. As a provider of cyber-security services, it tends to get called in to deal with the aftermath of hack attacks. In the past, it has been involved in investigations of several high-profile cyberattacks, such as when Sony Pictures had its computer system hacked in 2014. Forrester named CrowdStrike a "Leader" in The Forrester Wave: Endpoint Security, Q4 2023. Gartner ranked CrowdStrike #1 in the Market Share: Managed Security Services, Worldwide, 2022 report for Managed Detection and Response (MDR) Market Share for the second consecutive year. As at December 2023, CrowdStrike was the leader in Gartner's "Magic Quadrant."

To a degree this scenario has been predicted as early as in 2003 when, Dan Geer (whom I was privileged to meet at 1989 USENIX in San Diego, CA) and his colleagues (p. 48, https://www.tuhs.org/Archive/Documentation/AUUGN/AUUGN-V24.4.pdf) pointed to the negative impacts on cybersecurity caused by software monopolies like Microsoft.

Now, let's talk about "5 Whys" technique.

This technique has been developed in the 1930s by Sakichi Toyoda, the Japanese industrialist, inventor and founder of Toyota Industries, and became popular in the 1970s, and Toyota, as well as many other organisations (e.g., https://integratedcare.nnswlhd.health.nsw.gov.au/5-whys/), still use it to solve problems today.

Toyota has a "go and see" philosophy. This means that its decision-making is based on an in-depth understanding of what's actually happening on the shop floor, rather than on what someone in a boardroom thinks might be happening. The "5 Whys" technique is true to this tradition, and it is most effective when the answers come from people who have hands-on experience of the process or problem in question.

The method is remarkably simple: when a problem occurs, you drill down to its root cause by asking "Why?" five times (good example of using "5 Whys" is shown here https://en.wikipedia.org/wiki/Five_whys).

As recovery from this major world-wide IT outage continues, much of the discussion is focused on CrowdStrike, human error or inadequate/insufficient testing. And this is the 1st why that obviously does not give the reader the root cause.

The next question one may ask is why so many organisations suffered the consequences of this massive world-wide IT outage? The answer to this question is obvious: because they all are users of CrowdStrike. This is the 2nd why that obviously again does not give the reader the root cause.

The next question one may ask is why so many organisations use CrowdStrike? The answer to this question is also obvious: because CrowdStrike is the market leader, as confirmed by Gartner and Forrester. This is the 3rd why that obviously again does not give the reader the root cause.

The next question one may ask is why so many organisations need to use CrowdStrike? The answer to this question is obvious and has two aspects: because they all heavily rely on IT in conducting their business and because of exponential growth of cybersecurity threats and incidents (see Andrew Jenkinson "Ransomware and Cybercrime," CRC Press, 2022 for example). This is the 4th why that obviously again does not give the reader the root cause.

The next question one may ask is why modern organisations, both in private sector and various governments, are so heavily reliant on IT? And answer to this 5th why finally gives us the root cause: it is so-called **digital transformation**.

Now, let's explore where did the term "Digital Transformation" come from and what is it really supposed to mean? What is meant by digital transformation?

The term "Digital Transformation" was coined in 2011 by the consulting firm Capgemini (in partnership with the MIT). They defined the phrase as: **"the use of technology to radically improve performance or the reach of businesses."**

In short, digital transformation has emerged as the compass guiding organisations towards a more agile, efficient and customer-centric future. Digital transformation is a complex process that involves integrating new technologies into an organisation's operations and requires a holistic shift in how organisations work and communicate.

According to Deloitte (https://whatfix.com/digital-transformation/#:~: text=Digital%20transformation%20is%20the%20process,operated% 2C%20and%20how%20value%20is), "digital transformation is all about becoming a **digital enterprise - an organisation that uses technology to continuously evolve all aspects of its business models** (what it offers, how it interacts with customers and how it operates)."

Digital transformation is about evolving one's business by **experimenting with new technologies** and rethinking one's current approach to common issues. Because it's an evolution, digital transformation doesn't necessarily have a clear end point. The MIT Sloan Management Review, a publication that focuses on how management transforms in the digital age, says, "Digital transformation is better thought of as continual adaptation to a constantly changing environment." MIT Sloan Management Review highlights three key areas of digital transformation for enterprises:

1. Customer Experience – working to understand customers in more detail, using technology to fuel customer growth and creating more customer touchpoints.
2. Operational Processes – improving internal processes by leveraging digitisation and automation, enabling employees with digital tools and collecting data to monitor performance and make more strategic business decisions.
3. Business Models – transforming the business by augmenting physical offerings with digital tools and services, introducing digital products and using technology to provide global shared services

So, **digital transformation is the process of using digital technologies to transform existing traditional and non-digital business processes and services, or creating new ones,** to meet the evolving market and customer expectations, thus completely altering the way businesses are managed and operated, and how value is delivered to customers. Digital transformation is the **process by which organisations embed technologies across their businesses** to drive fundamental change.

Effectively, **digital transformation is the integration of digital technology into all areas of a business,** fundamentally changing how one operates and delivers value to customers. It's also a cultural change that requires organisations to continually challenge the status quo, experiment and get comfortable with failure. Digital transformation is the process of adoption and implementation of digital technology by an organisation in order to create new or modify existing products, services and operations by the means of translating business processes into a digital format (https://en.wikipedia.org/wiki/Digital_transformation).

As technology evolves, so should the business. At this point, it's not about organisations choosing to transform, it is more about deciding how to transform. **This approach is effectively based on ongoing escalation of organisation's reliance on IT.**

We often hear the saying "The Road to Hell is Paved with Good Intentions." According to The Phrase Finder (phrases.org.uk), the expression is often attributed to the Cistercian abbot Saint Bernard of Clairvaux (1090–1153), but that provenance is suspect given that the earliest reference to Saint Bernard saying this is in a work written almost 500 years later. This saying is more than true today, especially, for digital transformations.

There are no doubts that digital transformations aspire to achieve multiple things from improving customer experience and customer satisfaction, to minimising waste productivity and reducing costs, to increasing efficiency and profitability of organisations. While the ROI of digital transformation depends on a variety of factors, the right technology can greatly improve organisation's business functions and how customers engage with it:

1. Increases productivity while reducing labour costs – Using technology to work more efficiently is one of the most impactful ways to transform the business. For example, for enterprises, the time and money they spend training new employees and updating digital resources can quickly get out of hand. With the proper tools, organisation can keep costs down and productivity up.
2. Improves the customer experience – Tech-savvy customers want a great experience through multiple touchpoints – mobile apps, social media, email, live chat, etc. Digital transformations are the driving force behind improved customer experiences.
3. Drives innovation, keeping you ahead of your competition – competitors are looking into digital transformation regardless of whether or not you are. Choosing not to embrace digital transformation is essentially deciding that you don't mind being left behind.

But here the focus will be not on the positives, but rather than on negatives and unintended consequences of digital transformations.

So, what are the pitfalls and problems that digital transformation brings to organisations?

The original definition of digital modernisation emphasised a goal of **performance improvement or business development *enabled by technology, not*** *technology itself being the goal*. But in many cases (especially in public sector organisations), labelling a strategic plan as a digital transformation (or sometimes, digital modernisation) has diluted the concept of a clear and compelling goal, and has created a **focus on the wrong thing: the technology**. The term "Digital Transformation" is dangerous because it sends the wrong message.

Focus on technology is dangerous, especially if this comes from the CEO. In the past this has often created an "inflight magazine syndrome" when a CEO saw some add in an inflight magazine and then pushed CIO to implement this particular technology without any analysis of its actual suitability and impact on the organisation. Lately this has been more in the area of vendor (or lobbyist for public sector) pitching.

Even if the CEO is not pushing any particular technology, the overall mood of digital transformations creates desire for innovation within their direct reports and further down the food chain and this, in its own turn, has even higher probability of implementation of various technologies without full understanding of its impact.

As a result, one can often observe the following symptoms:

1. Lack of strategic approach and subsequently lack of centrally managed business and enterprise architecture resulting in a piecemeal.
2. Proliferation of shadow IT, when in many cases centralised IT does not have visibility and knowledge of all technology components (especially SaaS solutions bought on a credit card) used within the organisation.
3. Extremely high level of IT landscape complexity (see Chapter 4).

These symptoms create an untenable situation for any CISO that is not able under these circumstances to adequately assess all cybersecurity risks and subsequently develop an adequate risk mitigation strategy and approach. This becomes even more difficult in the case (which is almost always the case) of organisations that do not articulate and document their risk appetite.

If in the past these symptoms have been observed mainly in large multidivisional organisations (say, over 3,000 FTEs), today one can observe these symptoms even in smaller (just 200–300 FTEs) organisations.

Let's now use "5 Whys" technique again to find the roots of digital transformations and we will get to insatiable desire to grow the business by all possible (and impossible) means, or, effectively, greed, that is further exacerbated by short term focus (this quarter or this financial year), that has become prevalent over the last 15–20 years.

CrowdStrike incident, or better to say its consequences, prompted (at last) some rational thinking (https://medium.com/@matt.he.wanders.on/fragility-in-complexity-insights-from-the-bronze-age-collapse-and-the-2024-crowdstrike-outage-3a7faf91bb03).

Glasgow Caledonian University "smart technology" expert Matthew Anderson said: "We need to shift our focus from prioritising efficiency at all costs to balancing efficiency with resilience."

In particular, one of the consequences of digital transformations is the reality of supply chains becoming heavily dependent on technology (https:// theconversation.com/a-global-it-outage-brought-supply-chains-to-their-knees-we-need-to-be-better-prepared-next-time-235124).

Advanced IT systems now enable real-time tracking, inventory management and seamless communication across global supply chains. This has made them more efficient, transparent and responsive. But to achieve such precision and speed, they've also become highly interdependent. Making supply chains operate efficiently hinges on the timely success of everyone – and all the technology – involved. We've now seen just how quickly things can come undone and the question now is (and has actually been for a while!) **when, not if, the next global IT outage will occur.**

It is worth mentioning that another part of so-called digital revolution resulted in rapid proliferation of QR codes in general (from restaurant menus to registration at the entry point of many organisations) and QR codes embedded in PDF created additional attack vector that is not easy to defend against.

It is clear that IT became a victim of its own success.

The weak points are the millions of components, nodes, networks and pieces of software used today by the vast majority of organisations. Modern IT infrastructure is highly interconnected – think of possible network disruptions and DNS attacks (see Chapter 12) – and interdependent. If one component fails, it can lead to a situation where the failed component triggers a chain reaction that impacts other parts of the system (https://www. innovationaus.com/one-small-update-crippled-millions-of-it-systems-its-a-timely-warning/). "The catalogue of possible causes reads like the script of a disaster movie," said Tuffley.

Now, think of cyberattack. Terror attack. Sabotage. "Many organisations rely on the same cloud and SaaS (Software as a Service) providers and cyber security solutions. The result is a form of **digital monoculture,**" Tuffley explained. "Modern IT infrastructure is highly interconnected and interdependent. If one component fails, it can lead to a situation where the failed component triggers a chain reaction that impacts other parts of the system."

According to Zachary Ginsburg, Senior Director, Research of Gartner Risk & Audit Practice:

> Beyond politics, other global events, such as the July CrowdStrike outage, have raised questions about whether organisations **over-rely on their largest IT vendors.** For example, customers with a concentration of services with one vendor may face elevated risk in the event of outages, or they may face unanticipated changes in services depending on

new regulations or legal decisions in the EU, U.S. or elsewhere. Because third parties, like SaaS vendors, rely on other vendors, organisations may not realise the full extent of their exposure.

Unfortunately, people keep adding more and more extensions to the castle that has been built on quick sand of von Neumann architecture (see Chapter 3), TCP/IP (see Chapter 11, Figure 11.2), DNS and BGP (see chapter 12) using unsafe methods (see Chapters 6 and 9) and materials with not fully understood characteristics (see Chapters 7 and 8).

So, now that we had a wake-up call, it is the right time to urgently review how organisations rely (end-to-end) on technologies and how they can increase resilience of their operations in the wake of a major technology outage.

Chapter 6

Project management methodologies

A project is a temporary and unique endeavour designed to produce a product, service or result with a defined beginning and end (usually time-constrained, and often constrained by funding or staffing) undertaken to meet unique goals and objectives, typically to bring about beneficial change or added value. The temporary nature of projects stands in contrast with business as usual (BAU or operations), which are repetitive, permanent or semi-permanent functional activities to produce products or services. In practice, the management of such distinct production approaches requires the development of distinct technical skills and management strategies.

Project management is the process of supervising the work of a team to achieve all project goals within the given constraints. This information is usually described in project documentation, created at the beginning of the development process. The primary constraints are scope, time, and budget. The secondary challenge is to optimise the allocation of necessary inputs and apply them to meet pre-defined objectives.

Project management is the application of knowledge, skills, tools, and techniques to project activities to meet project requirements. It's the practice of planning, organising and executing the tasks needed to turn a brilliant idea into a tangible product, service or deliverable.

The objective of project management is to produce a complete project which complies with the organisation's objectives. In many cases, the objective of project management is also to shape or reform the project brief to feasibly address organisation's objectives. Once the organisation's objectives are established, they should influence all decisions made by people involved in the project – for example, project managers, designers, contractors and subcontractors. Ill-defined or too tightly prescribed project management objectives are usually detrimental to decision-making.

Project management has a very long history, going back to 2750–2550 BC and Egyptian pyramids, the ziggurats of Mesopotamia, Pharos of Alexandria, cities of the Indus Valley civilisation, the Acropolis and Parthenon in ancient Greece, the aqueducts, Via Appia and Colosseum in the Roman Empire,

DOI: 10.1201/9781032672601-6

Teotihuacán, the cities and pyramids of the Mayan, Inca and Aztec Empires. Up until 1800, projects were mainly in the civil engineering space and were generally managed by creative architects, engineers and master builders themselves. The first railway line in the world dates back to 1825, when George Stephenson connected the towns of Stockton and Darlington in England by rail.

As a discipline, project management developed from several fields of application including civil construction, engineering and heavy defence activities. Two forefathers of project management are Henry Gantt, called the father of planning and control techniques, who is famous for his use of the Gantt chart as a project management tool (alternative method of scheduling work named Harmonogram was proposed by Karol Adamiecki in 1903 and made a sensation) and Henri Fayol for his creation of the five management functions that form the foundation of the body of knowledge associated with project and program management.

The 1950s marked the beginning of the modern project management era, where core engineering fields came together to work as one. Project management became recognised as a distinct discipline arising from the management discipline with the engineering model. Prior to the 1950s projects in the United States were managed on an ad-hoc basis, using mostly Gantt charts and informal techniques and tools. At that time, two mathematical project-scheduling models were developed. The critical path method (CPM) was developed as a joint venture between DuPont Corporation and Remington Rand Corporation for managing plant maintenance projects. The program evaluation and review technique (PERT), was developed by the US Navy Special Projects Office in conjunction with the Lockheed Corporation and Booz Allen Hamilton as part of the Polaris missile submarine program.

At the same time, as project-scheduling models were being developed, technologies for project cost estimation, cost management and engineering economics were evolving. In 1956, the American Association of Cost Engineers (now AACE International – the Association for the Advancement of Cost Engineering) was formed by early practitioners of project management and the associated specialties of planning and scheduling, cost estimating and project control. AACE continued its pioneering work and in 2006, released the first integrated process for portfolio, program and project management (total cost management framework).

There's an old saying in software development that goes something like, "Fast, good or cheap - pick any two." Known as the iron triangle, project management triangle or triple constraint, this concept is familiar to anyone who has ever felt the pressure of weighing the opposing forces of quality, speed and cost against each other (Figure 6.1). Today, the basic precepts of project management are represented by the project triangle, a symbol popularised by Harold Kerzner in his landmark work: "Project Management: A Systems Approach to Planning, Scheduling, and Controlling."

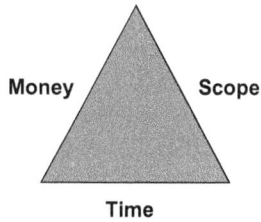

Figure 6.1 The Project Triangle.

When we talk about project management, we should remember that according to https://www.projectsmart.co.uk/history-of-project-management/brief-history-of-project-management.php, we should start back before 2570 BC – the year when the great pyramid of Giza was allegedly completed. Evolution of project management and complexity/scale of projects can be illustrated with this list of some major projects, events and methodologies mapped against the timeline:

- 2570–2550 BC – The Great Pyramid of Giza completed. The Pharaohs built the pyramids, and today, archaeologists still argue about how they achieved this feat. Ancient records show there were managers for each of the four faces of the Great Pyramid, responsible for overseeing their completion. We know there was some degree of planning, execution and control involved in managing this project.
- 208 BC – Construction of the Great Wall of China. Qin Shi Huang, the first emperor of a unified China under the Qin Dynasty (221–206 BC), built another wonder of the world. The emperor ordered millions of people to finish this project. According to historical data, the labour force was organised into three groups: soldiers, ordinary people and criminals. Unfortunately, a vast majority of the wall's ancient remains have long since disappeared due to materials being used for other constructions, erosion, theft and vandalism. The part we mainly see today is the Ming Dynasty section which lasted from 1368 to 1644. However, this section stretches for approximately 5,500 miles in itself, so it is easy to see why this is such a notable feat in engineering. The original structure stretched for 13,170 miles and is the world's longest wall and largest piece of ancient architecture. It stretches between the beaches of Qinhuangdao, the rugged mountains around Beijing, to a desert corridor between the mountain ranges at Jiayu Pass.
- 112 AD – The Aqueduct of Segovia. Aqueducts were watercourses designed to provide fresh drinking water and water for use in public baths. The Aqueduct of Segovia in Spain is an amazing structure built by the Roman Empire and is still, to this day, remarkably intact.
- 1883 – Brooklyn Bridge. New York City is one of the most iconic locations in the world with more famous landmarks. This suspension

roadway bridge connects Brooklyn and Manhattan and was one of the first of its kind to be constructed in the United States. It was designated a National Historic Civil Engineering Landmark in 1972. The Roeblings family are largely responsible for the conception, design, development and its execution. John Augustus Roebling had already overseen the construction of several other suspension bridges. However, it was his son Washington and his wife Emily who eventually brought that vision to fruition as sadly John passed away before the project was completed.

- 1903 – Harmonogram. One of the earliest known methods of scheduling work was invented in Poland by an engineer named Karol Adamiecki (1866–1933). His invention – the Harmonogram – led to increases in output between 100% and 400% in metal rolling mills, in machine shops, in chemical plants, in agriculture and in mining. The Harmonogram is known to have made a sensation in 1903 when Adamiecki first described it and the results of its application before the Society of Russian Engineers in Ekaterinoslaw.

- 1904 – Panama Canal. The Panama Canal is a 48-mile-long waterway that connects the Pacific and Atlantic Oceans. It was one of the most ambitious and complex civil engineering projects to have ever been completed. This valuable project still remains vital for passage of all types of vessels, including the important cargo ships. Interesting that the first proposal recorded for the construction of the Panama Canal was in 1534, when Charles V of Spain ordered surveys as he wanted to make it easier for travel between Spain and Peru. This had a tactical purpose as he knew he would gain military advantages over his Portuguese enemies.

- 1917 – The Gantt chart Developed by Henry Gantt (1861–1919). One of the forefathers of project management, Henry Gantt, is best known for creating his self-named scheduling diagram, the Gantt chart. It was a radical idea and an innovation of worldwide importance in the 1920s. Gantt charts are still in use today and form an essential part of the project managers' toolkit.

- 1917–1928 – The Ford River Rouge complex (commonly known as the Rouge complex, River Rouge or The Rouge) is a Ford Motor Company automobile factory complex located in Dearborn, Michigan, along the River Rouge, upstream from its confluence with the Detroit River at Zug Island. It measures 1.5 miles (2.4 km) wide by 1 mile (1.6 km) long, including 93 buildings with nearly 16 million square feet (1.5 km^2) of factory floor space. With its own docks in the dredged Rouge River, 100 miles (160 km) of interior railroad track, its own electricity plant and integrated steel mill, the titanic Rouge was able to turn raw materials into running vehicles within this single complex, a prime example of vertical-integration production. This complex turned coal, iron ore, limestone, rubber and sand into iron, steel, tires, glass and finished automobiles.

- 1931–1936 – Hoover Dam Project. This Project was one of first projects that used Gantt chart.
- 1942 – V-2 Rocket. V-2 rocket, German ballistic missile of World War II, the forerunner of modern space rockets and long-range missiles. Developed in Germany from 1936 through the efforts of scientists led by Wernher von Braun, it was first successfully launched on October 3, 1942.
- 1942–1946 – Manhattan Project. The Manhattan Project was a research and development program undertaken during World War II to produce the first nuclear weapons. It was led by the United States in collaboration with the United Kingdom and Canada. From 1942 to 1946, the project was directed by Major General Leslie Groves of the US Army Corps of Engineers. Nuclear physicist J. Robert Oppenheimer was the director of the Los Alamos Laboratory that designed the bombs. The Army program was designated the Manhattan District, as its first headquarters were in Manhattan. This name gradually superseded the official codename Development of Substitute Materials. The project absorbed its earlier British counterpart, Tube Alloys, and subsumed the program from the American civilian Office of Scientific Research and Development. The Manhattan Project employed nearly 130,000 people at its peak and cost nearly US$2 billion (equivalent to about $27 billion in 2023), over 80% of which was for building and operating the plants that produced the fissile material. Research and production took place at more than 30 sites across the United States, the United Kingdom and Canada.
- 1956 – The American Association of Cost Engineers (now AACE International) formed. Early practitioners of project management and the associated specialities of planning and scheduling, cost estimating, cost and schedule control formed the AACE in 1956. It has remained the leading professional society for cost estimators, cost engineers, schedulers, project managers and project control specialists since
- 1957 – The CPM invented by a joint venture between DuPont Corporation and Remington Rand Corporation. This technique was used to predict project duration by analysing which sequence of activities has the least amount of scheduling flexibility. It was designed to address the complex process of shutting down chemical plants for maintenance, and then with the maintenance completed, restarting them. The technique was so successful it saved the DuPont Corporation $1 million in the first year of its implementation.
- 1958 – PERT invented for the US Navy's Polaris Project. During the Cold War, the US Department of Defense's (DOD) US Navy Special Projects Office developed PERT as part of the Polaris mobile

submarine-launched ballistic missile project. PERT is a method for analysing the tasks involved in completing a project, especially the time needed to complete each task and identifying the minimum time required to complete the total project.

- 1962 – DOD mandated the work breakdown structure (WBS) approach. The DOD created the WBS concept as part of the Polaris mobile submarine-launched ballistic missile project. After completing the project, the DOD published the work breakdown structure it used and mandated the following of this procedure in future projects of this scope and size. WBS is an exhaustive, hierarchical tree structure of deliverables and tasks that need to be performed to complete a project. Later adopted by the private sector, the WBS remains one of the most common and valuable project management tools.
- 1964 – SR-71 "Blackbird." The Lockheed SR-71 "Blackbird" is a long-range, high-altitude, Mach 3+ strategic reconnaissance aircraft developed and manufactured by the American aerospace company Lockheed Corporation. First flight – December 22, 1964.
- 1965 – The International Project Management Association (IPMA) founded. IPMA was the world's first project management association, started in Vienna by a group as a forum for project managers to network and share information. Registered in Switzerland, the association is a federation of about 70 national and internationally oriented project management associations. Its vision is to promote project management and to lead the development of the profession.
- 1969 – Project Management Institute (PMI). PMI was launched to promote the Project Management Profession. PMI was founded by five volunteers as a non-profit professional organisation dedicated to advancing project management practice, science and profession. Articles of Incorporation for PMI were filed in Pennsylvania (signed by five persons, who are officially recognised as the founders of PMI - James Snyder, Eric Jenett, Gordon Davis, E.A. "Ned" Engman and Susan C. Gallagher) in 1969, which signified its official start. PMI held its first symposium in Atlanta, Georgia, attended by 83 people during that same year. Since then, the PMI has become best known as the publisher of "A Guide to the Project Management Body of Knowledge (PMBOK)," considered essential tool in today's project management profession. The PMI offers two levels of project management certification, Certified Associate in Project Management (CAPM) and Project Management Professional (PMP).
- 1969 – Concorde. Concorde is the first passenger supersonic airliner jointly developed and manufactured by Sud Aviation (later Aérospatiale) and the British Aircraft Corporation (BAC). Studies started in 1954, and France and the United Kingdom signed a treaty establishing the development project on November 29, 1962, as the programme cost was estimated at £70 million (£1.68 billion in 2023). Construction of the six prototypes began in February 1965, and the first flight took off from Toulouse on March 2, 1969.

- 1969 – First Moon landing. Apollo 11 was a spaceflight conducted by the United States from July 16 to July 24, 1969. It marked the first time in history that humans landed on the Moon. Commander Neil Armstrong and Lunar Module Pilot Buzz Aldrin landed the Apollo Lunar Module Eagle on July 20, 1969, at 20:17 UTC, and Armstrong became the first person to step onto the Moon's surface 6 h and 39 min later, on July 21 at 02:56 UTC. Aldrin joined him 19 min later. They spent about two and a quarter hours together exploring the site they had named Tranquility Base upon landing. Armstrong and Aldrin collected 47.5 pounds (21.5 kg) of lunar material to bring back to Earth as pilot Michael Collins flew the Command Module Columbia in lunar orbit, and were on the Moon's surface for 21 h, 36 min, before lifting off to rejoin Columbia. Apollo 11 was launched by a Saturn V rocket from Kennedy Space Center on Merritt Island, Florida
- 1975 – PROMPTII Method (Project Resource Organisation Management Planning Techniques) created by Simpact Systems Limited. Development of PROMPTII was in response to an outcry that computer projects were overrunning on time estimated for completion and original budgets as set out in feasibility studies. It was not unusual to experience factors of double, treble or even ten times the original estimates. PROMPTII was an attempt to set down guidelines for the stage flow of a computer project. In 1979, the UK Government's Central Computing and Telecommunications Agency (CCTA) adopted the method for all information systems projects.
- 1975 – "The Mythical Man-Month: Essays on Software Engineering" by Fred Brooks was published. In his classic book on software engineering and project management, Fred Brooks' central theme is that "Adding manpower to a late software project makes it later." This idea is called Brooks' law. The extra human communications needed to add another member to a programming team is more than anyone expects. It naturally depends on the experience and sophistication of the human programmers involved and the quality of available documentation. Nevertheless, no matter how much experience they have, the extra time discussing the assignment, commitments and technical details and evaluating the results becomes exponential as more people get added. These observations are from Brooks' experiences while managing the development of OS/360 at IBM.
- 1981 – Space Shuttle. The first mission of the space transport system (STS-1) or Space Shuttle, flew on April 12, 1981.
- 1984 – Theory of Constraints (TOC) introduced by Dr Eliyahu M. Goldratt in his novel "The Goal" (North River Press, 1984). TOC is an overall management philosophy that is geared to help organisations continually achieve their goals. The title comes from the view that any manageable system is limited in achieving more of its goal by a small number of constraints, and there is always, at least, one constraint.

The TOC process seeks to identify the constraint and restructure the rest of the organisation by using five focusing steps. The methods and algorithms from TOC went on to form the basis of Critical Chain Project Management (CCPM).

- 1986 – Scrum named as a project management style. Scrum is an agile software development model based on multiple small teams working in an intensive and interdependent manner. The use of the term scrum in software development came from a 1986 Harvard Business Review paper titled "The New New Product Development Game" by Hirotaka Takeuchi and Ikujiro Nonaka. Based on case studies from manufacturing firms in the automotive, photocopier and printer industries, the authors outlined a new approach to product development for increased speed and flexibility. They called this the rugby approach, as the process involves a single cross-functional team operating across multiple overlapping phases in which the team "tries to go the distance as a unit, passing the ball back and forth." In their paper, Takeuchi and Nonaka named Scrum as a project management style. Later they elaborated on it in "The Knowledge Creating Company" (Oxford University Press, 1995). Although Scrum is intended to manage software development projects, it can be used to run software maintenance teams or as a general project and programme management approach.
- 1987 – A Guide to the Project Management Body of Knowledge (PMBOK Guide) published by PMI. First published by the PMI as a white paper in 1987, the PMBOK Guide was an attempt to document and standardise accepted project management information and practices. The guide is one of the essential tools in the project management profession today and has become the global standard for the industry.
- 1988–1994 – The Channel Tunnel. An official agreement was reached between the United Kingdom and France in 1964 for the cross-channel connection. Due to political and economic issues, the first tunnelling didn't begin until 1988. This was a massive engineering undertaking and despite being two years behind original schedule and costing double the original estimate, it was still completed fairly swiftly. The first cross-Channel train journey took place on May 6, 1994. This monumental engineering achievement connects Folkestone Kent, England with Coquellles, Pas-de-Calais, Calais in northern France. It stretches 31.4 miles and its lowest point is 75 m deep.
- 1989 – Earned Value Management (EVM) Leadership elevated to Undersecretary of Defense for Acquisition, thus making EVM an essential part of programme management and procurement. The PMBOK Guide of 1987 outlines Earned Value Management (EVM) subsequently expanded on in later editions. Although the earned value concept has been around on factory floors since the early 1900s, it only came to prominence as a project management technique in the

late 1980s to early 1990s. In 1991, Secretary of Defense Dick Cheney cancelled the Navy A-12 Avenger II Programme because of performance problems detected by EVM.

- 1989 – PRINCE Method developed from PROMPTII in consultation with 150 European organisations. Published by the UK Government agency CCTA, PRojects IN Controlled Environments (PRINCE) became the UK standard for all government information systems projects. A feature of the original method, not seen in other methods, was the idea of "assuring progress" from three separate but linked perspectives. However, the PRINCE method developed a reputation for being too unwieldy, rigid and applicable only to large projects, leading to a revision in 1996.

- 1994 – CHAOS Report first published. The Standish Group collects information on project failures in the Information Technology (IT) industry, intending to make the industry more successful, showing ways to improve its success rates and increase the value of IT investments. The CHAOS report is its biennial publication about IT project failure.

- 1994–2003 – Three Gorges Dam. The Three Gorges Dam spans the Yangtze River near the town of Sandouping in China. It is officially the largest power station in the world producing huge amounts of electricity each year. Construction of the dam began in 1994 and it started producing electricity in 2003. Many more turbines have since been installed, the initial 6 has now risen to 32. The dam stretches 410 miles long and contains enough steel to build 63 Eiffel Towers.

- 1996 – PMBOK first edition was published. PMI saw a need to put together an official document and guide to advance the development of the project management profession.

- 1996 – PRINCE2 Published by CCTA. UK Office of Government Commerce (OGC) considered an upgrade to PRINCE in order, and the development was contracted out but assured by a virtual committee spread among 150 European organisations. Initially developed for Information Systems and Information Technology projects to reduce cost and time overruns, the second revision became more generic and applicable to any project type.

- 1997 – CCPM invented. CCPM was developed by Eliyahu M. Goldratt and is based on methods and algorithms drawn from his Theory of Constraints (TOC) introduced in his 1984 novel titled, "The Goal." A Critical Chain project network will keep the resources levelly loaded, but need them to be flexible in their start times and switch quickly between tasks and task chains to keep the whole project on schedule.

- 1998 – PMBOK becomes an ANSI standard. The American National Standards Institute (ANSI) recognised PMBOK as a standard in 1998, and later that year it was recognised as a standard by the Institute of Electrical and Electronics Engineers (IEEE).

- 1999 – Managing Successful Programmes (MSP) was developed in response to the requirement for organisations to have better connections between their projects and their long-term aims. MSP is a best practice method for programme management that enables organisations to coordinate, direct, implement and manage a portfolio of projects that deliver a desired outcome. And, no, MSP is not the same as PRINCE2. MSP is a programme management framework used to manage programmes of change within an organisation. PRINCE2 is a project management methodology used to manage projects.
- 2000 – PMBOK second edition was published. The second edition added new material reflecting the growth of practices and focuses on providing models for the project management profession in both plan-driven and change-driven adaptive (agile) life cycles. This practice standard describes the aspects of project estimating that are recognised as good practice on most projects most of the time and that are widely recognised and consistently applied.
- 2001 – The Agile Manifesto written. In February 2001, 17 software developers met at The Lodge, Snowbird, Utah resort to discuss lightweight software development methods. They published the Manifesto for Agile Software Development to define the approach now known by the same name. Some of the manifesto's authors formed the Agile Alliance, a non-profit organisation promoting software development according to the manifesto's 12 core principles.
- 2004 – PMBOK third edition was published. The third edition was restructured to emphasise the importance of the Process Groups and contains the fundamental, baseline practices that drive business results for any organisation – local, regional or global, updated to reflect the most current industry knowledge and practices.
- 2006 – Total Cost Management Framework released by AACE International. Total Cost Management is the name given by AACE International to a process for applying the skills and knowledge of cost engineering. It is also the first integrated process or method of portfolio, programme and project management. AACE first introduced the idea in the 1990s and published the whole process presentation in the "Total Cost Management Framework."
- 2008 – Fourth edition of PMBOK Guide released. The fourth edition continued the PMI tradition of excellence in project management with an easier-to-understand and implement the standard with improved consistency and greater clarification. The updated version has two new processes, absent in the previous versions.
- 2009 – Major PRINCE2 revision by the Office of Government Commerce (OGC). A significant revision has made the method more straightforward and easily customisable, based on frequent requests from users. The updated version has seven basic principles (not in the previous version) that contribute to project success. Overall, the

updated method aims to give project managers better tools to deliver projects on time, within budget and with the right quality.

- 2012 – ISO 21500:2012 Guidance on Project Management. In September 2012, the International Organisation for Standardisation published "ISO 21500:2012, Guidance on Project Management." The standard is designed for any organisation. It is the result of 5 years' work by experts from more than 50 countries. These include public, private or community groups and are applicable to any project, regardless of complexity, size and duration.
- 2012 – Fifth edition of PMBOK Guide released. The fifth edition provides guidelines, rules and characteristics for project management recognised as good practice in the profession. The updated version introduces a 10th knowledge area called "Project Stakeholder Management" and includes four new planning processes.
- 2013 – Ownership of the rights to PRINCE2 transferred from HM Cabinet Office to AXELOS Ltd, a joint venture by the Cabinet Office and Capita, with 49% and 51% stakes respectively
- 2017 – "PRINCE2 2017 Update" published. In 2013, ownership of PRINCE2 changed to AXELOS, which published the methods next major update in 2017. The new guidance focuses on scalability and flexibility. The 2017 update clarifies the bare minimum for a project to qualify as PRINCE2. It then shows examples, hints and tips about how to adjust these core principles to your project.
- 2017 – Sixth edition of PMBOK Guide released. This update reflects evolving good practices in project management. New to the sixth edition, each knowledge area contains a section entitled Approaches for Agile, Iterative and Adaptive Environments, describing how these practices integrate into project settings. It also emphasises strategic and business knowledge, including discussion of project management business documents, information on the PMI Talent Triangle and the essential skills for success today.
- 2018 – PRINCE2 Agile. PRINCE2 Agile is a tailored form of PRINCE2, suitable for Agile projects that use Kanban, Scrum or a similar Agile system in their delivery layer. It adds a management and governance layer to the relatively simple Agile methods focused on the delivery layer.
- 2021 – PRINCE2 was transferred to PeopleCert during their acquisition of AXELOS. PeopleCert became the exclusive Examination Institute (EI) for AXELOS in 2018, replacing the model with several competing EIs (such as APMG and EXIN). And now, PeopleCert is the custodian of best practice frameworks and methods such as PRINCE2 and ITIL.
- 2021 – Seventh edition of PMBOK Guide released. This latest edition addresses project practitioners' current and future needs and helps them be more proactive, innovative and agile in enabling desired

project outcomes. The critical change in this edition reflects the full range of development approaches, providing an entire section devoted to tailoring the development approach and processes.

- 2024 – MSP has been rebranded as PRINCE2 Programme Management.

Now that we have looked briefly at the evolving complexity of projects and growing maturity of project management methodologies and tools, we can discuss pros and cons of various approaches.

The "waterfall" methodology is one of the most popular, oldest and most traditional methodology in project management. This type of methodology is followed in a project where requirements are well-known and fixed and no further significant changes are expected. The "waterfall" methodology is a linear sequential design process, well established in software development processes. The "waterfall" development method originates in the manufacturing and construction industries. It gives highly regimented physical environments that are very difficult or impossible to change or adapt once work has begun.

Traditional "waterfall" methodology is a well-established project management workflow. Like a waterfall, each process phase cascades downward sequentially through five stages (requirements, design, implementation, verification and maintenance). The methodology comes from computer scientist Winston Royce's 1970 research paper on software development. Although Royce never named this model "waterfall", he gets credit for creating a linear, rigorous project management system. This methodology is a widely used project management method with a linear approach. In "waterfall" **each stage of the workflow needs to be completed before moving on to the next step.** While there are various types of project management methodologies, this one is well suited for projects where the objectives are clearly outlined from the beginning. It is important to note that this is a linear project management approach, where stakeholder and customer **requirements are gathered at the beginning of the project**, and then a sequential project plan is created to accommodate those requirements. Gantt charts are the preferred tool for project managers working in waterfall method. Using a Gantt chart allows mapping subtasks, dependencies and each phase of the project as it moves through the waterfall lifecycle.

The "waterfall" methodology is not exactly a method as much as it is an approach, however, the six distinct stages that make up this cycle are very common in most software development processes: Requirements, Design, Testing, Implementation, Verification/Integration, Deployment and Maintenance:

- **Requirements.** During this phase the big picture of the project's requirements is being outlined. These are, "high-level statements that could be implemented in many different ways," according to Dr Chris Mattmann, Chief Technology and Innovation Officer (CTIO) at NASA Jet Propulsion Laboratory. Requirements are typically fall into one of two categories: Functional Requirements (FR) and Non-functional Requirements (NFR).

- **Design.** Once the project requirements are understood, the next step is to come up with ways to design solutions that meet them. As such, design process outlines the end result and how it will be achieved.
- **Implementation.** During this phase, one of multiple possible designs is selected, as well as, technology to implement it. This could involve collecting data and inspecting whether the design is able to support the requirements.
- **Testing.** In this phase, all system components are tested. This includes an integration test, which makes sure that each part works properly with the others, a functional test, which guarantees that all functionality meets requirements and a test of performance, which ensures that the system can handle peak loads without crashing or slowing down significantly.
- **Verification/Integration.** During this phase implementation that has been built in the previous phase is tested whether it complies with and delivers against requirements.
- **Deployment and Maintenance.** The project isn't over once it has gone through validation and verification. The system still needs to be deployed and maintained. Maintenance also applies to adding new features or functionality.

Up until the late 1990s, the majority of software projects followed a simple "waterfall" life cycle. Requirements were gathered upfront, the solution was designed, built and tested. The release to users for acceptance testing (UAT) occurred, followed by bug fixes before the final production release. This approach worked well for many IT projects because they tended to be tightly scoped in both time and cost. It works with relatively fixed requirements that did not change much during the course of the project. Projects were small enough for management of changes, often by adding an extra week or two to the project timeline. This adjustment rarely caused significant problems.

Although the Waterfall methodology is one of the most stringent and planned out project management approaches, it is not without its set of advantages and disadvantages, as illustrated below (https://instituteproject management.com/blog/waterfall-methodology/) (Table 6.1).

Table 6.1 Advantages and disadvantages of the Waterfall approach

Advantages	Disadvantages
Presence of clear structure	Costly and inflexible
Smooth transfer of information	Doesn't prioritise client/end user
Easy to manage	Delayed testing
Early determination of goals	No scope for revision/reflection
Extremely stable	

One of the advantages of "waterfall" approach is that it has a fixed time-line and budget because the project goals are specific and delineated from the start. Once the goal of the project is established, the "waterfall" method-ology does not involve frequent feedback or collaboration from the client, apart from established milestones or deliverables for each phase. This makes it easier for project managers to plan and communicate with stakeholders or business partners. However, while this can help with planning, it is also only practical when a client has a clear and fixed end goal and does not need to be involved in the process of the project's development. One of the disad-vantages of this methodology is that addressing unexpected problems can be difficult and timely.

Having said this, it is important to remember about the cost of software bugs. Software bugs are more than just a minor inconvenience. According to the Consortium for Information and Software Quality (CISQ), poor soft-ware cost the US economy $2.08 trillion in 2020 alone. The costs of soft-ware bugs are not just limited to the direct costs that a software developer must make to fix software bugs. Another consequence is productivity loss because bugs contribute to worker downtime, disruptions and delays. Financial loss also occurs due to a loss of reputation – because buggy soft-ware can indicate to clients that developer(s) do not produce high-quality products. Furthermore, bugs can introduce security risks, which can have a large financial impact in the form of cyberattacks, data breaches and finan-cial theft. And there is significant difference in cost of fixing software bugs during various stages of so-called Software Development Life Cycle (SDLC).

SDLC is the standardised set of steps that a developer goes through when creating and maintaining software. These steps are very similar to the stages of "waterfall" methodology and include planning, analysing, designing, implementing, deploying and maintaining. Software bugs can occur at any stage of the SDLC. If software bug is introduced at an early stage of the SDLC and is not immediately addressed, then its costs will only increase as it progresses through the SDLC. For example, fixing a bug in the planning stage can cost $100 if found early, but that same bug can escalate to become a $10,000 problem if it is discovered later on in the production stage. That's because software bug can have a domino effect that leads to additional costs and delays. Historical observations show that the cost to fix software bug found after product release is then four to five times as much as one uncov-ered during design, and up to 100 times more than one identified during the maintenance phase. In other words, the cost of a bug grows exponentially as the software progresses through the SDLC. So, the least expensive place to address software bugs is during the design phase of SDLC. The biggest prob-lem in software development are misconceptions. If errors occur during this initial phase, it can significantly impact the entire development process. These are not problems of implementation but rather problems of misconception.

In 2004, NASA published a paper on error cost escalation through the project life cycle. The findings in this paper are astounding! The results show the degree to which costs escalate, as errors are discovered and fixed at later and later phases in the project life cycle. If the cost of fixing a requirements error discovered during the requirements phase is defined to be 1 unit, the cost to fix that error if found during the design phase increases to 3–8 units; at the manufacturing/build phase, the cost to fix the error is 7–16 units; at the integration and test phase, the cost to fix the error becomes 21–78 units; and at the operations phase, the cost to fix the requirements error ranged from 29 units to more than 1500 units. Discussion about various project management methodologies must pay attention to this aspect, but, unfortunately, it is often omitted.

Another extreme example of information on error-related cost factors can be found in a book on designing cost-effective space missions [Cloud, Giffen, Larson and Swan, "Designing Cost-Effective Space Missions: An Integrated, Systems Engineering Approach", Teaching Science and Technology, Inc., 1999]. These systems cost factors represent the costs of fixing errors in electronics hardware. Depending upon the phase that change occurs results in the costs are as follows:

Phase	Resulting cost
Product Design	$1,000
Product Testing	$10,000
Process Design	$100,000
Low-Rate Initial Production	$1,000,000
Final Production/Distribution	$10,000,000

One of the perceived disadvantages of "waterfall" approach are lengthy requirements and design stages that in the eyes of the customer do not produce any "tangible" outcomes, or better to say usable system or solution. This perceived disadvantage spawned agile approach discussed below. One can argue that this perceived disadvantage is also an advantage that allows earlier identification of errors. Also, before we dive into agile approach, it is important to note that focus on upfront requirements and design in "waterfall" approach means that security aspects can be dealt with at all six stages of the project.

Proliferation of Internet (and Internet applications) as well as desire to fast track "time to market" and Digital Transformations (see Chapter 5) trend resulted in development and subsequent proliferation of Agile methodology. Agile methodology focuses on creating working software quickly, collaborating with customers frequently and being able to adapt to changes easily. This methodology enables teams to follow a cycle of planning, executing and evaluating. Agile methodology is especially beneficial for projects that have uncertain requirements.

Agile methodology in project management is a structured approach that segments projects into manageable phases, focusing on continuous improvement.

It is an iterative process that involves planning, execution and evaluation and its roots can be found in Lean project management.

The birth of Lean Management can be tracked down to 1940, and in the past 80 years, it has become a universal management tool for work process optimisation. Toyota Founder Kiichiro Toyoda developed the Lean methodology after World War II to conserve resources and eliminate waste. After observing the purchasing and restocking of items at a supermarket, he conceived the just-in-time concept, which focuses on making products exactly when customers need them.

Toyoda's concept morphed into the Toyota Production System, which eventually became the Lean methodology. From these small beginnings, Lean evolved into the foundation of Agile project management – several industries, including software development, construction and healthcare, now use Lean methodology. Lean methodology aims to fully optimise team's process and output through continuous improvements. When done well, Lean allows teams to deliver customer value efficiently.

Lean project management is the application of lean manufacturing principles to the practice of project management. The goal of lean project management is to maximise value while minimising waste. Lean manufacturing principles were developed by Toyota in the 1950s and applied in the 1970s to combat the energy crisis. Lean project management relies on continuous improvement. That means that every process in the overall business value stream is improved by applying the principle of greater value and reduced waste. The term "lean" was coined in the late 1980s. The Project Management Institute sums it up: "To be Lean is to provide what is needed, when it is needed, with the minimum amount of materials, equipment, labour, and space."

Lean manufacturing identifies three types of waste: muda, muri and mura (known collectively as the 3M).

- Muda refers to activities that consume resources without providing additional value
- Muri refers to the overuse of equipment or employees
- Mura is operational "unevenness," which decreases efficiency and productivity in the long term

Lean Project Management is the evidence that the Lean principles can find beneficial application in many areas. Delivering value from organisation's perspective, cutting down waste and continuous improvement help project managers increase their projects efficiency and enable them to deliver more with less. Whereas traditional project management is structured in phases, which separate planning from execution, Lean project management enables teams to deliver faster by managing their workflow efficiently and focuses on delivering value from the organisation's perspective.

Lean methodology facilitates an ongoing process of incremental adjustment, significantly accelerating product delivery by optimising resources and effort and allowing teams to work efficiently and effectively.

Agile approach is about rapid testing, frequent delivery and continuous refinement. That is great for projects that require constant updates, like selling software or taking down a website when something goes wrong. However, it is not so good for projects that need to be built from scratch.

For this reason, agile methodology is focused on short-term projects that deliver value quickly, while long-term projects remain stuck in the planning and implementation phase. Agile development emphasises collaboration and communication throughout the project lifecycle. It involves iterative planning, and changing plans during implementation. This ensures all parties work toward a common goal.

The following table shows the main differences between "waterfall" and agile approaches (https://instituteprojectmanagement.com/blog/waterfall-methodology/) (Table 6.2).

Table 6.2 Waterfall vs Agile

Pros	Cons
Waterfall	
Works best when there are defined requirements	Requires investment to define scope and schedule before work begins
Best for stable environments	Scope changes can be slow and the adverse impact increases over the life cycle
The team is distributed and hence control can be managed by defined deliverables, milestones and dependencies	Risk of nothing to show for the money until the end
Best if scarce skills or resources have limited availability	Change adds effort and risk, so strict change control process must be in place to avoid scope creep
Plans are repeatable for similar projects	
Agile	
Works well when detailed requirements are unknown or subject to change	No advantage to projects when scope and detailed requirements are well understood and change can be controlled
Gives flexibility to "course correct"	Uncertainty about scope and schedules can make stakeholders nervous
Needs regular stakeholder feedback	Less effective if the team is "distributed"
The team is co-located, multi-functional and enables work in a collaborative way	Demands management and prioritisation of the backlog
Early return on investment by regular delivery	

Agile methodology is a project management framework that breaks projects down into several dynamic phases, commonly known as sprints. The Agile framework is an iterative methodology. After every sprint, teams reflect and look back to see if there was anything that could be improved so they can adjust their strategy for the next sprint. Agile process consists of four main stages: Preparation, Sprint planning, Sprint and Sprint retrospective:

- **Preparation.** In the preparation stage, the product owner creates a backlog of features they want to include in the final product. This is known as the product backlog. Then, the development team estimates how long each feature will take to build.
- **Sprint planning.** The sprint planning meeting is where the team decides which features from the product backlog they are going to work on during the sprint. A sprint is a set period (usually between two weeks and two months) during which the development team must achieve a specific goal. The team also decides how many of each type of task they can complete during the sprint. For example, the team may decide they can complete three coding tasks, two testing tasks and one documentation task during the sprint. This information is then added to the sprint backlog.
- **Sprint.** During the sprint, the team works on completing the tasks in the sprint backlog. They may also come across new issues to address or bugs that need fixing. If this happens, they will add these issues and bugs to the product backlog and prioritise them accordingly. At the end of the sprint, the development team should have completed all features in the sprint backlog. If not, the team will carry them over to the next sprint.
- **Sprint retrospective.** After each sprint team then holds a sprint review meeting where they demo completed features to the product owner and stakeholders. They also discuss what went well during the sprint and how they could improve their next one. Finally, the team holds a retrospective meeting, where they reflect on what went well and what didn't go so well during the sprint. They then create a plan of action for addressing these issues in future sprints. This feedback loop helps to ensure that each sprint is more successful than the last.

Not everything can be booted into a single sprint. It is time now to introduce epics. But before we do this, we need to introduce some other Agile terminology. Stories, also called "user stories," are short requirements or requests written from the perspective of an end user. An epic is too large to be completed in a single sprint but is a smaller representation of work than the highest-level goals and initiatives. An epic is typically completed over the course of several sprints or longer. Sometimes epics encompass multiple teams, on multiple projects. Initiatives are collections of epics that drive toward a common goal.

The Agile framework is an umbrella for several different variations. Here are some of the most common Agile methodologies:

- **Kanban.** Kanban is a visual approach to Agile. Teams use online Kanban board tools to represent where certain tasks are in the development process. Tasks are represented by cards on a board, and stages are represented in columns. As team members work on tasks, they move cards from the backlog column to the column that represents the stage the task is in. This method is a good way for teams to identify roadblocks and to visualise the amount of work that's getting done.
- **Scrum.** Scrum is a common Agile methodology for small teams and also involves sprints. The team is led by a Scrum master whose main job is to clear all obstacles for others executing the day-to-day work. Scrum teams meet daily to discuss active tasks, roadblocks and anything else that may affect the development team.
- **Extreme Programming (XP).** Extreme Programming is typically used in software development, it outlines values that will allow development team to work together more effectively. Similar to daily Scrum stand-ups, there are regular releases and iterations, yet XP is much more technical in its approach. If development team needs to quickly release and respond to customer requests, XP focuses on the "how" it will get done.
- **Adaptive Project Framework (APF).** Adaptive Project Framework, also known as Adaptive Project Management (APM) grew from the idea that unknown factors can show up at any time during a project. This technique is mainly used for IT projects where more traditional project management techniques don't apply. This framework is based on the idea that project resources can change at any time. For example, budgets can change, timelines can shift or team members working on the project may transition to different teams. APF focuses on the resources that a project has, as opposed to the resources a project needs.
- **Extreme Project Management (XPM).** This type of project management is often used for very complex projects with a high level of uncertainty. This approach involves constantly adapting processes until they lead to the desired result. This type of project involves many spontaneous changes and it's normal for teams to switch strategies from one week to the next. XPM requires a lot of flexibility. This is one of the reasons why each sprint is short – only a few weeks maximum. This methodology allows for frequent changes, trial-and-error approaches to problems and many iterations of self-correction.
- **Adaptive Software Development (ASD).** This Agile methodology enables teams to quickly adapt to changing requirements. The main focus of this process is continuous adaptation. The phases of this project type – speculate, collaborate and learn – allow for continuous learning as the project progresses. It's not uncommon for teams running ASD to be in all three phases of ASD at once. Because of its

non-linear structure, it's common for the phases to overlap. Because of the fluidity of this type of management, there's a higher likelihood that the constant repetition of the three phases helps team members identify and solve problems much quicker than standard project management methods.

- **Dynamic Systems Development Method (DSDM).** The Dynamic Systems Development Method is an Agile method that focuses on a full project lifecycle. Because of this, DSDM has a more rigorous structure and foundation, unlike other Agile methods. There are four main phases of DSDM:
  - Feasibility and business study
  - Functional mode or prototype iteration
  - Design and build iteration
  - Implementation
- **Feature Driven Development (FDD).** Feature Driven Development blends different Agile best practices. While still an iterative method of project management, this model focuses more on the exact features of a software that the team is working to develop. Feature-driven development relies heavily on customer input, as the features the team prioritises are the features that the customers need. This model also allows teams to update projects frequently. If there is an error, it's quick to cycle through and implement a fix as the phases of this framework are constantly moving.
- **Crystal Methods.** Crystal Methods approach was invented by Alistair Cockburn. He was one of the original, monumental persons in formulating the Agile manifesto for software development. Crystal is his latest iteration. Crystal is a group of smaller agile development methodologies: Crystal Yellow, Crystal Clear, Crystal Red, Crystal Orange and more. Each has its peculiar and exclusive framework that is characterised by factors such as system criticality, team size and project priorities. One chooses a framework depending on the nature of the project or system criticality. Some examples of the factors are Comfort (C), Essential Money (E), Discretionary Money (D) and Life (L). Similar to other Agile methodologies, Crystal also addresses prompt delivery of software, regularity, less administration with high involvement of users and high customer satisfaction. The Crystal family advocates that each system or project is inimitable. Therefore, each necessitates the solicitation of diverse practices, processes and policies to achieve the best results.
- **Disciplined Agile Delivery (DAD).** DAD is the software development portion of the Disciplined Agile Toolkit. DAD enables teams to make simplified process decisions around incremental and iterative solution delivery. DAD builds on the many practices espoused by advocates of agile software development, including scrum, agile modelling, lean

software development and others. The primary reference for disciplined agile delivery is the book "Choose Your WoW!," written by Scott Ambler and Mark Lines. WoW refers to "way of working" or "ways of working." In particular, DAD has been identified as a means of moving beyond scrum.

- **Scrum@Scale (S&S).** Scrum@Scale was developed by Dr Jeff Sutherland and is based on the fundamental principles of Scrum, Complex Adaptive Systems theory, game theory and object-oriented technology. Scrum, as originally outlined in the Scrum Guide, is a framework for developing, delivering and sustaining complex products by a single team. Since its inception, its usage has extended to the creation of products, processes, services and systems that require the efforts of multiple teams. Scrum@Scale was created to efficiently coordinate this new ecosystem of teams. It achieves this goal through setting up a "minimum viable bureaucracy" via a "scale-free" architecture.

- **Scaled Agile Framework (SAFe).** Scaled Agile Framework is a set of organisation and workflow patterns for implementing agile practices at enterprise scale. It was formed around three primary bodies of knowledge: agile software development, lean product development and systems thinking. Unlike other Agile methodologies that geared towards small to medium sized teams, SAFe is designed to scale agile practices across the entire organisation, including multiple teams, programs and portfolios. The SAFe is a set of organisation and workflow patterns intended to guide enterprises in scaling lean and agile practices. Along with DAD and S@S (Scrum@Scale), SAFe is one of a growing number of frameworks that seek to address the problems encountered when scaling beyond a single team.

As one can see there are multiple variations of agile methodology ranging from various "classical" pure Agile to "hybrid" models mixing Agile approach with traditional "waterfall" approach to requirements and design followed by actual delivery using one of the Agile variations.

No questions that Agile approach has succeeded in faster "time to market" and higher levels of flexibility that allow rapidly modernise functionality. It also gave business stakeholders stronger engagement and influence on what, when and how will be delivered. Today over 70% of US companies are using Agile approach.

However, there is (as always) a price to pay for this. One of the challenges, especially in larger organisations, is that development teams typically refine their backlog up to two to three iterations ahead, but in larger organisations the product marketing team needs to plan further ahead for their commitments to market and discussions with customers. They will often work with a very high level, 12- to 18-month roadmap, then plan collaboratively with the teams for three months of work. The development teams will still get into detailed refinement 2–3 iterations ahead, only getting into detailed task plans for the next iteration.

One of the biggest challenges that Agile approach poses to achieving required strength of cybersecurity posture is hiding in sprints, or to be more accurate, in sprints' duration. Sprints' duration (be it 2 weeks or 2 months) simply does not allow for proper security design, independent reviews and security testing, which creates high probability of security gaps being hidden in the delivered product. Business stakeholders usually do not understand cybersecurity (and potential implications of not following best practices in this space) and always press for delivery of functionality. The pressure to deliver functionality often even prevents bug fixing pushing this activity further down the track. And even if these gaps are discovered later at the security testing at UAT stage, the cost of closing them becomes very high. Even if the cost is tolerable (although typically by this stage project has already run out of money), organisation can rarely tolerate any additional delay to the product release.

One other interesting real-life observation is that Agile approach makes it difficult to define the end point, as focus is mainly on delivery of Minimum Viable Product (MVP), which is a product with enough features to attract early-adopter customers and validate a product idea early in the product development cycle and in industries such as software, the MVP can help the product team receive user feedback as quickly as possible to iterate and improve the product. However, multiple real-life observations show that projects very rarely (if at all) progress beyond MVP that is not clearly defined at the outset. More often than not, Agile projects run out either of money, or time, or both and stop at delivery of MVP.

Now, remember the saying: "Fast, good or cheap - pick any two"? Agile approach is clearly focused on "fast" (delivery of new functionality). Let's tick off one of these three. Intuitively (and this is arguably one of the selling points of Agile approach) the second one to tick off is "cheap." And now think about it – one can have any two out of three and we have ticked off two already. Can Agile approach bit this and deliver all three, including good in cybersecurity sense? The answer is – extremely unlikely.

It is worth mentioning that according to some sources Agile development is fading in popularity at large enterprises – and developer burnout is a key factor (https://www.itpro.com/software/agile-development-is-fading-in-popularity-at-large-enterprises-and-developer-burnout-is-a-key-factor#:~:text=Software-,Agile%20development%20is%20fading%20in%20popularity%20at%20large%20enterprises%20%2D%20and,burnout%20is%20a%20key%20factor&text=Agile%20development%20methodology%20is%20facing,and%20the%20rise%20of%20AI.) Agile development methodology is facing significant obstacles as the tech industry goes through a wave of changes including developer burnout, shifting working environments and the rise of AI. One of the studies found that while smaller organisations continue to consider Agile as a powerful productivity and organisational framework that exhibits "obvious benefits," medium-sized and larger organisations are less satisfied with what Agile can do for them, and are more likely to pick a software development strategy that uses a number of different frameworks.

Chapter 7

Head in the Cloud

The term "Cloud" is one of today's most frequently used buzzwords. Whether one is talking about storage, computing or security, people are always referencing the Cloud. But what is the reason behind using the term "Cloud computing?" The term "Cloud" refers to a symbol of an unknown domain. When network engineers were trying to understand what devices were on what network and how they intertwined with the Internet, they needed a way to illustrate and visualise it. Unfortunately, they didn't know every single detail when it came to these networks, but needed a way to show there was a network there, but were still unable to describe it. This is where the Cloud symbol was born (https://blog.icorps.com/the-history-of-cloud-computing#:~:text=Early%20Days%20of%20Cloud%20Computing,virtual%20machines%20in%20the%201970s.).

How big is Cloud usage today? According to Gartner, worldwide end-user spending on public cloud services is forecast to grow 20.4% to total $675.4 billion in 2024, up from $561 billion in 2023, (https://www.gartner.com/en/newsroom/press-releases/2024-05-20-gartner-forecasts-worldwide-public-cloud-end-user-spending-to-surpass-675-billion-in-2024#:~:text=Worldwide%20end%2Duser%20spending%20on,(GenAI)%20and%20application%20modernization). And the forecast is that it will keep growing at high pace (Table 7.1).

Table 7.1 Worldwide public Cloud services end-user spending forecast (US$ billion)

	2024 Spending	2024 Growth (%)	2025 Spending	2025 Growth (%)
Cloud Application Infrastructure Services (PaaS)	172.449	20.6	211.589	22.7
Cloud Application Services (SaaS)	247.203	20.0	295.083	19.4
Cloud Business Process Services (BPaaS)	72.675	9.8	82.262	13.2
Cloud Desktop-as-a-Service (DaaS)	3.062	13.1	3.437	12.3
Cloud System Infrastructure Services (IaaS)	180.044	25.6	232.391	29.1
Total Market	**675.433**	**20.4**	**824.763**	**22.1**

DOI: 10.1201/9781032672601-7

As of 2024, 46% of enterprises already have workloads in the public Cloud, with 8% planning to move additional workloads to the cloud in next 12 months; in addition to this, 48% of respondents reported having data stored on the public Cloud (https://www.statista.com/statistics/817316/worldwide-enterprise-workloads-by-cloud-type/#:~:text=Worldwide%20enterprise%20workload%2Fdata%20in%20public%20cloud%202024&text=As%20of%202024%2C%2046%20percent,cloud%20in%20next%2012%20months.).

As terms like XaaS and Public Cloud have been used, it is worthwhile to go back in time again and explore emergence of cloud technologies, types of Clouds and types of Cloud services.

The concept of Cloud computing can and should be traced back to the concept of service bureau and Service Bureau Division within IBM that was established in 1932 (in fact, IBM had operated service bureaus in major cities beginning in the 1920s allowing users to rent time on tabulating equipment, and later computing equipment to solve problems which couldn't justify a full-time equipment lease) and later was spun off as a wholly owned subsidiary (Service Bureau Corporation – SBC) in 1957 (this actually happened in 1956, as a result of a consent decree with the US Department of Justice) to operate IBM's burgeoning service bureau business (https://en.wikipedia.org/wiki/Service_Bureau_Corporation#:~:text=The%20Service%20Bureau%20Corporation%20(SBC,IBM's%20burgeoning%20service%20bureau%20businesses.). In 1968, IBM transferred its Information marketing Division to SBC. This included the CALL/360 time-sharing service, QUIKTRAN, BASIC and DATATEXT.

In 1955, John McCarthy (who originally coined the term "artificial intelligence"), created a theory of sharing computing time among a group of users. In 1961, he has suggested that one day "Computing can be sold as a Utility, like Water and Electricity." Getting the most out of computing time was an important consideration in the 1950s because it could cost upwards of several million dollars. It was a ridiculously expensive asset, and maximising it was a top priority among those who were shelling out the cash for the technology.

As one can see, Cloud computing has a rich history with the initial concepts of time-sharing computing that became popular in the 1960s via remote job entry (RJE). During this era users submitted jobs (typically decks of punched cards) to operators to run on mainframes – this was the era of "data centre" model. This was the time of exploration and experimentation with ways to make large-scale computing power available to more users through time-sharing, optimising utilisation of infrastructure, platform and applications, and increasing efficiency and turnaround time for end users (https://en.wikipedia.org/wiki/Cloud_computing#:~:text=History,-Main%20article%3A%20History&text=Cloud%20computing%20has%20a%20rich,predominantly%20used%20during%20this%20era).

When mainframe computers were originally launched in the 1950s and 1960s, the idea of Cloud computing was already in existence. Large-scale computing systems at the time were pricey and required significant infra-structure investments. As a result, businesses started looking into how to pool computing resources and maximise consumption. IBM toyed with operating system (OS) virtualisation (allowing for multiple users to time-share the same resource) since 1967. Virtual machines (VMs) were created in the 1970s (in 1972 IBM released an OS the VM operating system), enabling the simultaneous use of several OSs on a single physical machine. By enabling greater resource usage and multi-tenancy, this innovation set the groundwork for the future growth of cloud computing.

In 1963, DARPA (the Defense Advanced Research Projects Agency) pre-sented MIT with US$2 million for Project MAC. The funding included a requirement for MIT to develop technology allowing for a "computer to be used by two or more people, simultaneously." In this case, one of those gigantic, archaic computers using reels of magnetic tape for memory became the precursor to what has now become collectively known as Cloud com-puting. It acted as a primitive cloud with two or three people accessing it. The word "virtualisation" was used to describe this situation, though the word's meaning later expanded.

In the mid-1960s, an American computer scientist J.C.R. Licklider came up with an idea for an interconnected system of computers. In 1969, Licklider's revolutionary idea helped Bob Taylor and Larry Roberts develop something known as ARPANET (Advanced Research Projects Agency Network), a network relying on the TCP/IP protocol. JCR, or "Lick," was both a psychologist and a computer scientist, and promoted a vision called the "Intergalactic Computer Network," in which everyone on the planet would be interconnected by way of computers and is able to access informa-tion from anywhere. This Intergalactic Computer Network, otherwise known today as the Internet, is necessary for access to the Cloud.

The "Cloud" metaphor for virtualised services dates to 1994, when it was used by General Magic for the universe of "places" that mobile agents in the Telescript environment could "go." The metaphor is credited to David Hoffman, a General Magic communications specialist, based on its long-standing use in networking and telecom.

The expression Cloud computing became more widely known in 1996 when Compaq Computer Corporation drew up a business plan for future computing and the Internet. The company's ambition was to supercharge sales with "Cloud computing-enabled applications." The business plan foresaw that online consumer file storage would likely be commercially successful.

Salesforce introduced the concept of Software as a Service (SaaS) in 1999, offering customer relationship management software over the Internet. This marked the beginning of the commercialisation of Cloud computing and set the stage for the growth of SaaS.

The inception of the modern era Cloud was realised by Amazon Web Services (AWS) launching its public cloud in 2002. This was the beginning of the application of modern cloud computing services, which allowed developers to build applications independently. In 2006, Amazon Simple Storage Service, known as Amazon S3, and the Amazon Elastic Compute Cloud (EC2) and the beta version of Google Docs were released. Then in 2008 Google introduced its radically new pricing models.

The following decade saw the launch of various Cloud services. In 2010, Microsoft launched Microsoft Azure, and Rackspace Hosting and NASA initiated an open-source Cloud-software project, OpenStack. IBM introduced the IBM SmartCloud framework in 2011, and Oracle announced Oracle Cloud in 2012. In December 2019, Amazon launched AWS Outposts, a service that extends AWS infrastructure, services, APIs and tools to customer data centres, co-location spaces or on-premises facilities.

So, today one can define Cloud computing as a technology that allows users to store, access and manage data and applications over the Internet. Instead of storing data and running applications on a local computer or server, Cloud computing allows users to access these resources from remote servers, which are managed and maintained by Cloud service providers (CSPs).

As Cloud computing kept developing, new service models emerged. The following are the top four Cloud computing service models:

Infrastructure as a Service (IaaS): This computer resource delivery model gives customers access to virtualised computing resources, such as virtual machines, storage and networks, enabling them to deploy and manage their software and applications. IaaS delivers on-demand infrastructure resources, such as compute, storage, networking and virtualisation. With IaaS, the service provider owns and operates the infrastructure, but customers will need to purchase and manage software, such as OSs, middleware, data and applications.

Platform as a Service (PaaS): This model expands on IaaS by providing an all-inclusive environment for app development and deployment. Developers may build, test and deploy their product on this platform without having to worry about managing the underlying infrastructure. PaaS delivers and manages hardware and software resources for developing, testing, delivering and managing cloud applications. Providers typically offer middleware, development tools and cloud databases within their PaaS offerings.

SaaS: This model provides whole software programs that may be accessed online. These programs can be used by users without them having to locally install or maintain any software. SaaS provides a full application stack as a service that customers can access and use. SaaS solutions often come as ready-to-use applications, which are managed and maintained by the CSP.

Serverless computing: This is the latest model and is also called Function as a Service (FaaS). This is a relatively new cloud service model that provides solutions to build applications as simple, event-triggered functions without managing or scaling any infrastructure.

A simple analogy to help readers understanding the difference between **IaaS, PaaS, SaaS** and **Serverless computing** (FaaS) is to think of the models like eating fresh pasta. One could make their own from scratch (on-premises data centre), where one buys all the basic ingredients to make everything like the sauce and dough. However, most of us generally don't have enough time or skills or don't want to spend so much time and effort to eat a bowl of pasta. Thus, one might choose from the following options instead:

IaaS: Buying pre-packed ingredients like fresh pasta and sauce made by someone else that one uses to cook at home.

PaaS: Order takeaway (or takeout, as they call it in the United States) or delivery where one's meal is prepared and one doesn't have to worry about the ingredients or how to cook it, but one has to worry about where to eat, the utensils and cleaning up after your meal.

SaaS: Call ahead to the restaurant and order the exact meal one wants. The restaurant prepares everything ahead of time for you so that all one has to do is to show up and eat – without worrying about where to eat, utensils and cleaning.

FaaS: Go out to dinner and order pasta designed by you at a restaurant, alone or with friends. One eats and pays whatever one wants and the restaurant makes sure there's enough ingredients and staff to create the order without a long wait.

This analogy brings us to the discussion about prices associated with various models, but we will have this discussion a little bit later. At this stage it is worth mentioning that in the used earlier analogy one can easily notice that each step along the "line" (from buying prepacked ingredients to ordering takeaway to calling the restaurant ahead to ordering customer-designed pasta – **IaaS, PaaS, SaaS** and **FaaS**) comes with a higher price.

After we have discussed Cloud service models, it is important to talk about various types of Clouds: public, private and hybrid.

Public Clouds deliver resources, such as compute, storage, network, develop-and-deploy environments and applications over the Internet. They are owned and run by third-party CSPs like AWS, Microsoft Azure and Google Cloud.

Private Clouds are built, run and used by a single organisation. Sometimes, it is not easy to draw a line between Private Clouds and managed services. They provide greater control, customisation and data security but come with similar costs and resource limitations associated with traditional IT environments.

Environments that mix at least one private computing environment (traditional IT infrastructure or private Cloud, including edge) with one or more public Clouds are called hybrid Clouds. They allow leveraging the resources and services from different computing environments and choosing the most optimal for the workloads.

When talking about types of Cloud deployment, one may also hear the term Multicloud environment. In fact, industry research shows that a number of organisations are considering Multicloud approach, meaning they combine Cloud services from at least two different CSPs, whether public or private. Adopting a Multicloud approach gives greater flexibility to choose the solutions that best suit specific business needs and also reduces the risk of vendor lock-in, but comes with increased cost and complexity.

So, looking back, one can arrive to a conclusion that Cloud computing is a form of outsourcing, or, as per its definition, obtaining goods or services by contract from an outside supplier. Outsourcing is a business practice in which services or job functions are hired out to a third party on a contract or ongoing basis (https://www.cio.com/article/272355/outsourcing-outsourcing-definition-and-solutions.html).

There are multiple drivers for outsourcing, including:

- lower costs (due to economies of scale or lower labour rates),
- increased efficiency,
- variable capacity and scalability (up and down),
- increased focus on strategy/core competencies,
- access to skills or resources,
- increased flexibility to meet changing business and commercial conditions,
- accelerated time to market,
- lower investment in internal infrastructure,
- access to innovation, intellectual property and thought leadership,
- possible cash influx resulting from transfer of assets to the new provider, etc.

Pretty much all these reasons are behind the drive to utilise Cloud computing.

However, as everything, outsourcing in general and Cloud computing, in particular, have their drawbacks:

- lack of business or domain knowledge,
- language and cultural barriers,
- time zone differences,
- lack of direct control and reliance on a third party,
- contractual risks,
- Supply chain/network dependency, etc.

Among specific to Cloud computing benefits the following are usually mentioned (https://en.wikipedia.org/wiki/Cloud_computing#:~:text=History,-Main%20article%3A%20History&text=Cloud%20computing%20has %20a%20rich,predominantly%20used%20during%20this%20era.): **cost reduction, maintenance, skillset shortage avoidance, productivity,** performance, speed of provisioning, availability, scalability and elasticity, security. Let's have a quick look at each of these publicly declared benefits.

> **Cost reduction** is an interesting area and as promised earlier let's have a closer look at it. Public Cloud delivery model converts capital expenditure (e.g., buying servers) to operational expenditure. This was a great value proposition for those organisations that were cash-strapped, but needed technology refresh, as this was taking away the need for immediate capital outlay. Purportedly it was lowering entry barriers, as infrastructure is typically provided by a third party and need not be purchased for one-time or infrequent intensive computing tasks, especially for fixed-term projects. Pricing on a utility computing basis is "fine-grained," with usage-based billing options. As well, less in-house IT skills are required for implementation of projects that use Cloud computing.

There are numerous important factors that should be discussed here.

Firstly, capital avoidance looks very attractive at the point in time when it needs to be invoked. However, if one looks through the lenses of Total Cost of Ownership (TCO), say over 5 or 10 years, financial benefits becoming far less attractive. Moreover, clients of Cloud services provider enter a one-way door and become captive audience for CSP that can arbitrarily increase prices of various services. For example, Microsoft announced that as of September 1, 2023 price of Microsoft 365 subscriptions in Australia will increase by 9% (https://www.data3.com/knowledge-centre/blog/heres-everything-you-need-to-know-about-the-microsoft-price-increase/) which is significantly higher than CPI! For European countries price increases (effective April 1, 2023) for Microsoft Cloud ranged from 9% for United Kingdom to 11% for Denmark and Euro-zone to 15% for Norway (https://news.microsoft.com/europe/2023/01/05/consistent-global-pricing-for-the-microsoft.cloud/).

Another factor to consider in TCO analysis is network costs, like for example cost of Microsoft ExpressRoute. Public Cloud might look cheaper, but this is not always the case. Ingress-egress fees, data transfer fees, they add up. And as the workload grows, organisations might realise that it's actually cheaper to run these workloads in on-premises environments.

Thirdly, it is important to note that many large Cloud services providers used to charge their customers in US$ exposing their clients to currency exchange rate variations and thus making it difficult to do accurate financial forecasts.

Then, in the absence (which very often the case) of strong cloud governance clients' footprints in the Cloud tend to grow further increasing clients' costs. One of Australian universities faced a difficult dilemma during their first encounter with Cloud computing as actual cost eclipsed the projected one by a factor of three. Cloud costs can be far more unpredictable than on-premise equivalents, especially if configured incorrectly without spending controls in place.

Despite of the cost savings promise, Cloud has turned out to be expensive. As a result of this, quite a number of organisations now consider so-called Cloud repatriation (https://www.forbes.com/sites/forbestechcouncil/2023/04/18/the-rise-of-cloud-repatriation-why-companies-are-bringing-data-in-house/). Cloud repatriation involves moving applications or workloads or data that were once on a public Cloud environment either back on on-premises infrastructure, colocation infrastructure or to a different Cloud environment with an alternative cloud services provider. Unexpected costs are driving some data-heavy and legacy applications back from public Cloud to on-premises locations or private Clouds (https://journal.uptimeinstitute.com/high-costs-drive-cloud-repatriation-but-impact-is-overstated/).

For example, between 2013 and 2016, Dropbox pulled significant amount of data back from AWS and this worked out significantly cheaper and gave Dropbox more control over the data the company hosted (https://www.datacenterdynamics.com/en/analysis/how-dropbox-pulled-off-its-hybrid-cloud-transition/). More recently (in 2023) web company 37signals (which runs project management platform Basecamp and subscription-based email service Hey) announced the two services were migrating off of AWS and Google Cloud (https://www.datacenterdynamics.com/en/analysis/cloud-repatriation-and-the-death-of-cloud-only/). This move boosted 37signals profit by US$1 million and the expectation is to save US$7 million in 5 years (https://world.hey.com/dhh/we-stand-to-save-7m-over-five-years-from-our-cloud-exit-53996caa)! Details of what and why 37signals done can also be founed in: https://www.datacenterdynamics.com/en/analysis/how-dropbox-pulled-off-its-hybrid-cloud-transition/. Popular SEO tools developer, Ahrefs has revealed that its decision to use own hardware over the AWS cloud will save them $400 million over three years. The company calculated the costs of having its equivalent hardware and workloads entirely within AWS' Singapore region over the last 2 years, and estimates the cost would be $440m, vs. the $40m it actually paid for 850 on-premise servers during that time (https://tech.ahrefs.com/how-ahrefs-saved-us-400m-in-3-years-by-not-going-to-the-cloud-8939dd930af8).

The global proliferation of Cloud services in the last decade has brought significant advantages to businesses worldwide, offering scalability, agility and a pay-per-use model that often appears more cost-effective than traditional on-premises infrastructure. However, a closer analysis of the costs and benefits associated with cloud-based services suggests that this may not always be the case. A 2021 report from Andreessen Horowitz noted that cloud repatriation could drive a 50% reduction in Cloud spend, but notes it is a "major decision" to start moving workloads off of the Cloud.

Talking about Cloud repatriation it is important to note that cost is not the only driver for this. For example, latency, performance and management may be a driver – super latency-sensitive applications may not meet performance expectations in public cloud environments to the same degree they might on-premise or in colocation sites. Another driver may be data sovereignty demands that can be a driver in some markets. Countries with stricter data residency laws may force enterprises to keep data within their own borders – and in some cases out of the hands of certain companies under the purview of data-hungry governments. Many Cloud providers are looking to offer "sovereign Cloud" solutions that hand over controls to a trusted domestic partner to overcome some of these issues.

Finally, I had personal experience in doing a study for one insurance company showing that Cloud repatriation would have saved this company 45% of their costs on the same footprint. Managed Service Providers (MSPs), particularly those with their own data centres, are vital in facilitating Cloud repatriation. They don't just provide technical know-how but also the infrastructure necessary for a successful move. By collaborating with MSPs, companies can enjoy a controlled and secure environment akin to a private Cloud. This customised setting offers better flexibility, scalability and security.

From the cost point of view, public Cloud is great for projects where infrastructure is required for a finite duration of time, especially if one can stand it up in the morning and shut down in the evening maximising benefits of "pay per use" model, but for workloads that require 24×7 operation public Cloud is often not the cheapest option.

> **Maintenance** of Cloud environments is perceived being easier because of servers and storage maintained by the Cloud services provider(s). Although this may be the case in many situations and the benefit of not struggling with the search for skilled personnel, but the price one pays for this is threefold: loss of full control, loss of full visibility (and often – understanding) of the entire environment end-to-end, as well as potential security issues that will be discussed later in this book.
>
> **Skillset shortage avoidance** has been covered earlier. Having said this, it is important to remember that in the fact that organisations inevitably need as many (if not more) people to manage increasingly complex and intertwined Cloud environments as they did on-premise, and some companies may prefer to just keep things in-house or use an MSP.
>
> **Productivity, performance and availability** are questionable benefits as they can be achieved both within and outside Cloud services and are significantly linked to the overall applications, data and network architectures and implementations. When talking about availability one must be mindful that replication and/or hot-hot data replication

is a double-edged sword and if primary database becomes poisoned, the same is true for the secondary database. The author of these lines about 10 years ago had a questionable privilege to manage one of such incidents that closed motor registries across the state for 3 days in a row.

Speed of provisioning is an unquestionable benefit of using Cloud computing as one can stand up a number of new environments in the matter of several (1–8) hours and sometimes, even minutes. This is especially useful for projects and experimentation.

Scalability and elasticity are great benefits, as long as one remembers that applications have to be written with these in mind and that "lift-and-shift" of legacy applications into the Cloud is unlikely to result in such benefit. It is also important to remember that dynamic provisioning of resources is not "free for all," but comes with reservation price.

Security is the most interesting and controversial aspect that deserves deeper discussion. Typical argument one can here is that public Cloud providers (and they the biggest players like Microsoft, AWS, Google and Oracle) can offer much better levels of protection than organisation that 1/1000 of their size or even smaller, as the big players are able to attract and keep professionals of the highest qualifications. This is true, but, unfortunately, this is not the full picture due to two factors: the concept of shared responsibility and the fact that complexity of security is greatly increased when data is distributed over a wider area or over a greater number of devices, as well as in multi-tenant systems shared by unrelated users. In addition, user access to security audit logs may be difficult or impossible. Private Clouds are in part driven by clients' desire to retain control over the infrastructure and avoid losing control of information security.

But let's talk about shared responsibility concept first. What does this concept actually mean?

The shared responsibility model is a security framework that outlines the roles and responsibilities shared between CSPs and their clients to ensure security. It establishes the Cloud security obligations of a CSP and of the organisation which uses those services. It aims to determine accountability and responsibility, so that all aspects of cloud security are covered. Under the shared responsibility model:

- The CSP responsible for the **security of the Cloud**
- The client responsible for **security in the Cloud**

CSPs are responsible for safeguarding the integrity of their infrastructure. This includes all elements associated with the **security of the Cloud**, such as maintaining network devices, updating server firmware, managing virtualisation hypervisors and securing physical facilities like data centres. This helps building trust among clients that their mission-critical data stored on CSPs' servers are safe and protected against potential loss and cyberthreats.

On the other hand, clients are responsible for the safety and security of operations within their own business systems, often referred to as **security in the Cloud**. Clients should ensure critical security elements, such as user access controls, data encryption at rest and in transit, firewall configurations and endpoint protection, alignment with established cybersecurity guidelines. However, it's important to note that data protection responsibilities vary depending on where the workloads are hosted – for example, on SaaS, PaaS, IaaS or in an on-premises data centre.

Both parties also must comply with industry standards and regulations. Depending on the type of service and the specific CSP, there may also be some overlapping responsibilities between the client organisation and the CSP.

The shared responsibility model for Cloud security is one of those things that seems simple enough on the surface but is actually very complex when putting it into practice. Security will tend to be an afterthought for a large portion of users deploying workloads to the Cloud. Adhering to a shared responsibility model, means clients' security team maintains responsibilities for security as they move applications, data, containers and workloads to the Cloud, while the CSP takes some responsibility, but not much.

These responsibilities are not widely understood. Although 98% of organisations reported Cloud data breach within the past eighteen months, only 13% understand their own Cloud security responsibilities (https://www.wiz.io/academy/shared-responsibility-model). Explaining this complex landscape is especially difficult when one needs to explain it to non-technical senior executives, like, for example, CFOs and CEOs. Many organisations erroneously rely on their CSPs for data protection and application security. Closing this knowledge gap is an essential step toward fulfilling cloud security obligations.

Unfortunately, this notion of shared responsibility can be (and often is) misunderstood, leading to the assumption that Cloud workloads – as well as any applications, data or activity associated with them – are fully protected by the CSP. This can result in clients unknowingly running not fully protected workloads in a public Cloud, making them vulnerable to attacks that target the OS, data or applications. Even securely configured workloads can become a target at runtime, as they are vulnerable to zero-day exploits.

The level of a CSP client's shared responsibility depends on service type: SaaS, PaaS or IaaS.

In the **SaaS** model, CSPs bear most security responsibilities. They secure the software application, including infrastructure and networks, and they are responsible for application-level security. Client's responsibilities often include managing user access and ensuring data is protected and accounts are secure. In short, customers rely heavily on their CSP for security, uptime and system performance.

In the **PaaS** model, CSPs manage infrastructure and underlying platform components, such as runtime, libraries and OSs. Clients are responsible for developing, maintaining and managing data and user access within their applications.

Of the three models, **IaaS** clients have the highest level of responsibility. The CSP secures the foundational infrastructure, including virtual machines, storage and networks – while clients are responsible for securing everything built on the infrastructure, such as the OS, runtime, applications and data.

Complexity of using shared responsibility model in practice stems from three main sources. Firstly, CSPs' marketing machines created impression that Cloud is inherently secure. Secondly, complexity of the contracts (written in the most complex form of "legalise" to protect CSPs) makes the author wonder how many of CSPs' clients actually fully understand contracts they are signing. Thirdly, due to "grey areas" of responsibility that Cloud computing comes with variation of CSPs' obligations depending upon Cloud computing model that is being used (https://blog.r2ut.com/shared-responsibility-model-what-is-it and https://www.wiz.io/academy/shared-responsibility-model) (Table 7.2 and Figure 7.1).

One of the challenges of the shared responsibility model is clear (and, sorry for the pun, shared!) understanding of who is responsible for what. And if it is reasonably easy to achieve in any of the "fully-shaded" areas, it is not easy to achieve in "half-shaded" areas.

This situation is further exacerbated by continuously growing complexity and size of IT infrastructure. And situation becomes even murkier when other MSPs enter the scene (https://vmc.techzone.vmware.com/vmcwaf/avs-shared-responsibility-model#vmware-cloud-shared-responsibility). Despite of all claims about enhanced security clients gain in the Cloud, according to Gartner "by 2025, 99% of cloud-security failures are forecast to come from customers."

Table 7.2 Complexity and vagueness of shared responsibility model, where CC means Cloud Customer and CP means Cloud Provider

Responsibility	On-premises	IaaS	PaaS	SaaS	FaaS
Data classification	CC	CC	CC	CC	CC
Client and end-point protection	CC	CC	CC	CC/CP	CC/CP
Identity and access management	CC	CC	CC/CP	CC/CP	CC/CP
Application-level controls	CC	CC	CP	CP	CP
Network controls	CC	CC/CP	CP	CP	CP
Host infrastructure	CC	CC/CP	CP	CP	CP
Physical security	CC	CP	CP	CP	CP

SHARED RESPONSIBILITY MODEL

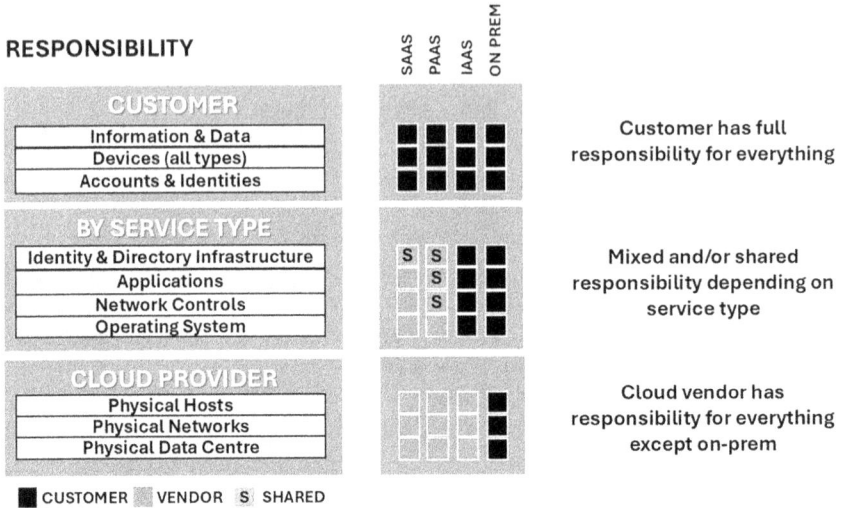

Figure 7.1 Another illustration of shared responsibility model.

Source: https://blog.r2ut.com/shared-responsibility-model-what-is-it.

Every CSP has its own definition of shared responsibility model, like for example MS Azure (https://learn.microsoft.com/en-us/azure/security/fundamentals/shared-responsibility) and AWS (https://docs.aws.amazon.com/wellarchitected/latest/security-pillar/shared-responsibility.html). Some responsibilities are shared by both parties based on service type (SaaS, PaaS or IaaS) and have overlaps, adding to confusion.

For example, within the Microsoft Azure shared responsibility model, CSPs and SaaS and PaaS clients share responsibility for securing identity and directory infrastructure. Alternatively, application security and network controls are shared under the PaaS model. CSPs clearly define the boundaries of responsibilities through service level agreements (SLAs). Overlaps usually exist in the following areas:

Operating systems: Whether a client brings their own OS or deploys an OS provided by the CSP, the responsibility of choosing the appropriate OS to meet an organisation's security requirements lies with the user. If the client chooses the CSP's OS, then CSP is responsible for its security. However, if a client brings their own OS, then client is responsible for its security.

Native vs. third-party tools: Service providers are responsible for deploying, managing, maintaining and updating services. Though when deploying a third-party tool or application as a workload, the customer is charged with securing the application and its data, while the CSP's responsibility is limited to the infrastructure and virtualisation layer.

Server-based vs. serverless computing: In server-based computing, client is responsible for choosing the OS, deploying the workload and configuring the necessary security settings. On the other hand, in serverless or event-based computing, client is accountable for the deployed code and user-defined security or configuration options provided by the CSP.

Network controls: Whether deploying their own firewall or using the CSP's, clients are responsible for configuring firewall rules and ensuring proper security standard configuration.

This earlier discussion raises a lot of questions about practicality and multiple interpretations of shared responsibility model and as a result - about practical levels of security achievable (especially for clients with complex IT landscape) with Cloud computing.

While we are talking about Cloud it is important to remember that CSPs are not immune against Distributed Denial of Service (DDoS) attacks. In February 2020, AWS was hit by a gigantic DDoS attack that used a technique called Connectionless Lightweight Directory Access Protocol (CLDAP) reflection. This technique relies on vulnerable third-party CLDAP servers and amplifies the amount of data sent to the victim's IP address by 56 to 70 times. The attack lasted for three days and peaked at an astounding 2.3 terabytes per second. While the disruption caused by the AWS DDoS attack was far less severe than it could have been, the sheer scale of the attack and the implications for AWS hosting customers potentially losing revenue and suffering brand damage are significant.

In the meantime, Cloud computing technology continues to accelerate digital transformations (see Chapter 5), providing organisations with everything from compute and storage to cloud databases and development tools to advanced data analytics and AI/ML capabilities. In its 2024 report, Crowdstrike noted 75% increase in Cloud intrusions.

Chapter 8

SaaS solutions

The notion of Software as a Service (**SaaS**) has been introduced earlier in Chapter 7 and one of the very earliest examples of **SaaS** applications were introduced in mid-late 1990s for certain verticals with Salesforce launching in 1999 web-based Customer Relationship Management (CRM) solution that allowed very wide spectrum of organisations to manage their customer relationships more efficiently and with greater scalability. Other early **SaaS** applications are NetSuite (launched in 1998), which provided accounting and ERP software, as well as WebEx (launched in 1995), which provided an early example of web conferencing software.

The idea of a **SaaS** model is not new. Software vendors have been selling subscription-based software for years allowing customers to pay only for the services they use. The main difference with **SaaS** is that this model reduces or eliminates the costs of upfront purchasing, installing and maintaining of software. In theory, **SaaS** can greatly reduce organisation's IT budget with a subscription plan that meets the current and future needs of the organisation with flexibility and scalability. This includes using **SaaS** to build mobile and web applications through the Cloud, which has great appeal to organisations by eliminating expensive equipment and ongoing maintenance costs normally associated with software on desktop PCs.

Software as a Service is one on the top four cloud computing service models. So, **SaaS** is a web-based software delivery model that has become an industry standard worldwide. It encompasses an array of applications, spanning accounting, applicant tracking, CRM, document creation/editing and management, email management and photo editing/design. This particular model provides whole software programs that may be accessed online. These programs can be used without users having to locally install or maintain any software. **SaaS** provides a full application stack as a service that customers can access and use. **SaaS** solutions often come as ready-to-use applications, which are managed and maintained by the CSP.

Typically, **SaaS** offerings are hosted on the Cloud 24/7 via the independent software vendor (ISV) or, more frequently now, via a third-party Cloud provider such as AWS (Amazon Web Services), Microsoft Azure or Google. So,

DOI: 10.1201/9781032672601-8

SaaS is a software deployment model in which a third-party provider builds applications on Cloud infrastructure and makes them available to customers via the Internet. This means software can be accessed from any device with an Internet connection and web browser rather than just on the local machine where it is installed, as is the case with traditional software.

Proliferation of **SaaS** started with verticals. Construction collaboration was one of the first areas (as well as CRM) for proliferation of **SaaS** solutions, as **SaaS** is very well suited for collaboration between multiple geographically distributed organisations. The terms "construction collaboration" and "construction collaboration software" were coined in Australia by Aconex in 2001. It was later adopted in 2003 in the United Kingdom when seven UK-based vendors joined together to form the Network for Construction Collaboration Technology Providers (NCCTP), to promote the benefits and use of collaborative technologies in the architecture, engineering, construction (AEC) and related industries.

But before this occurred, two Australian companies have come up with two **SaaS** solutions for construction collaboration. Both of them later have been acquired by the US-based companies.

In 1995, former engineers Russell Mortimer and Steven Joustraand founded Australian company QA Software. The company developed **SaaS** solution for construction collaboration called TeamBinder. TeamBinder is used by many organisations, including, but not limited to, Alcoa, Laing O'Rourke, Transport for NSW, Brookfield Multiplex and others in Australia as well as overseas. In 2018, QA Software was acquired by US construction project management software firm InEight in a deal the US firm declared will help it grow in the local market and begin selling QA's product globally.

TeamBinder is the second local construction technology product to be acquired by larger US firms in recent times, after Oracle's $1.6 billion buyout of ASX-listed Aconex was finalised in 2018 a week earlier. Another Australian company with focus on construction collaboration was founded in 2000 by Leigh Jasper and Robert Phillpot to offer construction collaboration and procurement management services (its name Aconex is a concatenation of Australian Construction Exchange).

SaaS provides a complete software solution that one typically purchases on a pay-as-you-go basis from a service provider. Having said this, there are multiple implementations of charging models for **SaaS** offered by various service providers and sometimes some of them are either in the form of an obscured license agreement or a combination of license agreement and pay-as-you-go, like for example Service Now offering. In the past, TeamBinder charging model was based on the dollar value of the project it was being used for.

In the early 2000s, advancements in technology made it possible for Cloud-based software services to become a reality for businesses and initial pitch was to smaller businesses that did not have a "critical mass" (or, better to say, enough either money, or expertise, or time, or all of the above) to build, implement, deploy and maintain either in-house developed or COTS (commercial-off-the-shelf) applications. Larger organisations can be attracted to SaaS technology for short-term projects or applications that aren't needed after the project has been delivered.

Over the next two decades, SaaS applications started progressive replacement of on-premise software – it was seen as more cost-effective, easier to deploy and maintain, and all around more flexible.

Global circumstances have helped accelerate adoption of SaaS tools, with the GFC of 2007–2009 driving businesses to reduce expenses and the COVID-19 pandemic making Cloud-hosted applications essential in a remote-first environment since 2020. The COVID-19 pandemic impacted global industry in a number of significant ways. One major development was how it created workflows that promoted safety and flexibility with remote access for employees. Pandemic shutdowns forced organisations' leaders to seek digital transformation solutions to protect and adapt their businesses operations and SaaS is one such solution that emerged where businesses gained a significant competitive advantage amid the chaos of the changing global economy.

In 2021, the average organisation used 110 SaaS applications, up from 80 in 2010 (+38%). This marks a nearly 7 times increase since 2017, and a nearly 14 times increase since 2015. It is expected that by 2025 SaaS tools will account for 85% of a organisations' tech stack. Gartner notes that, in 2024 "organisations maintain an average of over 125 different SaaS applications totalling $1,040 per employee annually and that IT typically is aware of only a third of those due to decentralised ownership and sourcing."

Today smaller organisations usually don't have to worry about bringing on a big IT team to manage their technology needs. Tapping into SaaS solutions is seen as an efficient and easy way for organisations to access the prebuilt tools they need for everything from graphic design to project management – and beyond.

The value proposition was perceived so strong, that, for example, since 2013, the Australian Government has made no secret of its desire to pursue a Cloud-first strategy. As part of its 2021 Secure Cloud Strategy, the Government has stated that moving to the Cloud will "generate a faster pace of delivery, continuous improvement cycles and broad access to services." In August 2015, NSW Government followed this trend and in introduced Cloud policy, that has been updated in 2020 and further complemented by Cloud strategy that works hand-in-hand with NSW Digital Strategy and the Federal Digital Transformation Strategy.

Some of the most well-known examples of SaaS are email, calendaring and office tools (such as Microsoft Office 365). Microsoft launched Office 365 in 2011 and this was the real tipping point for SaaS.

Microsoft Office 365 has its roots in two separate earlier products.

The first one might be obvious: Microsoft Office. Introduced in 1988, Office bundled together three core productivity applications: Word, Excel and PowerPoint. Later editions expanded this list to include tools like Visio, OneNote, Outlook, Publisher, SharePoint and Access.

Despite its name, however, Office 365 is not simply the Office productivity applications moved to the Cloud. Office 365 offers the Microsoft Office applications using a subscription model, constantly providing users with fixes, updates and new features, including functionality not included in the traditional desktop Office suites.

But Office 365 is actually more the successor of Microsoft's Business Productivity Online Suite (BPOS) – a set of enterprise products delivered as a subscription service hosted by Microsoft as part of its Online Services. Chief among those products were the 2007 versions of Exchange, SharePoint and Lync (which later became Skype for Business). Launched in 2008, BPOS was primarily targeting smaller businesses.

With Office 365, Microsoft leapt into the **SaaS** world with a solution built for organisations of all sizes. Indeed, the first release of Office 365 was based on the Cloud-centric 2010 versions of the enterprise products in BPOS. For example, in BPOS, SharePoint was little more than a repository for document sharing, while in Office 365, it became a true collaboration tool.

Over the years Microsoft dramatically updated and enhanced its Office 365 platform. For example, it ventured into social networking with Yammer; into business intelligence and data mining with Power BI; and into teamwork organisation with Planner. In 2017, it launched Microsoft Teams, which has quickly become the collaboration platform of choice for organisations around the globe. It's really no surprise that by 2017, revenue from Office 365 was exceeding conventional license sales.

It's also no surprise that the moniker "Office 365" started to feel inadequate for the vastly expanded platform. In July 2017, Microsoft launched the brand Microsoft 365. Initially, it was little more than a marketing or licensing exercise, establishing a bundle that allowed enterprise customers to buy Office 365 Enterprise (E3 and E5), Enterprise Mobility and Security, and Windows 10 Enterprise. As that bundle gained popularity, Microsoft began applying the new brand more liberally. In particular, in 2020, Office 365 plans designed for consumer and small business use were rebranded as Microsoft 365. Office 365 Personal became Microsoft 365 Personal, Office 365 Home became Microsoft 365 Family and so on.

Moreover, since the launch of the brand, Microsoft has transformed Microsoft 365 from a simple bundle of products into a coherent and comprehensive Cloud productivity platform with not just discrete products but broader functionality like information governance, information protection and compliance.

It is logical now to cover real and perceived pros associated with **SaaS**, as well as cons.

There are many pros to using a SaaS development in your business, including:

Focus on core capabilities – Organisation runs its business and lets the SaaS provider to focus on delivery of SaaS solution supporting organisation's core business – this is especially important for smaller organisations that often can't attract and retain the right talent and are always strapped for everything from cash to resources.

Shorter implementation time or, in other words, reduced time to benefit – this is one of the main benefits of using SaaS, as it differs from the traditional model because the software (application) is already installed and configured. Service provider can simply provision the server in Cloud, and in a couple hours and have the application ready for use. This reduces the time that is spent on hardware acquisition, installation and configuration, as well as, on installation and configuration of software and can reduce the issues that get in the way of the software deployment.

Effectively SaaS is ready to go, which is one of the major operational benefits of SaaS as it offers out-of-the-box functionality. Put simply, one can get started almost immediately, as the software is installed and configured ahead of time on the Cloud, and one won't need to go through a lengthy deployment process, as is often the case with on-premises software. Faster deployment times mean organisations can benefit from SaaS product new features and functionality sooner and realise the ROI of their investment faster. For this reason, shorter deployment times are a crucial selling point for many SaaS service providers.

Speaking about shorter implementation time, one needs to remember about another factor that has significant impact on the implementation timeline. Majority (especially vertical) SaaS solutions are designed and built along Henry Ford's saying "Any customer can have a car painted any colour that he wants so as long as it is black" (Henry Ford, "My Life and Work", 1922). Yes, it is possible to configure SaaS solutions to reflect customer's organisation org structure, cost centres, users, direction, origination and destination of various workflows and etc., but all these do not require any changes in the code or any new coding. However, some of SaaS solution allow customisations, sometimes very heavy customisations, by introducing new code developed for the particular customer (or changes to the core code of SaaS solution) and this effectively kills any thoughts about shorter implementation time and moves implementation in the space of development of custom-built software. There are numerous examples of initiatives (especially around Salesforce platform) that take years to develop, build and implement – on par with any in-house or outsourced software development. It is worth noting that heavy customisation also kills the benefit of automated updates: for example, one organisation that uses heavily customised

Salesforce **SaaS** solution based on Process Builder faces 18 months and $6mln project to migrate this to Flow Builder (as Salesforce will no longer be supporting Workflow Rules and Process Builder on December 31, 2025), as well as, it is often has negative impact on security, especially when complex real-time integration is required (see Chapter 4).

Scalability – One of the most significant **SaaS** benefits for customers is its scalability. Compared with traditional software models, **SaaS** can scale up and down to meet fast-changing business requirements. So, if organisation suddenly receives an influx of new users, **SaaS** offers the flexibility to immediately boost organisation's capacity without the need to acquire more hardware that may be required only for a short-term use; and dialling capacity down is similarly easy. With traditional software deployment models, adding or removing users can be a difficult and time-consuming process, whilst **SaaS** solutions allow organisations to easily add or remove users as needed. This makes it easy software resources based on changing demand.

This benefit is especially important for small organisations focused on growth, **SaaS** programs can be a way to accommodate current needs and budgets while considering expectations for the future, as well as, for organisations running campaigns. For many **SaaS** solutions, pricing is determined on a per-user basis so organisations only pay for the users or "seats" they need. There are no space constraints, enabling organisations with fluctuating demands to expand or downsize efficiently without infrastructure worries. Scalability is another notable feature of **SaaS** and is another crucial selling point for many **SaaS** service providers.

Reduced management/maintenance cost/effort and automated updates – another very attractive feature of **SaaS** is automatic access to patches and updates. Use of **SaaS** requires no installation, equipment updates or traditional licensing management. In addition to lower upfront costs, **SaaS** also minimises maintenance costs. With traditional software deployment models, organisations are responsible for maintaining their own servers and updating their software. This can be time-consuming and costly, especially for smaller organisations that don't have dedicated IT staff. On the other hand, **SaaS** vendors take care of server maintenance and software updates, freeing up time and organisation's resources offering 24 × 7 × 365 support. Providers of **SaaS** deal with hardware and software updates, deploying upgrades centrally to the hosted applications and removing this workload and responsibility from their customers.

Use of **SaaS** also allows organisations to access new features and updates easily, as providers of **SaaS** routinely update licenses with new versions.

Software vendors typically roll out new features and updates regularly, and these updates are usually already included in the subscription fee. This means that customers always have access to the latest version of the software, without having to invest in expensive upgrades or add-ons. Outdated tools are eliminated, and organisations no longer shoulder expenses related to licenses, hardware upgrades or infrastructure. Additionally, **SaaS** providers bear the responsibility of addressing bugs and errors. Providers of **SaaS** solutions always look for ways to improve their software and customer experience by providing automatic updates, meaning customers will be automatically updated to the latest version when a new release appears.

It is an excellent benefit because it means that customers will always have the latest features and bug fixes. It also saves customers from manually updating their software, which can be time-consuming and frustrating. As such, automatic updates are a crucial **SaaS** benefit that can save customers time, money and frustration. Automatic updates ensure that customers have access to the safest and most recent version of **SaaS** solution. Although automated updates are typically seen as a benefit of **SaaS**, though in some cases when customer(s) for whatever reason don't want this to happen, automated updates become a drawback of using **SaaS**. Mandatory upgrades may even cause damage due to a lack of compatibility with existing older software that organisation may need to use for whatever reason.

Research by GoCardless has shown that as many as 37% of all organisations choose **SaaS** because of the regular product updates, making this one of the key operational benefits of **SaaS** subscriptions. Instead of buying the upgrade package and going through the lengthy installation process, **SaaS** provider will simply upgrade the solution and make it available to their customers. It's easier to update software continuously using the **SaaS** model because these updates are the provider's responsibility. Some vendors create new versions of products as frequently as every week or every few months to ensure a product remains useful to current customers' needs. For software vendors, this is an opportunity to refine products over time, take advantage of shifts and trends in consumer markets and keep customers satisfied.

Pay-per-use or subscription pricing model is another often-cited benefit of using **SaaS** solutions. Typically, this benefit is centred around several aspects: no capital investment required, leveraging costs over a large user base, avoiding capital expenditure by moving to pay-per-use model that takes away the need for capital expenditure every 3–5–7 years when hardware upgrade is required and moving to recurrent (e.g., perceived more predictable) cost model, and automated software updates. As much as this is true in many cases the picture is not as "black-and-white," as it is often portrayed and costs should be looked at through the prism of Total Cost of Ownership (TCO) over the lifetime of the solution (say, 10–15 years). When looked through these lenses, picture often becomes less obvious, especially considering recent price hikes by **SaaS** providers like, for example, Microsoft.

It is probably right to bring here an analogy of using renting not buying, which is an attractive business model but ultimately costs more, ties companies in for long-term contracts and has dire consequences on security as rarely is any diligence taken by vendor or customer as assumptions are made in the absence of good understanding of shared responsibility model.

Whilst the subscription-based pricing model of **SaaS** can be a benefit, it can also be a disadvantage. Over the long term, the cost of using **SaaS** may be higher than traditional licensing models, especially for businesses that require a large number of users or specific features.

> **Ubiquitous access and business continuity in times of crisis** – Access to **SaaS** solutions via Internet makes it accessible from anywhere, as long as Internet connectivity is available. However, business continuity support in times of crisis is another non "black-and-white" benefit as it can be true and can be false. As **SaaS** solutions are reliant on Internet connectivity, this is definitely true if crisis at hand does not impact Internet connectivity. But if Internet connectivity is gone – **SaaS** solution becomes useless, like cordless power tool without charged battery.
>
> **Duplication in Cloud and automated backups** – typically **SaaS** providers ensure duplication in the Cloud (protecting information across the servers located in different geographies) and perform automated backups. Having said this, one must remember that duplication does not protect against database poisoning (as poisoning happens in real time to all instances) and automated backups are not equal to storing backups off site.

Apart from benefits associated with use of **SaaS** solutions, there are also disadvantages to relying on **SaaS solutions,** and organisations should be aware of potential limiting factors, security concerns and cost issues. While leveraging the "pros" of **SaaS** can open up capabilities and boost efficiencies, every organisation should also review the list of potential "cons" to using SaaS, as there are also a number of cons.

The problem with outsourcing technology or buying it as a service (**SaaS**) is that development and upgrading remain in the hands of another party. If anything unforeseen happens to them – such as an M&A, a disaster, receivership or other event – customer's organisation is vulnerable. It is important to remember that with a **SaaS** model customer's organisation never actually owns the service it is using, leaving the control of important business functions with a third party.

Transition to using **SaaS** is, in a sense, a one-way door as the cost of reversal of such a transition is typically very high. Organisation moving to use **SaaS** solution becomes heavily dependent on this solution and in case of something going wrong – be it an automated update/patch (see Chapter 5), or Internet outage, or **SaaS** service provider going out of business (or selling it to another company) – organisation's business operation may be in jeopardy.

It is also important to mention, that organisations using **SaaS** solutions are a captive audience and has no ability to accurately predict it costs over time, for example, after the 1st of September 2023, Microsoft 365 subscriptions renewals increased by 9%, which is a significant number. These are some obvious disadvantages, associated with the use of **SaaS** solutions:

- Vendor lock-in is the limited ability to negotiate terms or change vendors. Since the vendor controls the software, they also control the terms of the contract, leaving customers with limited bargaining power. This is especially true for small companies, that often face a dilemma of "take it or leave it." In addition, organisations that use **SaaS** solutions are often dependent on a single vendor for their software needs, which can make it difficult to switch to another provider if needed. This can be a problem if the vendor's service is poor, or if the business needs to migrate to another solution.
- Opaqueness of the **SaaS** solutions, as customers get only vague idea about infrastructure and software implementation architectures and maintenance processes – the main issue here is lack of transparency, as there tend to be contradictions between the sales pitch and achievable, practical results. Lack of control is a major disadvantage of **SaaS** solutions, since organisation relies on the **SaaS** provider for updates and new features – software is hosted in the Cloud, updates and new features are controlled by the **SaaS** provider, and customers have little control over when or how those updates and features are implemented.
- There is no guarantee about how long the **SaaS** will be available and if needed cost of migration to elsewhere or cost of re-development can be more than building organisation's own solution in the first place.
- Despite of or in line with various marketing claims about using the best practice, in reality use of **SaaS** solutions (especially in "out of the box" mode without customisations) is likely to result in staying with the pack, instead of getting ahead of the curve), as it's not easy to stay ahead of the curve with **SaaS** – majority of **SaaS** solutions are typically designed to be one-size-fits-all, and users have limited ability to customise or configure the software to their specific needs – **SaaS** empowers, but it is also an equaliser and organisation may lose its competitive advantage/distinction by conforming to "standard (allegedly the best) practice" and faces "customisation dilemma."
  - No customisation potentially leading to loss of competitive advantage/distinction, but maximising numerous benefits of using **SaaS** solution.
  - Customisation resulting in loss of numerous benefits of using **SaaS** solution and, most likely, to high costs, but offering ability differentiate from others.
- System features can change and may cause dysfunction in customer's organisation and customer's organisation has a very limited ability to

manage **SaaS** provider's priorities for new/changed functionality or around decisions to stay with an older version of the software because organisation is not ready to retrain its personnel (or for any other reason), similarly customer may end up waiting for a long time to get new functionality (important for their organisation) implemented in **SaaS** solution.

- Typical organisation needs more than one **SaaS** solution to address all organisation's needs which often increases architectural and integration complexity (see Chapter 4) and creates additional challenges in securing organisation's data.
- One of the biggest drawbacks of **SaaS** is its dependence on Internet connectivity. Since the software is hosted in the Cloud, users need fast, stable and reliable Internet connection to access it – degraded or downed Internet connection significantly impacts **SaaS** solution and subsequently organisation's ability to operate; bad Internet connection causes slow response times, service disruptions or a total outage.
- Providers of **SaaS** solutions typically offer service level agreements (SLAs) for uptime and performance. These SLAs guarantee a certain level of service, and offer compensation if the vendor fails to meet their obligations. However, as **SaaS** solutions are totally dependent upon Internet connectivity and **SaaS** service providers are accountable only for their Cloud-based solutions, one should take their SLAs with a grain of salt – they are necessary to have in place, but, unfortunately, are not sufficient to give customers full peace of mind. For example, Microsoft Office 365 guarantees a 99.9% uptime SLA, but this is meaningful only in case if Internet connectivity is fast, stable and reliable.
- Loss of control as **SaaS** model turns much of that control over to the **SaaS** provider (whilst with traditional perpetual software licence sales model, applications were largely controlled by the organisation that used them) and customer (unless they have very strong **SaaS** governance processes) may not be aware of the increased usage (be it number of users, storage or something else).
- Performance of **SaaS** solution may deteriorate over time – many of **SaaS** solutions get slower over time, especially if they get a lot of adoption and a high number of users – they just get too loaded and heavy. There are likely going through optimisation cycles on the **SaaS** provider's side, but even so, these cycles typically do not improve the performance all that much.
- It can be difficult to terminate services organisation doesn't want any more as some of **SaaS** providers – especially at the enterprise level – have long lead times for termination or allow organisation to terminate only once per year.
- One of the major unintended consequences of **SaaS** proliferation is so-called "**SaaS sprawl**" that can be defined as uncontrollable proliferation of **SaaS** solutions across organisation and resulting in increased

cost, disjointed user experience, difficulties in managing and organising data and increased security concerns – this topic deserves more detailed discussion below.
- Security is a big topic when it is concerned **SaaS**. Marketeers are often claiming higher security offered by **SaaS** solutions, but is this really the case? Let's talk about it in more detail below.

Let's have a closer look at **SaaS** sprawl. Let's deduce the common misconception: **SaaS** sprawl is not similar to shadow IT.

According to Wikipedia, shadow IT refers to information technology (IT) systems, devices, software, applications, services and other IT resources, deployed by departments other than the central IT department, to bypass limitations and restrictions that have been imposed by central information systems and without explicit IT department approval. It has grown exponentially in recent years with the adoption of Cloud-based applications and services. While shadow IT can promote innovation and improve employee productivity, it often introduces security risks to organisation through data leaks and potential compliance violations, especially when such systems are not aligned with corporate governance. Cloud services, and especially **SaaS**, have become the biggest category of shadow IT. The number of services and apps has increased, and staff members routinely purchase them using credit cards and install and use them without involving central IT department. But shadow IT isn't always the result of employees acting alone – according to Gartner, 38% of technology purchases are managed, defined and controlled by business leaders of departments rather than by IT department.

The use of shadow IT has become increasingly prevalent in recent years because of digital transformation efforts (see Chapter 5). In 2019, Everest Group study estimated that nearly half of all IT spend "lurks in the shadows." Notably, these figures were pre-pandemic. It is likely that a sudden influx of remote workers due to COVID-19 restrictions has further increased the use of shadow IT as workers struggle to maintain productivity in a new environment with limited resources.

It is worth noting that digital transformation initiatives brought with them adoption of DevOps, which became a major driver for the proliferation of shadow IT. Cloud and DevOps teams like to run fast and without friction. However, obtaining the visibility and management levels that security teams require often leads to setbacks and delays within the development cycle. When a developer spawns a Cloud workload using their personal credentials, they do so not as a matter of preference or out of malice but because going through the proper internal channels may delay work and cause the entire team to miss a deadline.

So, **SaaS** sprawl is a consequence of shadow IT. The common factor between the two phenomena is that both reveal that the internal departments are not communicating or coordinating, and there is no ideal process to keep these unverified activities manageable.

In essence, **SaaS** sprawl refers to the proliferation of multiple unapproved, duplicate, redundant and poorly integrated **SaaS** solutions. Lack of strong governance, coordination or oversight from IT department, results in **SaaS** sprawl, or in other words, fragmented **SaaS** ecosystem. It's like a garden where each plant grows independently, without any planned arrangement or guidance. Without a clear strategy or framework in place, teams are left to their own devices, making individual decisions about which SaaS applications to adopt.

Scary picture has been painted by Zylo (https://zylo.com/blog/too-many-apps/). Zylo's **SaaS** Management Index report showed that organisations often underestimate the number of applications they use by two to three times the actual number. This is in part because IT now only manages, on average, 17% of an organisation's **SaaS**. This means the remaining 83% are being managed and purchased by business units and individual employees. The repercussion of this is that it becomes difficult (if possible, at all) for IT department to manage and optimise their organisation's **SaaS** assets. The average organisation has approximately 269 **SaaS** applications. For large organisations, this number skyrockets to an average of 650. What's even more scary, is that Zylo's research has shown that the average organisation adds six new **SaaS** applications every 30 days.

One may ask, what are the consequences of **SaaS** sprawl? It is likely to result in inefficiencies and duplicated efforts, as different departments or teams use different applications to perform similar tasks. This can lead to data silos and confusion, negatively impacting productivity and collaboration.

- A recent survey found that over 30% of businesses reported duplicated work due to multiple **SaaS** applications.
- The average company wastes more than $135,000 on unused, underused or duplicate **SaaS** tools.
- According to a RingCentral and CITE Research survey of 2000 knowledge professionals, 7 out of 10 employees waste up to an hour of their workday just hopping between business tools.

Moreover, 50% of employees claim that they find it challenging to look for resources in the jungle of applications in their office, and 46% of workers report that their job is chaotic while juggling between apps, a phenomenon currently represented by the expression "app fatigue." As a result, up to 80% of productivity is lost, as are various other workplace difficulties such as stress, knowledge loss and a drop in output quality. Add to this integration complexities and vendor management challenges:

- Integrating multiple **SaaS** applications can be challenging, especially when they are not designed to work together. This can hinder data flow and collaboration among different systems.

- Managing relationships with multiple **SaaS** providers, including contract negotiations, support and updates, can become complex and time-consuming.

Apart from the above-mentioned issues **SaaS** sprawl brings with it numerous data security and compliance risks, as each **SaaS** application may have its own security protocols and compliance requirements. Without proper oversight, organisations may expose sensitive data or violate regulatory standards. Uncontrolled **SaaS** applications often increase security risks, as data can be stored in unsecured systems or shared with unauthorised individuals and stats below support this:

- 51% organisations experienced a ransomware attack that targeted their **SaaS** data, and 52% of these attacks were successful.
- 43% organisations agree that security misconfigurations lead to security incidents. To minimise these risks, companies must implement proper security measures and monitor SaaS usage within their organisations.

According to a recent survey, 75% of respondents believe that the most significant risk of **SaaS** sprawl is security. It is given that **SaaS** applications hold a large amount of confidential data, consumer financial data, records and other information.

One of the side effects of **SaaS** sprawl can be erosion of trust in IT department when other departments outside IT department influence and control outcomes, and the IT department is often left picking up the pieces without having contributed to the fragmentation. Security is another area that is heavily impacted by this process as **SaaS** sprawl dramatically increases attack surface. When it comes to integration, user experience suffers as employees navigate through a maze of systems, and data quality takes a hit with redundant, sometimes conflicting inputs.

When organisation starts looking at **SaaS** solution it should remember that **SaaS** provider (despite of all marketing claims) is not responsible for ensuring data security of organisation's data. Provider of **SaaS** services and the **SaaS** customer share responsibility for data security in a third-party **SaaS** product (see Chapter 7). Provider of **SaaS** services is primarily responsible for the security of the infrastructure and the application itself, while **SaaS** customer is responsible for configuring the application's security settings, managing user access and educating users about security practices. Whilst **SaaS** provider secures the application itself, strict measures should be taken by the customer in regard to sensitive data. One thing customers of **SaaS** (and especially smaller organisations) need to be aware of is that they cannot assume that their data is secure or retrievable in case of ransomware, disasters or misconfiguration of user privileges. These are situations when having off site backup copy is absolutely important. Organisations that take data security seriously should think about customer-specific encryption both of data at rest and data in transmission.

One of the risks associated with SaaS data security (especially with no data encryption at rest) is a scenario when an employee of SaaS provider with right level of access is incentivised (or threatened with harm to their family) to harvest and farm out sensitive customer's data that can be monetised, or used in the interest of a nation state actors.

As discussed earlier in this chapter, SaaS solutions have a lot of benefits, but there are some unique risks that exist as well. A big one is the risk of credential theft. With stolen credentials, cybercriminals can gain access to employees' Microsoft 365 accounts, databases in AWS and apps with sensitive data such as Salesforce. Once inside the corporate network, they spread malware to steal valuable data and hold it for ransom. Another risk associated with SaaS solutions is that SaaS providers rely on shared infrastructure to deliver their services to multiple customers and this can be very attractive to cybercriminals or nation state actors. The bigger the service and the more sensitive the data, the more attractive it becomes to attackers, who can gain access to multiple customers' data in one place.

As we touched on data security, it is important to mention some obvious considerations, such as data residency and privacy – residency of customers' data and compliance with privacy laws is important for some organisations that prefer to maintain data sovereignty. Concerns around privacy and security are one of the major disadvantages of SaaS for many customers. Whilst many SaaS providers offer an excellent standard of data security, data breaches and hacks are still possible, and some organisations simply aren't comfortable handing over confidential company data to a third party. Similar situation may occur with regulatory compliance requirements – some organisations may find it difficult to meet regulatory requirements while using SaaS solutions. Organisations using SaaS solutions have limited control over the software and SaaS provider's operational processes and practices. This makes it difficult for organisations to conduct audits and ensure that the SaaS provider is following best practices as statements like "Trust me, I am a Doctor!" do not fly in this case. Additionally, SaaS providers most likely will not provide detailed logs or information about the software operation, which can make it difficult for organisations to troubleshoot issues or identify potential security breaches.

Chapter 9

Understanding supply chain challenges

As these lines are being written, the world is still reeling over a double attack on Hezbollah. During the afternoon of Tuesday, September 17, 2024 (approximately at 15:30 local time) pagers belonging to Hezbollah fighters, collaborators and civilian supporters exploded across Lebanon. According to the New York Times, the pagers received a message at 3.30 pm local time that appeared to have come from the group's leadership. It was this message that is believed to have activated the explosions. This attack targeted approximately 5,000 pagers and resulted in at least 12 people killed and nearly 3,000 injured. The following day Wednesday, September 18, 2024 was marked by another round of blasts, when exploding walkie-talkies killed at least 25 and injured more than 600 people.

Pagers are wireless telecommunications devices that receive and display alphanumeric or voice messages. Pagers are wireless devices that can send messages without an Internet connection. Though they've lost popularity to mobile phones, some fields like healthcare still depend on them. Hezbollah has relied heavily on pagers as a low-tech means of communications trying to evade location-tracking by Israel. Pagers are much harder to track than mobile phones, which have long since been abandoned as simply too vulnerable. Israel's assassination of the Hamas bomb-maker Yahya Ayyash demonstrated this as long ago as 1996, when his phone exploded in his hand. In February 2024, Hassan Nasrallah directed Hezbollah fighters to get rid of their phones, saying they had been infiltrated by Israeli intelligence. He told his forces to break, bury or lock their phones in an iron box.

Subsequently, Hezbollah ordered 5,000 AR-924 pagers marketed by the Taiwan-based company Gold Apollo, according to the Lebanon official, and it was these new devices that exploded. Two firms based in Taiwan and Hungary accused in media reports of manufacturing the pagers have both denied responsibility, with the Taiwanese government saying the different parts of the pagers were not from Taiwan. Taiwanese company Gold Apollo said it had authorised use of its brand on the AR-924 pager model – and that a Budapest-based, Hungary company called BAC Consulting KFT produced and sold the pagers. According to one Hezbollah operative, the pagers that

DOI: 10.1201/9781032672601-9

exploded on Tuesday were a new brand that the group had not used before. Elijah Magnier, a Brussels-based security analyst, told AFP: "For Israel to embed an explosive trigger within the new batch of pagers, they would have likely needed access to the supply chain of these devices."

The IC-V82 walkie-talkies are understood to have been bought around the same time as the pagers, and images of the devices examined by Reuters showed an inside panel labelled "ICOM" and "made in Japan." But the devices may not even be from Icom, as the statement from Icom describes the IC-V82 model as a handheld radio which was exported to the Middle East from 2004 to 2014 and has not been shipped since then. Icom said production of that model stopped 10 years ago. The manufacturing of the batteries has also stopped, it says. Icom also said that it is not possible to confirm whether the IC-V82s that exploded in the attack were shipped directly from Icom, or via a distributor. It said any products for overseas markets were sold only to the company's authorised distributors. A sales executive at the US subsidiary of Icom told AP news agency that the devices which exploded in Lebanon appeared to be a knock-off product – adding that it was easy to find counterfeit versions of the product online. Lebanon's communications ministry said that exploded walkie-talkie devices were a discontinued model made by the Japanese firm ICOM. The IC-V82 radios were not supplied by a recognised agent, were not officially licensed and had not been vetted by the security services, the ministry said.

Why are we talking about this in a book on cybersecurity? Because these explosions were the result of supply chain interference. An American official, who spoke on the condition of anonymity, said Israel briefed the United States on the operation – where small amounts of explosives hidden in the pagers were detonated. The Lebanese government and Iran-backed Hezbollah also blamed Israel for the deadly explosions. The Israeli military declined to comment. Whether devices have been rigged with explosives and a switch that could remotely detonate the device, or hacked to cause batteries' thermal run away – in either case we are seeing a supply chain attack. And this is arguably the most impactful and most widely publicised so far supply chain attack that opened a new chapter in what can be used as a warfare.

Going back to cybersecurity the most well-known supply chain attack was SolarWinds attack of 2020. In December 2020, the SolarWinds attack sent shockwaves around the world. Attackers gained unauthorised access to SolarWinds' software development environment, injected malicious code into Orion platform updates, and created a Sunburst malware, potentially compromising national security. The attack affected 18,000+ organisations, including government agencies and major corporations (those impacted included a significant number of the US federal agencies such as the Department of Homeland Security, the State Department, Department of Energy, the National Nuclear Security Administration, Department of Commerce and the Department of the Treasury and private companies like

AT&T, Microsoft, Cisco and Deloitte), and the malicious actors responsible for the breach may have been preparing to carry out the attack since 2019.

SolarWinds Orion is an IT management platform that many government and private organisations use. In April 2020 SolarWinds was recognised in Gartner Magic Quadrant. All software vendors generally provide periodic updates to their systems. Sometimes the update is a critical security patch for a newly discovered zero-day vulnerability, but most often we see general bug and security fixes, as well, as roll out of new features. In March 2020, SolarWinds released a general update for their Orion platform. Unbeknownst to SolarWinds and their thousands of customers, hackers who are now believed to have been nation-state sponsored had inserted malicious code (see Chapter 2) into the update itself, which allowed them access to the many thousands of organisations using the Orion platform. All the client organisations had to do to enable hackers to gain access was to download their legitimate update from their trusted vendor. Given the push to make sure patching cadence is quick, it's no surprise 18,000+ organisations installed the update quickly.

SolarWinds attack was so extensive that communications at the US Treasury and Commerce Departments were reportedly compromised, as it was reported by Krebs. More than 425 of the Fortune 500 use SolarWinds, and some 18,000 SolarWinds Orion customers have downloaded the software with the trojan. This attack allowed attackers to penetrate FireEye and steal tools used by the company's Red Team, the team simulating the attacker during penetration testing.

In April 2023, it was disclosed that the US Department of Justice detected the SolarWinds breach in May 2020, 6 months before the official announcement, and informed SolarWinds of the anomaly. During the same period, Volexity traced a data breach at a US think tank to the organisation's Orion server. In September 2020, Palo Alto Networks identified anomalous activity related to Orion. In each case, SolarWinds was notified but found nothing suspicious.

Some other well-known supply chain attacks include Target USA (2014 – 70 million customers impacted, 40 million debit and credit cards details stolen), Panama Papers (2016 – tax evasion tactics of over 214,000 companies and high-ranking politicians exposed), Equifax (2017 – 147 million customers impacted, sensitive data, like, social security numbers, driver's license numbers, DOBs, addresses stolen), Paradise Papers (2017 – 13.4 million investment records of the wealthy 1% including, Donald Trump, Justin Trudeau, Vladimir Putin's son-in-law and even Queen Elizabeth exposed).

But here's the thing: one of the reasons this happens is because of the widespread adoption of best-of-breed tools, such as the SolarWinds Orion management platform or Microsoft suite (see Chapter 5). With more tools comes an increased attack surface for threat actors to exploit (see Chapter 4).

So, what is a supply chain attack? A supply chain attack is a cyberattack that seeks to damage an organisation by targeting less secure elements in the supply chain. A supply chain attack can occur in any industry, from the financial sector, oil industry, to a government sector. A supply chain attack can happen in software or hardware. Cybercriminals typically tamper with the manufacturing or distribution of a product by installing malware or hardware-based spying components. Symantec's 2019 Internet Security Threat Report states that in 2018 supply chain attacks increased by 78 percent. Although supply chain attack is a broad term without a universally agreed upon definition, in reference to cybersecurity, a supply chain attack can involve physically tampering with electronics (computers, ATMs, power systems, factory data networks) in order to install undetectable malware for the purpose of bringing harm to a player further down the supply chain network. Alternatively, the term can be used to describe attacks exploiting the software supply chain, in which an apparently low-level or unimportant software component used by other software can be used to inject malicious code into the larger software that depends on the component.

Supply chain attack is a type of cyberattack that targets trusted third-party vendors and/or suppliers who offer services or software instead of directly targeting a specific organisation. Supply chain attacks exploit the trust relationships between different organisations. All organisations have some degree of trust in other organisations when they install and use software on their networks, or collaborate as part or vendor or contractor agreements. Supply chain attacks target the weakest link in the chain of trust. Even if your organisation is well-defended and has a strong cyber-security posture, if a trusted vendor is not secure, attackers will target that vendor to bypass whatever security is in place in the primary organisation. By gaining a foothold in the provider's network or provider's software, an attacker can exploit this trust to gain access to a more secure network.

A supply chain attack, which is also known as a third-party attack, value-chain attack or backdoor breach, is when a cybercriminal accesses organisation's network via third-party vendors or through the supply chain. Supply chains can be massive and complex, which is why some attacks are so difficult to trace. Many organisations work with dozens of suppliers for everything from ingredients or production materials to outsourced work and technology. This is why it's so important to protect the supply chain and ensure the organisations one's organisation is working with are as committed to security as you are. Supply chain attacks are a type of cyberattack that is often overlooked. This type of attack can cause catastrophic damage over time and can be more difficult to detect and prevent if one's vendors aren't maintaining strong cybersecurity posture (which is difficult to control). Supply chain attacks work by delivering malicious software via a supplier or vendor. For example, via a software update, like in SolarWinds case, or a keylogger placed on a USB drive can make its way into a large retail organisation, which then logs keystrokes to determine passwords to employee accounts. Cybercriminals can then gain access to sensitive information, customer records, payment information and more.

Considering huge size and complexity of today's IT ecosystem, one should remember that supply chain attack may be of several types (go back the story about pagers and walkie-talkies):

Software Supply Chain Attack: A software supply chain attack only requires one compromised application or piece of software to deliver malware across the entire supply chain. Attacks will often target an application's source code, delivering malicious code into a trusted app or software system. Cybercriminals often target software or application updates as entry points. Software supply chain attacks are incredibly difficult to trace, with cybercriminals often using stolen certificates to "sign" the code to make it look legitimate.

Firmware Supply Chain Attack: Inserting malware into a computer's booting code is an attack that only takes a second to unfold. Once a computer boots up, the malware is executed, jeopardising the entire system. Firmware attacks are quick, often undetectable if you're not looking for them and incredibly damaging.

Hardware Supply Chain Attack: Hardware attacks depend on physical devices, much like the USB keylogger or pagers mentioned earlier. Cybercriminals will target a device that makes its way through the entire supply chain to maximise its reach and damage.

Let's take another trip back in time, as the most serious (and the most denied) case of hardware supply chain attack happened in 2015. On October 4, 2018, Bloomberg Businessweek published a story, which is the culmination of years of investigative work and cites nearly 20 anonymous sources from both the US government and private companies reportedly involved in the affair. The piece says that American authorities first became aware of the existence of the chips in 2015, that the classified probe is still ongoing, and that US officials have identified an unspecified unit of the People's Liberation Army (PLA) as being responsible for sneaking the malicious hardware into the servers. This report alleged that the Chinese government directly interceded to insert small microchips into motherboards from a company called Supermicro, that are in use in servers everywhere from the adult film industry to US military and US Intelligence Community data centres, which make them vulnerable and open them up to remote hacks. The report actually said the following: "Nested on the servers' motherboards, the testers found a tiny microchip, not much bigger than a grain of rice, that wasn't part of the boards' original design. Amazon reported the discovery to US authorities, sending a shudder through the intelligence community." According to unnamed US officials cited in the report, the spying hardware was designed by a unit of the People's Liberation Army and was inserted on equipment manufactured in China for US-based Super Micro Computer Inc.

"Think of Supermicro as the Microsoft of the hardware world," a former US intelligence official told Bloomberg. "Attacking Supermicro motherboards is like attacking Windows. It's like attacking the whole world" he said. By 2015, the San Jose-based firm had sold thousands of servers to more than 900 customers in around 100 countries. That the customer base includes the Central Intelligence Agency, various elements of the US military, the Department of Homeland Security, NASA and the US Congress, as well as big-name tech firms such as Apple. The basic concept behind the alleged plan was pretty straightforward. The PLA unit in question allegedly infiltrated Supermicro's China-based subcontractors who actually make the motherboards and added its own hardware, reportedly no bigger than a grain of rice or the tip of a pencil. These chips themselves don't do much on their own, but what they do is immensely important. The small amount of computer code they contain instructs the completed servers to be open to outside modifications and to be ready to receive further code from other computers remotely, creating a backdoor for hackers to access the information they contain. It could potentially have other functions, as well, including acting as a remotely-operated kill-switch to just shut down a system entirely on command. Hackers could also potentially use it as a gateway to feed false or confusing information into a target system, as well.

The issue reportedly only became apparent in 2015 after Amazon sent systems produced by a company called Elemental, which included Supermicro servers, for a deep security inspection, according to Bloomberg. Elemental manufactured equipment for Department of Defense data centres, the CIA's drone operations and onboard networks of Navy warships. Report said that "Elemental also started working with American spy agencies. In 2009 the company announced a development partnership with In-Q-Tel Inc., the CIA's investment arm, a deal that paved the way for Elemental servers to be used in national security missions across the U.S. government."

Amazon Web Services was looking to acquire Elemental, which specialised in hardware to support online video-streaming services, to help with its own projects, such as Amazon Prime Video. The unnamed third-party security firm located the chips, after which Amazon reportedly informed the Federal Bureau of Investigation, prompting the still ongoing investigation. One of Bloomberg's anonymous sources said that US officials identified at least 30 private companies, including Apple, that had the sabotaged servers. It is important to note, however, that Amazon, Apple and Supermicro have all vociferously and publicly denied Bloomberg's reporting categorically. The three companies say they have never located a piece of malicious hardware in the servers, contacted the US government about such an issue, or are aware of any investigation. The Chinese government, not surprisingly, issued a vague and indirect response when the outlet asked for comment. That being said, in 2016, Apple did stop buying products from Supermicro entirely, citing a security incident it said was unrelated to any hardware tampering.

Software supply chain attacks inject malicious code into an application in order to infect all users of an app, while hardware supply chain attacks compromise physical components for the same purpose. Some examples of software supply chain attacks are:

- **Browser-based attacks** run malicious code on end-user browsers. Attackers may target JavaScript libraries or browser extensions that automatically execute code on user devices. Alternatively, they may also steal sensitive user information that is stored in the browser (via cookies, session storage and so on).
- **Software attacks** disguise malware in software updates. As in the SolarWinds attack, systems may download these updates automatically, inadvertently allowing attackers to infect devices and carry out further actions.
- **Open-source attacks** exploit vulnerabilities in open-source code. Open-source code packages can help organisations accelerate application and software development, but they may also allow attackers to tamper with known vulnerabilities or conceal malware that is then used to infiltrate in the future the target system or device.
- **JavaScript attacks** exploit existing vulnerabilities in JavaScript code or embed malicious scripts in webpages that automatically execute when loaded by a user.
- **Magecart** attacks use malicious JavaScript code to skim credit card information from website checkout forms, which are often managed by third parties. This is also known as "formjacking."
- **Watering hole attacks** identify websites that are commonly used by a large number of users (e.g., a website builder or government website). Attackers may use a number of tactics to identify security vulnerabilities within the site, then use those vulnerabilities to deliver malware to unsuspecting users.
- **Cryptojacking** allows attackers to steal computational resources needed to mine cryptocurrency. They can do this in several ways: by injecting malicious code or ads into a website, embedding cryptomining scripts into open-source code repositories, or using phishing tactics to deliver malware-infected links to unsuspecting users.

Supply chain attacks typically piggyback legitimate processes to gain uninhibited access into an organisation's ecosystem. This attack begins with infiltrating a vendor's security defences. This process is usually much simpler than attacking a victim directly due to the unfortunate myopic cybersecurity practices of many vendors. Penetration could occur via multiple attack vectors. Once injected into a vendor's ecosystem, the malicious code needs to embed itself into a digitally signed process of its host. This is the key to gaining access to a vendor's client network. A digital signature verifies that a piece of software is authentic to the manufacturer, which permits the transmission of the software to all networked parties. By hiding behind this digital signature, malicious code is free to ride the steady stream of software update traffic between a compromised vendor and its client network. For example, the malicious payload that compromised the US government was injected into a SolarWinds Dynamic Link Library file (.dll file). This file was a digitally signed asset of SolarWinds Orion software, the disguise nation-state hackers needed to gain access to SolarWinds' client base.

Compromised vendors unknowingly distribute malware to their entire client network. The software patches that facilitate the hostile payload contain a backdoor that communicates with all third-party servers, this is the distribution point for the malware. A popular service provider could infect thousands of businesses (18,000+ in case of SolarWinds) with a single update, helping threat actors achieve a higher magnitude of impact with a lot less effort. When a victim installs a compromised software update from a service provider, the malicious code is also installed with the same permissions as the digitally signed software, and the cyberattack is initiated. Once installed, a remote access trojan (RAT) is usually activated to give cybercriminals access to each infected host for sensitive data exfiltration.

The SolarWinds supply chain attack was unique in that the hackers didn't initiate remote control immediately. Rather, the malware lay dormant for two weeks before initiating contact with a command-and-control server (a remote session manager for compromised systems also known as C2) via a backdoor.

Supply chain attacks are cyberattacks against third-party vendors in an organisation's supply chain. Historically, supply chain attacks have referred to attacks against trusted relationships, in which an unsecure supplier in a chain is attacked in order to gain access to their larger trading partners. This is what happened in the 2013 attack against Target, where the threat actor gained access to an HVAC contractor in order to enter Target's systems.

While traditional supply chain attacks are still a concern, an even bigger threat facing organisations today is the software supply chain. Software supply chains are highly susceptible to attack, because in modern development organisations, software is not created from scratch, and uses many off-the-shelf components such as third-party APIs, open-source code and proprietary code from software vendors. Any of these could be exposed to security threats and vulnerabilities.

So, today the greater concern is a software supply chain attack. Software supply chains are particularly vulnerable because modern software is not written from scratch: rather, it involves many off-the-shelf components, such as libraries, third-party APIs, open-source code and proprietary code from software vendors. Today, the average software project has 203 dependencies. If a popular app includes one compromised dependency, every organisation that downloads from the vendor is compromised as well, so the number of victims can grow exponentially. And the risk growth with the growth of IT ecosystem (see Chapter 4).

Another interesting case to explore is October 2023 Okta breach as it demonstrates the chain reaction mechanism of theft, rather than direct compromise of organisation that uses certain software through malware covertly pushed into this organisation's environment(s). As discussed earlier, supply chain attacks involve a compromised vendor that's been breached, and the data stolen is then used to compromise the vendors' customers. In this case, attackers breached Okta's support ticket system using a compromised service account. From there the attackers stole HAR files uploaded by Okta's customers, which contained Okta's customers' credentials.

Cloudflare, being an Okta customer, responded to the initial breach by rotating 5000 exposed credentials. Sadly, their efforts fell short. In an extensive report, Cloudflare described how a few weeks after the incident, the Okta attackers used two credentials that were not rotated to compromise their Atlassian suite: a token and service account credentials, both belonging to integrations within Cloudflare's Atlassian environment, and were used to gain administrative access to Cloudflare's Jira, Confluence and Bitbucket. The compromised production Atlassian suite contained Cloudflare's internal Confluence wiki (14,099 pages), Jira bug tracking (2M tickets) and Bitbucket source code (11,904 repositories), all of which the attackers had access to. This was a devastating attack on one of the largest SaaS (see Chapter 8) companies, and severely highlighted the risks of supply chain attacks. Although not initially their fault, Cloudflare's most sensitive data was leaked. This attack demonstrated again how attackers abuse non-human access, which usually goes unmonitored, to achieve high-privilege access to internal systems. Another noteworthy point is that the attackers targeted Cloud, SaaS and also on-prem solutions to expand their access. This emphasises the growing need for a holistic approach to securing non-human identities across the entire organisation.

A supply chain attack is a highly effective way of breaching security by injecting malicious libraries or components into a product without the developer, manufacturer or end client realising it. It's an effective way to steal sensitive data, gain access to highly sensitive environments, or gain remote control over specific systems.

Also, software is reused, so vulnerability in one application can live on beyond the original software's lifecycle. Software that lacks a large user community is particularly vulnerable, because a large community is more likely to expose vulnerability faster than a project with few followers.

As we started to talk about software supply chain attacks it is important to discuss Software Bill of Materials or SBOM. Today's software is complex and consists of multiple components including newly written code, historical and re-used code, libraries and etc. According to the US Cybersecurity & Infrastructure Security Agency (CISA) software bill of materials (SBOM) is a nested inventory, a list of ingredients that make up software components. Effectively, SBOM is a list of all the components, libraries, metadata and other dependencies used in a software application. SBOM also lists the licenses that govern those components, the versions of the components used in the codebase, and their patch status, which allows security teams to quickly identify any associated security or license risks. The idea of SBOM is derived from the term "Bill of Materials" (BOM) that has its origins in the world of manufacturing, where a BOM is an inventory detailing all the items included in a product. In the automotive industry, for example, manufacturers maintain a detailed Bill of Materials for each vehicle.

Numerous high-profile security breaches like Codecov, Kaseya, Apache Log4j – are all supply chain attacks. These attacks prompted President Biden to issue a cybersecurity executive order (EO) detailing guidelines for how federal departments, agencies and contractors doing business with the government must secure their software. Among the recommendations was a requirement for SBOMs, to ensure the safety and integrity of software applications used by the federal government. Although the EO is directed toward organisations doing business with the government, these guidelines, including SBOMs, are likely to become a de facto baseline for how all organisations build, test, secure and operate their software applications. CISA recently released their secure software development attestation form and any providers of software for use in critical infrastructure have 90 days to provide a completed form, and all other providers of software to the US government have 180 days to do the same.

As mentioned earlier, any organisation that builds software needs to maintain an SBOM for their codebases. Organisations typically use a mix of custom-built code, commercial off-the-shelf code, and open-source components to create software. As one principal architect of a leading software supply chain provider noted, "We have over a hundred products, with each of those products having hundreds to thousands of different third-party and open source components." The idea of SBOM is that it allows organisations to track all the components in their codebases (Figure 9.1).

There are various standards that are used for SBOM representation. For example, FDA requires SBOMs to be generated in a format and structure that aligns with the NTIA's (the US Government's National Telecommunications

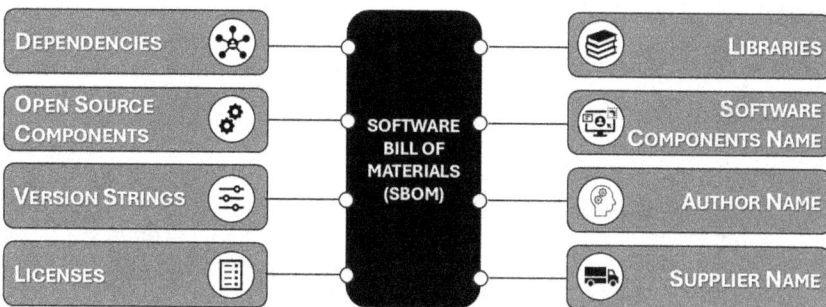

Figure 9.1 What is SBOM.

Source:https://scribesecurity.com/sbom/#definition-of-software-bill-of-materials.

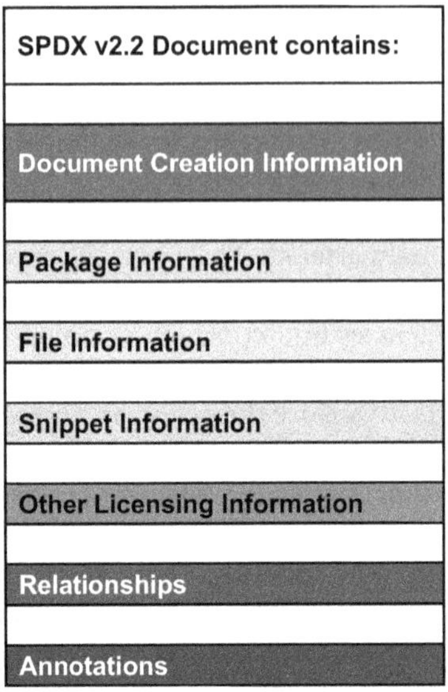

Figure 9.2 Software Package Data Exchange v2.2.

and Information Administration) minimum elements. One of the early standards is Software Package Data Exchange (SPDX) – the primary open standard for Software Bill of Materials formats developed by the Linux Foundation in 2010 (https://scribesecurity.com/sbom/#definition-of-software-bill-of-materials.) (Figure 9.2).

The International Organisation for Standards (ISO) began establishing a standard for marking software components with machine-readable IDs. Software Identification tags (SWID tags), as they're now known, are structured embedded metadata in software that contains information such as the name of the software product, version, developers, relationships and more. SWID Tags can aid in automating patch management, software integrity validation, vulnerability detection, and permitting or prohibiting software installs, similar to software asset management. In 2012, ISO/IEC 19770-2 was confirmed, and it was modified in 2015. There are four main types of SWID tags that are used at various stages of the software development lifecycle:

- **Corpus Tags:** These are used to identify and characterise software components that aren't ready to be installed. According to NIST, corpus tags are "designed to be utilised as inputs to software installation tools and procedures."
- **Primary Tags:** A primary tag's purpose is to identify and contextualise software items once they've been installed.
- **Patch Tags:** Patch tags identify and describe the patch (as opposed to the core product itself). Patch tags can also, and often do, incorporate

contextual information about the patch's relationship to other goods or patches.

- **Supplemental Tags:** Supplemental tags allow software users and software management tools to add useful local utility context information like licensing keys and contact information for relevant parties.

In 2017, the OWASP Foundation released CycloneDX as part of Dependency-Track, an open-source software component analysis tool. CycloneDX is a lightweight standard for multi-industry use, with use cases like vulnerability detection, licensing compliance and assessing old components. CycloneDX 1.4 was launched in January 2022. Cyclone DX can handle four different types of data:

- **Material Flow Chart Metadata:** It contains information on the application/product itself, such as the supplier, manufacturer and components described in the SBOM, as well as any tools used to create an SBOM.
- **Components:** This is a comprehensive list of both proprietary and open-source components, together with license information.
- **Services:** Endpoint URIs, authentication requirements and trust boundary traversals are all examples of external APIs that software can use.
- **Dependencies:** include both direct and indirect connections (Figure 9.3).

It is not enough just to prepare (and maintain) SBOM. It needs to be secured with a digital signature. A digital signature is exactly what it sounds like:

HIGH LEVEL OBJECT MODEL

METADATA	– Supplier – Authors –Component – Manufacturer – Tools
COMPONENTS	– Application – Framework – Operating System – Firmware – Library – Container – Device – File
SERVICES	– Provider – Data – Endpoints – Trust Boundary
DEPENDENCIES	– Components – Services
COMPOSITIONS	– Components – Services – Dependencies
VULNERABILITIES	– Details – Source – Advisories – Risk Ratings – Exploitability
EXTENSIONS	– Properties – Per Organization – Per Team – Per Industry – Formal Taxonomy

Figure 9.3 High Level Object Model.

Source: https://scribesecurity.com/sbom/#definition-of-software-bill-of-materials.

an electronic version of the traditional paper and pen signature. It checks the validity and integrity of digital communications and documents using a sophisticated mathematical approach. It ensures that a message's contents are not tampered with while in transit, assisting us in overcoming the problem of impersonation and tampering in digital communications. Digital signatures have increased in adoption over time and are a cryptographic standard. A signed SBOM provides a checksum, which is a long string of letters and numbers that represent the sum of a piece of digital data's accurate digits and can be compared to find faults or changes. A checksum is similar to a digital fingerprint. On a regular basis, it checks for redundancy (CRC). Changes to raw data in digital networks and storage devices are detected using an error-detecting code and verification function. As a digital signature is meant to serve as a validated and secure way of proving authenticity in transactions – that is, once signed, a person cannot claim otherwise – it holds all signatories to the procedures and actions laid out in the bill.

As one of the core purposes of digital signatures is verification, an unsigned SBOM is not verifiable. One can think of it as a contract: if a contract hasn't been signed by participating parties, there's no real way to enforce it. Similarly, an unsigned SBOM is just an unsigned document: customer cannot hold supplier accountable. This can also lead to further problems down the road, as an unsigned SBOM can also pose risks for organisation's security. Anything that might have otherwise been protected by a signed SBOM is now not protected, and therefore data and information can be sent or replicated anywhere. One of the main purposes of signed SBOMs – accountability – is lost when an SBOM is unsigned as changes can then be made to it without consequences from the creator's or client's sides.

As one can easily see SBOM is very complex, whichever model for its representation is chosen. More importantly, as software evolves (and it always evolves, unless one uses software that is beyond its end of life (EOL) date) SBOM needs regular reviews and updates. Now, to the practicalities. Do you think you will be able to see SBOM for any of Microsoft products? Or SAP? Or any other major software vendor? The answer is 99,999% no, unless your organisation is something like CIA or NSA or FBI. More importantly, even if you could, SBOM that you would see is a point in time (most likely time of the original deployment) SBOM. Now, with quarterly (monthly?) patches and twice a year updates how much effort would it take to validate each next iteration of SBOM? Now, think about organisation with 400 people that has more applications than bums on seats (see Chapter 4) – how much effort would it take to analyse each SBOM update? Or organisation with 4000 applications (see Chapter 4). The effort required to analyse each SBOM change is simply astronomical and thus cannot be implemented in practice as no organisation can afford to do it every time software vendor provides a patch or an update. What does this mean? This means that at no point in time no organisation, especially one of those who embarked on the digital journey (see Chapter 5), can be confident in the third-party software it uses.

Another aspect to consider is a possibility of an insider threat looked through a potential supply chain attack lens. Is it possible to incentivise a software developer to inbuilt malware in their code (or data)? Of course it is possible. And if the money will not be strong enough motivation (though sufficient enough financial inducement that will set the person for life will typically, do it), then a threat to harm next of kin will definitely do this. Again, the question – what does this mean? This means that at no point in time no one can be confident that software used by an organisation does not contain a timebomb malware that can go off at some point in time. And this is before we start talking about vulnerabilities that slipped through code reviews, testing and etc.

In recent years, software supply chain attacks have moved from the periphery of concerns to the forefront. According to Verizon's "2024 Data Breach Investigations Report," the use of vulnerabilities to initiate breaches surged by 180% in 2023, compared to 2022. Of those breaches, 15% involved a third party or supplier, such as software supply chains, hosting partner infrastructures, or data custodians. So far, in 2024, approximately 183 thousand organisations were affected by supply chain cyberattacks worldwide (compared with the annual peak of over 263 million impacted customers in 2019). In the first quarter of 2023, over 60 thousand organisations reported being impacted by supply chain attacks. In June 2024 BlackBerry revealed that more than 75 percent of software supply chains have experienced cyberattacks in the last 12 months. A comparison to a similar study conducted in 2022 reveals persistent challenges in securing software supply chains. In today's interconnected digital landscape, the security of the software supply chain has become a paramount concern for cybersecurity professionals and their organisations. Rapidly increasing reliance on third-party vendors and suppliers introduces numerous vulnerabilities and keeps raising risks associated with supply chain attacks.

Supply chain attacks such as those perpetrated on Blackbaud, Accellion, Microsoft Exchange servers and – most notably – SolarWinds, represent a unique challenge and a key shift in attack vectors for threat actors around the world. The SolarWinds attack demonstrated to organisations that they must have their guard up at all times when it comes to their supply chains. It displayed particular vulnerabilities of manufacturing a software supply chain and how they can pose a risk for high-profile, highly protected companies such as Cisco, Intel and Microsoft. It also shows IT security leaders that once a bad actor has infiltrated one part of the chain, they've infiltrated the whole thing. Security researchers state that supply chain attacks are some of the most difficult threats to prevent because they take advantage of inherent trust. Beyond that, they're difficult to detect, and they can have longer lasting residual effects. Mitigating and remediating a supply chain attack isn't as simple as installing an antivirus or resetting an operating system. As Lucian Constantin said, supply chain attacks "are some of the hardest types of threats to prevent because they take advantage of trust relationships between vendors and customers and machine-to-machine communication channels, such as software update mechanisms that are inherently trusted by users." This was echoed by Jake Williams (SANS Institute): "Supply chain compromises will continue. They are extremely difficult to protect against, highlighting the need for security to be considered as part of the vendor selection process." As Ken Thompson said in his Turing Award Lecture (https://www.cs.cmu.edu/~rdriley/487/papers/Thompson_1984_ ReflectionsonTrustingTrust.pdf): "You can't trust code that you did not totally create yourself."

Chapter 10

Before TCP/IP

From switching circuits to switching packets

We live in the Internet era and Internet is based on TCP/IP protocol. But before we discuss TCP/IP let's again go back in time.

Humans used many types of information communication over a distance since before recorded history. The earliest methods of communication at a distance made use of such media as drums, smoke, fire and reflected rays of the Sun.

Today this is called a telegraph (device or system that allows the transmission of information by coded signal over distance). The word telegraph is derived from the Greek words tele, meaning "distant," and graphein, meaning "to write."

It came into use towards the end of the 18th century to describe an optical semaphore system developed in France. Before the development of the electric telegraph, visual systems were used to convey messages over distances by means of variable displays. One of the most successful of the visual telegraphs was the semaphore developed in France by the two brothers Claude and Ignace Chappe in 1791. Visual signals given by flags and torches were used for short-range communication and continued to be utilised well into the 20th century, when the two-flag semaphore system was widely used, particularly by the world's navies. This system consisted of a pair of movable arms mounted at the ends of a crossbeam on hilltop towers. Each arm of the semaphore could assume seven angular positions 45° apart, and the horizontal beam could tilt 45° clockwise or counterclockwise. In this manner, it was possible to represent numbers and the letters of the alphabet. Multiple towers spaced at intervals of 5 to 10 km (3 to 6 miles) were built to permit transmission over long distances. Signalling rate of three symbols per minute could be achieved, allowing to relay messages across the country in minutes (Figure 10.1).

Electrical telegraphy was the first electrical telecommunications system and the most widely used of a number of early messaging systems called telegraphs, which were devised to send text messages more quickly than physically carrying them. Electrical telegraphy is a point-to-point text messaging system, primarily used from the 1840s until the late 20th century

DOI: 10.1201/9781032672601-10

Figure 10.1 Optical telegraph tower on the Litermont (Liter Mountain), near Nalbach in Saarland, Germany.

Source: https://openverse.org/image/b79c2c62-2bdb-4c94-83fe-62b288 3f234b?q=Litermont+tower.

Electrical telegraphy can be considered the first example of electrical engineering.

Text telegraphy consisted of two or more geographically separated stations, called telegraph offices. The offices were connected by wires, usually supported overhead on utility poles. Many electrical telegraph systems were invented that operated in different ways, but the ones that became widespread fit into two broad categories. First are the needle telegraphs, in which electric current sent down the telegraph line produces electromagnetic force to move a needle-shaped pointer into position over a printed list. Early needle telegraph models used multiple needles, thus requiring multiple wires to be installed between stations. The first commercial needle telegraph system and the most widely used of its type was the Cooke and Wheatstone telegraph, invented in 1837. The second category are armature systems, in which the current activates a telegraph sounder that makes a click; communication on this type of system relies on sending clicks in coded rhythmic patterns.

The electric telegraph did not burst suddenly upon the scene but was a result of a scientific evolution that had been taking place since the 18th century in the field of electricity. One of the key developments was the invention of the voltaic cell in 1800 by Alessandro Volta of Italy. This made it possible to power electric devices in a more effective manner using relatively low voltages and high currents. Previous methods of producing electricity employed frictional generation of static electricity, which led to high voltages and low currents. Many devices incorporating high-voltage static electricity and various detectors such as pith balls and sparks were proposed for use in telegraphic systems. All were unsuccessful, however, due to the severe losses in the transmission wires (especially, in bad weather) and being limited to relatively short distances. Application of the battery to telegraphy was made possible by several further developments in the new science of electromagnetism. In 1820, Hans Christian Ørsted of Denmark discovered that a magnetic needle could be deflected by a wire carrying an electric current. In 1825, William Sturgeon in Britain discovered the multiturn electromagnet, and in 1831, Michael Faraday again in Britain and Joseph Henry in the United States refined the science of electromagnetism sufficiently to make it possible to design practical electromagnetic devices.

The first two practical electric telegraphs appeared at almost the same time. In 1837, two British inventors Sir William Fothergill Cooke and Sir Charles Wheatstone obtained a patent on a telegraph system that employed six wires and actuated five needle pointers attached to five galvanoscopes at the receiver. If current was sent through the proper wires, the needles could be made to point to specific letters and numbers on their mounting plate.

In 1832, Samuel F.B. Morse, a professor of painting and sculpture at the University of the City of New York (later New York University), became interested in the possibility of electric telegraphy and made sketches of ideas for such a system. In 1835, he devised a system of dots and dashes to represent letters and numbers. In 1837, he was granted a patent on an electromagnetic telegraph. Importance of this invention was recognised in an 1838 letter from him to Francis O. J. Smith in 1838, in which Morse wrote: "This mode of instantaneous communication must inevitably become an instrument of immense power, to be wielded for good or for evil, as it shall be properly or improperly directed." (Figures 10.2 and 10.3).

Morse's original transmitter incorporated a device called a portarule, which employed moulded type with built-in dots and dashes. The type could be moved through a mechanism in such a manner that the dots and dashes would make and break the contact between the battery and the wire to the receiver. The receiver, or register, embossed the dots and dashes on an unwinding strip of paper that passed under a stylus. The stylus was actuated

Figure 10.2 Samuel Finley Breese Morse.

Source: https://openverse.org/image/4766de4d-1dfb-4125-a6b4-3947b 6d8e5bd?q=samuel+morse.

Figure 10.3 Key-type Morse telegraph transmitter from the 1840s.

Source: https://openverse.org/image/96a58ce1-f8b2-4e2b-b83a-41d56 a1f1188?q=Morse+telegraph+transmitter+1840s.

by an electromagnet turned on and off by the signals from the transmitter. In 1865, the Morse system became the standard for international communication, using a modified form of Morse's code that had been developed for German railways. The register received a transmitted signal and transcribed the Morse Code symbols onto a strip of paper wound from the spools.

Morse had formed a partnership with Alfred Vail, who was a clever mechanic and is credited with many contributions to the Morse system. Among them are the replacement of the portarule transmitter by a simple make-and-break key, the refinement of the Morse Code so that the shortest code sequences were assigned to the most frequently occurring letters and the improvement of the mechanical design of all the system components. The first demonstration of the system by Morse was conducted for his friends at his workplace in 1837. In 1843, Morse obtained financial support from the US government to build a demonstration telegraph system 60 km (35 miles) long between Washington, DC, and Baltimore, MD. Wires were attached by glass insulators to poles alongside a railroad. The system was completed and public use initiated on May 24, 1844, with transmission of the message, "What hath God wrought!," a translation of a phrase from the Book of Numbers (Numbers 23:23). This inaugurated the telegraph era in the United States, which was to last more than 100 years. Beginning in 1850, submarine telegraph cables allowed for the first rapid communication between people on different continents

Ironically, when Morse offered to sell his telegraph to the US government for $100,000, the postmaster general rejected the offer and as explained by James D. Reid in his 1879 book "The Telegraph in America" the reason for rejection was: "... the operation of the telegraph between Washington and Baltimore had not satisfied him that under any rate of postage that could be adopted, its revenues could be made equal to its expenditures."

Railroad traffic control was one of the earliest applications of the telegraph. Electrical telegraphs were used by the emerging railway companies to provide signals for train control systems, minimising the chances of trains colliding with each other. This was built around the signalling block system in which signal boxes along the line communicate with neighbouring boxes by telegraphic sounding of single-stroke bells and three-position needle telegraph instruments.

But almost immediately this technology became a vital tool for the transmission of news around the country. In 1848, the Associated Press was formed in the United States to pool telegraph expenses, and in 1849, Paul Julius Reuters in Paris initiated telegraphic press service (using pigeons to cover sections where lines were incomplete). By 1851 more than 50 telegraph companies were in operation in the United States. One of the most significant was the New York and Mississippi Printing Telegraph Company formed by Hiram Sibley, which in 1856 was consolidated with a number of other start-up telegraph companies into the Western Union Telegraph Company. Western Union became the dominant telegraph company in the

United States. In 1861, it completed the first transcontinental telegraph line, connecting San Francisco to the Midwest and then on to the East Coast. After the Union Pacific Railroad was finished in 1869, much of the line was relocated to run along the railroad right-of-way to facilitate maintenance.

In 1845, the Electric Telegraph Company was formed in Britain to promote development of the needle telegraph system. As in the United States, development of the telegraph was carried out by highly competitive private companies, but a movement towards monopoly was strong. In 1870, the telegraph industry was nationalised and became part of the British Post Office.

Because of worldwide interest in applications of the telegraph in 1865 the French government at the initiative of Napoleon III, invited international participants to a conference in Paris to facilitate and regulate international telegraph services. A result of the conference was the founding of the forerunner of the modern ITU. In the following year, the first successful transatlantic cables were completed. Soon after its introduction in Europe, it became apparent that the American Morse Code was inadequate for the transmission of the majority of non-English texts because it lacked letters with diacritical marks. A variant that ultimately became known as the International Morse Code was adopted in 1851 for use on cables, for land telegraph lines except in North America, and later for wireless telegraphy. Except for some minor improvements in 1938, the International Morse Code has remained unchanged. It is no longer a major means of commercial or maritime communications, but it is still used by amateur radio operators.

At the 1925 Paris conference, the ITU created two consultative committees to deal with the complexities of the international telephone services, known as CCIF (as the French acronym) and with long-distance telegraphy CCIT (Comité Consultatif International des Communications Téléphoniques à grande distance). Due the basic similarity of many of the technical problems faced by the CCIF and CCIT, a decision was taken in 1956 to merge them into a single entity, the International Telegraph and Telephone Consultative Committee (CCITT, in French: Comité Consultatif International Téléphonique et Télégraphique). The first Plenary Assembly of the new organisation was held in Geneva, Switzerland in December 1956. In 1992, the Plenipotentiary Conference (the top policy-making conference of ITU) saw a reform of ITU, giving the Union greater flexibility to adapt to an increasingly complex, interactive and competitive environment. The CCITT was renamed the Telecommunication Standardisation Sector (ITU-T), as one of three Sectors of the Union alongside the Radiocommunication Sector (ITU-R) and the Telecommunication Development Sector (ITU-D)

New technology and devices kept appearing and led to a continual evolution of the telegraph industry during the latter half of the 19th century and the first half of the 20th century. By 1856 the register in the Morse system was replaced by a sounder, and the code was transcribed directly from the sounds by the operator. During this time telegraph message traffic was rapidly expanding and, in the words of Western Union President William Orton, had become "the nervous system of commerce." Orton contacted inventors Thomas Alva Edison and Elisha Gray to find a way to send multiple telegraph messages on each telegraph line to avoid the great cost of constructing new lines – this was actually the starting point for packet switching. In 1871, J.B. Stearns of the United States completed refinement of the duplex transmission system that originated in Germany by Wilhelm Gintl, which allowed the same line to be used simultaneously for sending and receiving, thus doubling its capacity. This system was further improved by the American inventor Thomas Alva Edison, who patented a quadraplex telegraph system in 1874 that permitted the simultaneous transmission of two signals in each direction on a single line.

A major new concept was introduced in 1871 by Jean-Maurice-Émile Baudot in France. Baudot devised a system for multiplexing (switching) a single line among a number of simultaneous users. The heart of the system was a distributor consisting of a stationary face plate containing concentric circular copper rings that were swept by brushes mounted on a rotating assembly. The face plate was divided into sectors depending on the number of users. Each sector could produce a sequence of five on or off connections that represented a transmitted letter or symbol. The on/off connections were referred to as marks or spaces – in modern terminology, binary digits or bits, consisting of ones or zeros – and the 32 possible symbols that they encoded came to be known as the Baudot Code. In the Baudot system, the transmitter and receiver had to be synchronised so that the correct transmitter and receiver were connected at the same time. The first systems used manual transmission, but this was soon replaced with perforated tape. Variations of this system were used well into the 20th century. This was the forerunner of what is now known as time-division multiplexing.

During this time of rapid change in the telegraph industry a new device, the telephone, was patented by Alexander Graham Bell in 1876. Alexander Graham Bell (March 3, 1847 – August 2, 1922) was a Scottish-born Canadian–American inventor, scientist and engineer who is credited with patenting the first practical telephone. At the age of 12 he built a homemade device that combined rotating paddles with sets of nail brushes, creating a simple dehusking machine that was put into operation at the mill and used steadily for a number of years. With no formal training, he mastered the piano and became the family's pianist. Bell was also deeply affected by his mother's gradual deafness (she began to lose her hearing when he was 12) and this is from where his interest in sound originated. History of his work in this area is fascinating and worth reading about (https://en.wikipedia.org/wiki/Alexander_Graham_Bell). Alexander Graham Bell's telephone system needed a method for connecting calls, leading to the creation of circuit switched telephone network made of manual switchboards. Operators would physically connect wires to establish a circuit between callers.

Although the telephone was originally expected to replace the telegraph completely, this turned out not to be the case: both industries thrived side by side for many decades. Much of the technology developed for telephony had parallel applications in telegraphy. A number of systems were developed that allowed simultaneous transmission of telegraph and telephone signals on the same lines. In 1882, the Western Electric Company was acquired from Western Union by the American Bell Telephone Company. Western Electric had started as a telegraph manufacturing company but later became a major contributor to both the telephone and telegraph industries.

The vacuum tube, patented in the United States in 1907 by Lee De Forest, led to several improvements in telegraph performance and greatly intensified research efforts in telegraphy, telephony and the emerging field of wireless communication. In 1918, modulated carriers with frequency-division multiplexing, in which several different frequencies are transmitted simultaneously over the same line, were introduced. At the receiving end the different signals were separated from one another by frequency-selective filters and sent to separate decoding units, thus allowing as many as 24 telegraph signals to be transmitted over a single telephone channel. Vacuum tube circuits were used to amplify and regenerate weak signals in a manner not previously possible. The development of new magnetic materials enabled more effective loading of transmission lines, thereby improving transmission speeds. In 1928, loading was first successfully applied to submarine cables to allow duplex operation, but it was not until 1950 that Western Union installed the first successful underwater vacuum tube repeater.

In 1903, the British inventor Donald Murray, following the ideas of Baudot, devised a time-division multiplex system for the British Post Office. The transmitter used a typewriter keyboard that punched tape, and the receiver printed text. He modified the Baudot Code by assigning code combinations with the fewest punched holes to the most frequently encountered letters and symbols. Murray sold the patent rights to Western Union and Western Electric in 1912, and this formed the basis of the printing telegraph systems that came into use in the 1920s.

In 1909, Guglielmo Giovanni Maria Marconi, 1st Marquis of Marconi and Karl Ferdinand Braun won a Nobel Prize "in recognition of their contributions to the development of wireless telegraphy." Marconi's work laid the foundation for the development of radio, television and all modern wireless communication systems. He was an Italian inventor, electrical engineer, physicist and politician, known for his creation of a practical radio wave-based wireless telegraph system. Marconi was also an entrepreneur, businessman and founder of The Wireless Telegraph & Signal Company in the United Kingdom in 1897 (which became the Marconi Company).

In 1924, the American Telephone & Telegraph Company (AT&T) introduced a printing telegraph system called the Teletype, which became widely used for business communication. The unit consisted of a typewriter keyboard and a simplex printer. Each keystroke generated a series of coded electric impulses that were then sent over the transmission line to the receiving system. The receiver decoded the pulses and printed the message on a paper tape or other medium.

For many years teleprinters used the five-bit Baudot Code and (in some cases, other specialised codes). With the advent of computers, however, it became apparent that the Baudot Code was no longer adequate, and in 1966, the American Standard Code for Information Interchange (ASCII) was established. ASCII consisted of seven bits, compared with five bits for the Baudot Code. This allowed 128 different coded letters or symbols, as compared with 32 for the Baudot Code. Code speeds of 150 words per minute were possible with teleprinter systems using the ASCII code, as compared with 75 words per minute for those using the Baudot Code.

In today's digital age, it is essential to understand how data travels across networks and circuit switching was initially a key concept in this realm. There are two types of switching – circuit switching and packet switching and historically it all started with circuit switching. Circuit switching, a method of communication where a dedicated path is established for the duration of a transmission, plays a vital role in traditional telephony systems and has influenced modern network designs. This concept is most familiar in traditional telephone systems, where a physical circuit is created for each call. This straightforward approach ensures a consistent and reliable connection between communicating parties, making it an integral part of how information is exchanged. Traditional telephone networks use circuit switching. When one makes a phone call, a dedicated circuit is established between the caller and the receiver for the duration of the call. Three characteristics of circuit switching, are important.

Firstly, before the two parties can talk the circuit between them has to be created, and it takes time for a switch to check if a connection can be made and then to make the connection.

Secondly, when a connection has been made, it creates a dedicated connection. No other party can reach either party of a dedicated connection until that connection has ended.

Thirdly, since switches are very expensive, telephone companies implemented an accounting policy to recover their investment by instituting a minimum charge for every telephone call, generally three minutes.

For voice calls that lasted many minutes, a minimum charge did not represent a problem. But communications between computers often last less than seconds, much less than minutes. It was difficult to imagine how circuit switching could work efficiently for computer communications when such a system took minutes to make a connection, created dedicated connections so only one person, or party, could be in connection with another party, and had a prohibitive cost structure.

The vast majority of people attribute invention of the telephone to Alexander Graham Bell. However, to be fair, it is important to note that invention of the telephone was the culmination of work done by more than one individual, and led to an array of lawsuits related to the patent claims of several individuals and numerous companies. Notable people included in this were Antonio Meucci, Philipp Reis, Simon Alles, Elisha Gray and Alexander Graham Bell.

The concept of the telephone dates back to the string telephone or lover's telephone that has been known for centuries, comprising two diaphragms connected by a taut string or wire. Sound waves are carried as mechanical vibrations along the string or wire from one diaphragm to the other. The classic example is the tin can telephone, a children's toy made by connecting the two ends of a string to the bottoms of two metal cans, paper cups or similar items. The essential idea of this toy was that a diaphragm can collect voice sounds for reproduction at a distance. One precursor to the development of the electromagnetic telephone originated in 1833 when Carl Friedrich Gauss and Wilhelm Eduard Weber invented an electromagnetic device for the transmission of telegraphic signals at the University of Göttingen, in Lower Saxony, helping to create the fundamental basis for the technology that was later used in similar telecommunication devices. Gauss's and Weber's invention is purported to be the world's first electromagnetic telegraph.

In 1840, American Charles Grafton Page passed an electric current through a coil of wire placed between the poles of a horseshoe magnet. He observed that connecting and disconnecting the current caused a ringing sound in the magnet. He called this effect "galvanic music." Innocenzo Manzetti considered the idea of a telephone as early as 1844, and may have made one in 1864, as an enhancement to an automaton built by him in 1849. In 1854, Charles Bourseul, a French telegraph engineer, proposed (but did not build) the first design of a "make-and-break" telephone. That is about the same time that Meucci later claimed to have created his first attempt at the telephone in Italy. An early communicating device was invented around 1854 by Antonio Meucci, who called it a telettrofono ("electrophone"). In 1871, Meucci filed a patent caveat at the US Patent Office. His caveat describes his invention, but does not mention a diaphragm, electromagnet, conversion of sound into electrical waves, conversion of electrical waves into sound or other essential features of an electromagnetic telephone. The Reis telephone was developed by Johann Philipp Reis from 1857 onwards. Allegedly, the transmitter was difficult to operate, since the relative position of the needle and the contact were critical to the device's operation. Thus, it can be called a "telephone," since it did transmit voice sounds electrically over distance, but was hardly a commercially practical telephone in the modern sense. In 1874, the Reis device was tested by the British company Standard Telephones and Cables (STC). The results also confirmed it could transmit and receive speech with good quality (fidelity), but relatively low intensity. In February 1878, Cyrille Duquet invented the handset and obtained a patent for a number of modifications "giving more facility for the transmission of sound and adding to its acoustic properties," and in particular for the design of a new apparatus combining the speaker and receiver in a single unit. Elisha Gray, of Highland Park, IL, also devised a tone telegraph of this kind about the same time as La Cour. In Gray's tone telegraph, several vibrating steel reeds tuned to different frequencies interrupted the current, which at the other end of the line passed through electromagnets and vibrated matching tuned steel reeds near the electromagnet poles. Gray's "harmonic telegraph," with vibrating reeds, was used by the Western Union Telegraph Company. On February 14, 1876, at the US Patent Office, Gray's lawyer filed a patent caveat for a telephone on the very same day that Bell's lawyer filed Bell's patent application for a telephone. The water transmitter described in Gray's caveat was strikingly similar to the experimental telephone transmitter tested by Bell on 10 March 1876, a fact which raised questions about whether Bell (who knew of Gray) was inspired by Gray's design or vice versa. Although Bell did not use Gray's water transmitter in later telephones, evidence suggests that Bell's lawyers may have obtained an unfair advantage over Gray.

When Bell mentioned to Gardiner Hubbard and Thomas Sanders that he was working on a method of sending multiple tones on a telegraph wire using a multi-reed device, the two began to financially support Bell's experiments. Patent matters were handled by Hubbard's patent attorney, Anthony Pollok. In 1876, Bell became the first to obtain a patent for an "apparatus for transmitting vocal or other sounds telegraphically," after experimenting with many primitive sound transmitters and receivers. Because of illness and other commitments, Bell made little or no telephone improvements or experiments for eight months until after his US patent was published, but within a year the first telephone exchange was built in Connecticut (it went into operation in January 1878) and the Bell Telephone Company was created in 1877, with Bell the owner of a third of the shares, quickly making him a wealthy man. In 1880, Bell was awarded the French Volta Prize for his invention and with this money, founded the Volta Laboratory in Washington. In 1885, he co-founded the American Telephone and Telegraph Company (AT&T).

Alexander Bell's telephone system needed a method for connecting calls, leading to the creation of circuit switched telephone network made of manual switchboards. Operators would physically connect wires to establish a circuit between callers. Over time, these manual systems evolved into automated switchboards, significantly improving efficiency and reliability. By the mid-20th century, electronic switches replaced mechanical ones, marking a significant advancement in telephony. Circuit switching provided the backbone for the global telephone network, enabling millions of people to communicate reliably.

Both telegraph and telephone used circuit switching, which is a form of a point-to-point communication that allows communication between two not directly connected points via a number of relay points. This approach is expensive (as one needs a lot of links or cables), does not allow to fully utilise available bandwidth (link is busy no matter if it is used or not during the session) and often has a single point of failure (if one of the links or relay points is disrupted, end-to-end communication becomes impossible). The latest version of circuit switching standard Signalling System No. 7 (SS7) was published by the International Telecommunication Union Telecommunication Standardisation Sector (ITU-T) in March 1993.

SS7 is a set of telephony signalling protocols developed in the 1970s that is used to set up and tear down telephone calls on most parts of the global public switched telephone network (PSTN). The protocol also performs number translation, local number portability, prepaid billing, Short Message Service (SMS) and other services. The protocol was introduced in the Bell System in the United States by the name Common Channel Interoffice Signaling in the 1970s for signalling between No. 4ESS switch and No. 4A crossbar toll offices. The SS7 protocol is defined for international use by the Q.700-series recommendations of 1988 by the ITU-T. Of the many national variants of the SS7 protocols, most are based on variants standardised by the American National Standards Institute (ANSI) and the European Telecommunications Standards Institute (ETSI). National variants with striking characteristics are the Chinese and Japanese Telecommunication Technology Committee (TTC) national variants. It is well known that SS7 has numerous security vulnerabilities, allowing location tracking of callers, interception of voice data, intercept two-factor authentication keys and possibly the delivery of spyware to phones.

First attempts to connect computers were based on the use of telephone networks. And some may also remember dial-up modems used for remote access to computers via telephone lines, with transmission speed evolving from 110 baud/s (1958, Bell 101 modems used in SAGE) to 56 kbit/s (1998, V.90); a baud is one symbol per second; each symbol may encode one or more data bits. Dial-up modems could attach in two different ways: with an acoustic coupler or with a direct electrical connection. Quick evolution of these devices (that is an interesting story in itself) is shown in the table below:

Technology	Kbit/s	Year released
Bell 101	0.1	1958
Bell 103 or V.21	0.3	1962
Bell 202	1.2	1976
Bell 212A or V.22	1.2	1980
Bell 201A	2.0	1962
V.22bis	2.4	1984
V.27ter	4.8	1976
V.32	9.6	1984
V.32bis	14.4	1991

Technology	Kbit/s	Year released
V.32terbo	19.2	1993
V.34	28.8	1994
V.34	33.6	1996
V.90	56.0/33.6	1998
V.92	56.0/48.0	2000

When we talk about early computer networks we should talk about Advanced Research Projects Agency Network (ARPANET) that was an end-product of a decade of computer-communications developments spurred by military concerns that the Soviets might use their jet bombers to launch surprise nuclear attacks against the United States.

By the 1960s, a system called SAGE (Semi-Automatic Ground Environment) had already been built and was using computers to track incoming enemy aircraft and to coordinate military response. The system included 23 "direction centres," each with a massive mainframe computer that could track 400 planes, distinguishing friendly aircraft from enemy bombers. The system required six years and $61 billion to implement.

At the height of the Cold War, military commanders were seeking a computer communications system without a central core, with no headquarters or base of operations that could be attacked and destroyed by enemies thus blacking out the entire network in a single hit. It is important to remember that despite claims that ARPANET's purpose was always more academic than military, at the time it was not about general-purpose network, but about special purpose military/academic network with the main focus on Availability (survivability) and then on Integrity (guaranteed (or almost guaranteed) delivery) with significantly less (if any at all) focus on Confidentiality (as it was assumed that multiple packet delivery routes will prevent interception of the whole message). As David D. Clark, an MIT scientist whose air of genial wisdom earned him the nickname "Albus Dumbledore," mentioned much later: "It's not that we didn't think about security." "We knew that there were untrustworthy people out there, and we thought we could exclude them." Vinton Cerf, who in 1970s and '80s designed key building blocks of the Internet later said: "We didn't focus on how you could wreck this system intentionally. You could argue with hindsight that we should have, but getting this thing to work at all was non-trivial."

How wrong was this judgement! What started as an online community for a few dozen researchers now is accessible to an estimated 3 billion people. That's roughly the population of the entire planet in the early 1960s, when talk of building a revolutionary new computer network began.

ARPANET was the first wide-area packet-switched network with distributed control and one of the first computer networks to implement TCP/IP protocol suite. Both technologies became the technical foundation of the Internet.

ARPANET was established by the Advanced Research Projects Agency (now DARPA) of the United States Department of Defense. Interestingly it has changed its name several times. The Advanced Research Projects Agency (ARPA) gained a "D" when it was renamed the Defense Advanced Research Projects Agency (DARPA) in 1972. The Agency's name briefly reverted to ARPA in 1993, only to have the "D" restored in 1996. It is estimated that 70 percent of all US computer-science research was funded by ARPA. However, many of those involved said that the agency was far from being a restrictive militaristic environment and that it gave them free rein to try out radical ideas. As a result, ARPA was the birthplace not only of computer networks and the Internet but also of computer graphics, parallel processing, computer flight simulation and many other key achievements.

In 1962, Joseph Carl Robnett Licklider joined ARPA and became the first director of ARPA's Information Processing Techniques Office (IPTO). Ivan Sutherland succeeded Licklider as IPTO director in 1964, and two years later Robert Taylor became IPTO director. Taylor would become a key figure in ARPANET's development, partly because of his observational abilities. In his room at the Pentagon, he had access to three teletype terminals, each hooked up to one of three remote ARPA-supported time-sharing mainframe computers – at Systems Development Corp. in Santa Monica, at UC Berkeley's Genie Project and at MIT's Compatible Time-Sharing System project (later known as Multics). Taylor decided that it made no sense to require three teletype machines just to communicate with three incompatible computer systems. It would be much more efficient if the three were merged into one, with a single computer-language protocol that could allow any terminal to communicate with any other terminal. These insights led Taylor to propose and subsequently to secure funding for ARPANET.

ARPANET arose from a desire to share information over great distances without the need for dedicated phone connections between each computer on a network. As it turned out, fulfilling this desire would require packet switching.

In the early 1960's, existing communication networks were made from dedicated, analogue circuits mainly used for voice telephone connections which were always on once activated. In the 1960s, there were few specialist data networks. Communications between computers were usually carried across the public switched telephone network (PSTN) or through dedicated private circuits. The private circuit solution required point-to-point connections on a one-to-one basis. This was expensive and networks were somewhat complex to manage. One solution to this was packet switching that completely changed this perspective by viewing networks as discontinuous, digital systems that transmit data in small packets only when required. At first glance, this looks like it introduces two big changes:

- **Need for conversions**: Analogue communications like voice have to undergo analogue-to-digital encoding to get onto the network and then digital-to-analogue decoding at the destination to be read which means extra work.
- **Losing always-on**: It gives up the advantage of an always-on, continuous connection.

However, it turned out that packet switching introduces four practical advantages that far outweigh any hypothetical disadvantages:

- **Redundancy**: It eliminates dependence on any one communication link (and any relay centre), enabling the network to survive considerable damage.
- **Efficiency**: It enables more than one communication to share a given link at the same time, greatly increasing the number of total communications the network can simultaneously support.
- **Processing**: It moves the computer into the network by placing software systems at each node, which can then be upgraded and improved to enable the network to continually get better.
- **Digital**: It makes communications digital, which means they can be made error free. It also means that communications between digital computers have no conversion overhead or transformation error.

Packet switching represented a paradigm shift in communications technology. Packet switching contrasts with traditional principal networking paradigm, circuit switching, a method which pre-allocates dedicated network bandwidth specifically for each communication session, each having a constant bit rate and latency between nodes. In cases of billable services, such as cellular communication services, circuit switching is characterised by a fee per unit of connection time, even when no data is transferred, while packet switching may be characterised by a fee per unit of information transmitted, such as characters, packets or messages.

A plan for the network was first made available publicly in October 1967, at an Association for Computing Machinery (ACM) symposium in Gatlinburg, Tennessee. There, plans were announced for building a computer network that would link 16 ARPA-sponsored universities and research centres across the United States. In the summer of 1968, the Defense Department put out a call for competitive bids to build the network, and in January 1969, Bolt, Beranek and Newman (BBN) of Cambridge, Massachusetts, won the $1 million contract.

The packet switching concept was first invented by Paul Baran in the early 1960's, and then independently a few years later by Donald Davies. Leonard Kleinrock conducted early research in the related field of digital message switching, and helped build the ARPANET, the world's first packet switching network. Packet switching is a method of grouping data into short messages in fixed format, that is, packets, which are transmitted over a digital network. Packets are made of a header and a payload. Data in the header is used by networking hardware to direct the packet to its destination, where the payload is extracted and used by an operating system, application software or higher layer protocols. Today packet switching is the primary basis for data communications in computer networks worldwide.

Image of UCLA scientist Leonard Kleinrock stands next to a specialised computer – a forerunner to today's routers – that sent the first message over the Internet in 1969 from his original laboratory on the school's campus can be seen at: https://blog.knowbe4.com/how-the-nsa-killed-internet-security-in-1978).

Paul Baran was a researcher at the RAND Corporation think tank who introduced the idea first. He was instructed to come up with a plan for a computer communications network that could survive nuclear attack and continue functioning. He came up with a process that he called "hot-potato routing," which later became known as packet switching. Packets are small chunks of digital information broken up from larger messages. To illustrate in more recent terms: an e-mail might be split into numerous electronic packets of information and transmitted almost at random across the labyrinth of the nation's communication lines. They do not all follow the same route and do not even need to travel in proper sequential order. They are precisely reassembled at the receiver's end, because each packet contains an identifying "header," revealing which part of the larger message it represents, along with instructions for reconstituting the intended message. As a further safeguard, packets contain mathematical verification schemes that ensure data does not get lost in transit. The network on which they travel, meanwhile, consists of computerised switches that automatically forward packets on to their destination. Data packets made computer communications more workable within existing telephone infrastructure by allowing all those packets to flow following paths of least resistance, thereby preventing logjams of digital data over direct, dedicated telephone lines.

Baran invented the concept of packet switching while a young electrical engineer at RAND when he was asked to perform an investigation into survivable communications networks for the US Air Force, building on one of the first wide area computer networks created for the SAGE radar defence system with the goal of providing a fault-tolerant, efficient routing method for telecommunication messages. At the time his ideas contradicted then-established principles of pre-allocation of network bandwidth, exemplified by the development of telecommunications in the Bell System. His results were first presented to the Air Force in the summer of 1961 as briefing B-265, then as paper P-2626, and then in 1964 as a series of eleven amazingly thorough, comprehensive papers titled "On Distributed Communications." Baran's 1964 papers went well beyond documenting the breakthrough concept of packet switching and described a detailed architecture for a large-scale, distributed, survivable communications network designed to withstand almost any degree of destruction of individual components without loss of end-to-end communications. Baran also assumed that any link of the network could fail at any time, and so the network was designed with no central control or administration. Baran's groundbreaking work helped to convince the US Military that wide area digital computer networks were a promising technology. Baran also talked to Bob Taylor and J.C.R. Licklider at the IPTO about the concept since they were also working to build a wide area communications network. Baran's papers then influenced Roberts and Kleinrock to adopt the technology when they joined the IPTO for development of the ARPANET, laying the groundwork that led to its incorporation into the TCP/IP network protocol used on the Internet today.

A packet switching network follows networking protocols that divide messages into packets before sending them. Packet switching is the primary basis for data communications in computer networks worldwide. A simple definition of packet switching is: the routing and transferring of data by means of addressed packets so that a channel is occupied during the transmission of the packet only, and upon completion of the transmission the channel is made available for the transfer of other traffic. Packet switching is a method of grouping data transmitted over a digital network into packets which are composed of a header and a payload. Data in the header is used by networking hardware to direct the packet to its destination where the payload is extracted and used by application software. A network packet can hold about 1500 bytes, but this can be changed. The Maximum Transmission Unit (MTU) for Ethernet, for instance, is 1500 bytes. This might include all the information in a packet (including header and footer information). The data size for a packet might be around 536 bytes.

Packet switching allows delivery of variable bit rate data streams, realised as sequences of short messages in fixed format, that is packets, over a computer network which allocates transmission resources as needed using statistical multiplexing or dynamic bandwidth allocation techniques. As they traverse networking hardware, such as switches and routers, packets are received, buffered, queued and retransmitted (stored and forwarded), resulting in variable latency and throughput depending on the link capacity and the traffic load on the network. Packets are normally forwarded by intermediate network nodes asynchronously using first-in, first-out buffering, but may be forwarded according to some scheduling discipline for fair queuing, traffic shaping or for differentiated or guaranteed quality of service, such as weighted fair queuing or leaky bucket. Packet-based communication may be implemented with or without intermediate forwarding nodes (switches and routers). In case of a shared physical medium, the packets may be delivered according to a multiple access scheme.

As it often happens in the history of innovations, Baran's packet switching work was strikingly similar to the work performed independently a few years later by Donald Davies at the UK National Physical Laboratory, including common details like a packet size of 1024 bits. The term "packet switching" itself was taken from Davies' work, since Baran had called the concept the bit less catching "distributed adaptive message block switching." Davies coined the modern term packet switching and inspired numerous packet switching networks in the decade following, including the incorporation of the concept into the design of the ARPANET in the United States and the CYCLADES network (built by Louis Pouzin) in France. The ARPANET and CYCLADES were the primary precursor networks of the modern Internet.

Established communications – primarily telecommunications - companies expressed scepticism about the idea at first. However, it was quickly shown that a packet switching network typically worked better, faster and cheaper than a dedicated circuit switching network. Since the network shared all of the available bandwidth on a packetised basis, many communications could occur simultaneously. This was a major discovery, and the key concept that made wide-area communication networks and the Internet itself possible and cost-effective.

The first successful test of ARPANET occurred in October 1969. ARPANET was showcased in October 1972 at the first International Conference on Computer Communications (ICCC). The presentation showed that it was possible to create a practical and usable packet-switching network, but much of the communications industry in the United States remained uninterested, if not outright hostile. Seeing an opportunity, BBN and Roberts founded the commercial network company Telenet that year.

There were numerous other initiatives associated with packet switching. Let's have a look at some of them.

In 1975, five nations (Canada, France, Japan, the United Kingdom and the United States) began discussions about a standardising host-network interface and building public packet networks. The result was the protocol CCITT Recommendation X.25, which was adopted by all of the nations involved. In March 1976, X.25 ushered in the next phase of packet switching: the rise of interconnected public service networks. Additional agreements soon followed, among them X.75, a standard protocol for connecting international networks. X.25 is an ITU-T standard protocol suite for packet-switched data communication in wide area networks (WAN). It was originally defined by the International Telegraph and Telephone Consultative Committee (CCITT, now ITU-T) in a series of drafts and finalised in a publication known as The Orange Book in 1976. In December 1978, French semi-public Transpac society opened domestic packet network named TRANSPAC based on the use of X.25. In 1987, Transpac was the world's largest public packet-switched network with revenues of nearly $400m. In 1990, Minitel videotex services accounted for 45% of its data and 20% of its $678m revenue. By 1991, it was operating in fifteen European countries. France Télécom closed the Minitel service, and the Transpac network via which it was available, in June 2012. X.25 was popular with telecommunications companies for their public data networks from the late 1970s to 1990s, which provided worldwide coverage. It was also used in financial transaction systems, such as automated teller machines, and by the credit card payment industry. However, most users have since moved to the Internet Protocol Suite (TCP/IP). It still has a niche use, for example, by the aviation industry. However, neither X.25, nor X.75 did not withstand the test of time.

In Japan, the Nippon Telegraph and Telephone Corporation (NTT) has been carrying out developmental research on a new packet switched data network since 1971 with focus on X.25 protocol.

The Trans-Canada Telephone System announced its own DATAPAC in October 1974. DATAPAC, or Datapac in some documents, was Canada's packet switched X.25 data network. Initial work on a data-only network started in 1972 and was announced by Bell Canada in 1974 as Dataroute. DATAPAC was implemented by adding packet switching to the existing Dataroute networks. It opened for use in 1976 as the world's first public data network designed specifically for X.25. It was operated first by Trans-Canada Telephone System, then Telecom Canada, then the Stentor Alliance, it finally reverted to Bell Canada when the Stentor Alliance was dissolved in 1999. Like most X.25 networks in the western world, DATAPAC services were largely replaced by TCP/IP in the 1990s and 2000s. Bell phased out the service on 31 December 2009.

Whether this was the case of X.25 being a victim of the "camel syndrome" ("a camel is a horse designed by a committee"), or a failed competition with TCP/IP, or all of the above, or something else can be debated.

In 1974, IBM created its proprietary network architecture called SNA. It was designed in the era when the computer industry had not fully adopted the concept of layered communication. Applications, databases and communication functions were mingled into the same protocol or product, which made it difficult to maintain and manage. Systems Network Architecture (SNA). It is a complete protocol stack for interconnecting computers and their resources. SNA describes formats and protocols but, in itself, is not a piece of software. IBM describes SNA as a data communication architecture established by IBM to specify common conventions for communication among the wide array of IBM hardware and software data communication products and other platforms. Among the platforms that implement SNA in addition to mainframes are IBM's Communications Server on Windows, AIX and Linux, Microsoft's Host Integration Server (HIS) for Windows, and many more. The implementation of SNA takes the form of various communications packages, most notably Virtual Telecommunications Access Method (VTAM), the mainframe software package for SNA communications. SNA is still used in banks and other financial transaction networks, as well as in many government agencies. In 1999, there were an estimated 3,500 companies "with 11,000 SNA mainframes." One of the primary pieces of hardware, the 3745/3746 communications controller, has been withdrawn from the market by IBM. IBM continues to provide hardware maintenance service and microcode features to support users. A robust market of smaller companies continues to provide the 3745/3746, features, parts and service. VTAM is also supported by IBM, as is the NCP required by the 3745/3746 controllers. In 2008, an IBM publication said: "with the popularity and growth of TCP/IP, SNA is changing from being a true network architecture to being what could be termed an 'application and application access architecture.' In other words, there are many applications that still need to communicate in SNA, but the required SNA protocols are carried over the network by IP." SNA at its core was designed with the ability to wrap different layers of connections with a blanket of security. To communicate within an SNA environment, one would first have to connect to a node and establish and maintain a link connection into the network. Then one has to negotiate a proper session and then handle the flows within the session itself. At each level there are different security controls that can govern the connections and protect the session information.

Also, in 1975 Digital Equipment Corporation (DEC) created DECnet – a suite of network protocols, originally in order to connect two PDP-11 minicomputers. Later DECnet evolved into one of the first peer-to-peer network architectures, thus transforming DEC into a networking powerhouse in the 1980s. Initially built with three layers, it later (1982) evolved into a seven-layer OSI-compliant networking protocol. DECnet was inbuilt right into the DEC flagship operating system OpenVMS since its inception. Later DEC ported it to Ultrix, OSF/1 (later Tru64) as well as Apple Macintosh and IBM PC running variants of DOS, OS/2 and Microsoft Windows under the name PATHWORKS, allowing these systems to connect to DECnet networks of VAX machines as terminal nodes. While the DECnet protocols were designed entirely by DEC, DECnet Phase II (and later) were open standards with published specifications, and several implementations were developed outside DEC, including ones for FreeBSD and Linux. DEC was a major American company in the computer industry from the 1960s to the 1990s. It was co-founded by Ken Olsen, Harlan Anderson and H. Edward Roberts in 1957. Olsen was president until he was forced to resign in 1992, after the company had gone into precipitous decline. Eventually June 1998 DEC was acquired by Compaq, then in 2001, Compaq was acquired by Hewlett-Packard. The DEC line was discontinued and faded from the market. Subsequently DECnet code in the Linux kernel was marked as orphaned on February 18, 2010 and removed August 22, 2022.

Honeywell Bull also developed its proprietary networking architecture for Honeywell Bull mainframes called Distributed Systems Architecture (DSA). DSA is also no longer supported for client access. Honeywell Bull mainframes are fitted with Mainway for translating DSA to TCP/IP.

The networking architecture for Univac mainframes was the Distributed Computing Architecture (DCA), and the networking architecture for Burroughs mainframes was the Burroughs Network Architecture (BNA).

After the two merged to form Unisys, both were provided by the merged company. Both were obsolete by 2012.

Frame relay was a packet-switching telecommunications service designed for cost-efficient data transmission for intermittent traffic between local area networks (LANs) and between endpoints in wide area networks (WANs). Today it has been largely replaced by other technologies such as Ethernet over copper or fibre optic cable.

Chapter 11

TCP/IP

Now, back to ARPANET.

Building on the ideas of J.C.R. Licklider, Robert Taylor initiated the ARPANET project in 1966 to enable resource sharing between remote computers. Taylor appointed Larry Roberts as program manager. Roberts made the key decisions about the request for proposal to build the network. He incorporated Donald Davies' concepts and designs for packet switching, and sought input from Paul Baran on dynamic routing. In 1969, ARPA awarded the contract to build the Interface Message Processors (IMPs) for the network to Bolt Beranek & Newman (BBN).

The first packet-switched message between two computers was sent in late 1969 by a team of UCLA graduate students under the leadership of Professor Leonard Kleinrock. A member of Kleinrock's team, Charley Kline, had the honour of being first to send it, but anecdotally it was not a successful start. As Charley Kline at UCLA tried logging into the Stanford Research Institute's computer for the first time, the system crashed just as he was typing the letter "G" in "LOGIN." The first computers were connected in 1969 and the Network Control Protocol was implemented in 1970, development of which was led by Steve Crocker at UCLA and other graduate students, including Jon Postel and Vinton Cerf.

By the end of 1969, academic institutions were scrambling to connect to ARPANET. The University of California Santa Barbara and the University of Utah linked up that year. The network was declared operational in 1971. Further software development enabled remote login and file transfer, which was used to provide an early form of email. By April 1971, there were 15 nodes and 23 host terminals in the network (Figure 11.1).

In addition to the four initial schools, contractor BBN had joined, along with MIT, the RAND Corporation and NASA, among others. By January 1973 there were 35 connected nodes, by 1976 there were 63 connected hosts.

During its first 10 years, ARPANET was a test bed for networking innovations. New applications and protocols like Telnet, file transfer protocol (FTP) and network control protocol (NCP) were constantly being devised, tested and deployed on the network. In 1971, BBN's Ray Tomlinson wrote

DOI: 10.1201/9781032672601-11

ARPA COMPUTER NETWORK

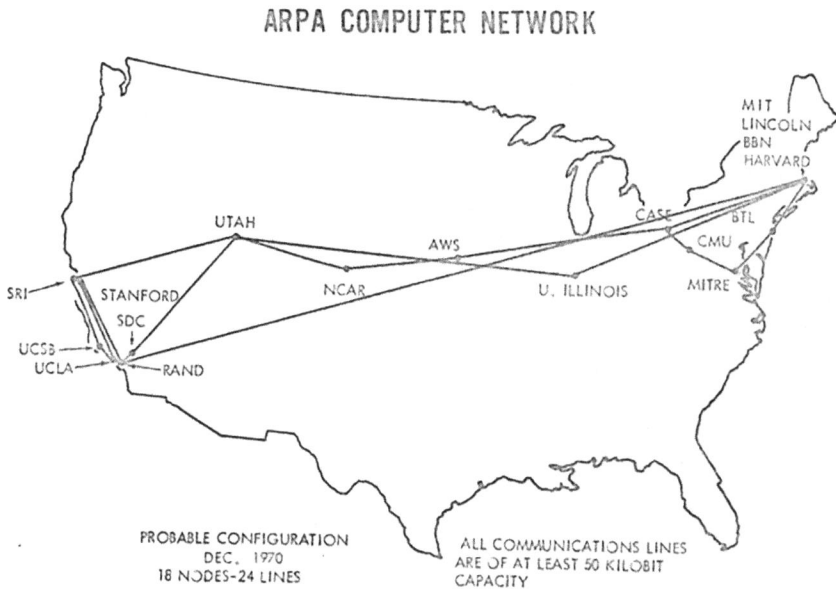

Figure 11.1 ARPANET, 1970.

Source: https://commons.wikimedia.org/w/index.php?curid=61887207.

the first e-mail program, and the ARPANET community took to it instantly. "Mailing lists," which eventually became known as "LISTSERVs," followed e-mail almost immediately, creating virtual discussion groups.

However, ARPANET could not talk to any of the other computer networks that inevitably sprang up in its wake. So in the spring of 1973, Vinton Cerf and Robert Kahn began considering ways of connecting ARPANET with two other networks that had emerged, specifically SATNET (satellite networking) and a Hawaii-based packet radio system called ALOHANET. One day, waiting in a hotel lobby, Vinton Cerf dreamed up a new computer communications protocol, a gateway between networks, which eventually became known as transmission-control protocol/Internet protocol (TCP/IP). TCP/IP, which was first tested on ARPANET in 1977, was a way that one network could hand off data packets to another, then another, and another. As this work progressed, a protocol was developed by which multiple separate networks could be joined into a network of networks. This incorporated concepts pioneered in the French CYCLADES project directed by Louis Pouzin.

By the summer of 1973, Robert Kahn and Vinton Cerf had worked out a fundamental reformulation, in which the differences between local network protocols were hidden by using a common internetwork protocol, and, instead of the network being responsible for reliability, as in the existing ARPANET protocols, this function was delegated to the hosts. Vinton Cerf credited Louis Pouzin and Hubert Zimmermann, designers of the CYCLADES network, with important influences on this design. The new protocol was implemented as the Transmission Control Program in 1974 by Vinton Cerf, Yogen Dalal and Carl Sunshine. Initially, the Transmission Control Program (the Internet Protocol did not then exist as a separate protocol) provided only a reliable byte stream service to its users, not datagrams. Several experimental versions were developed. As experience with the protocol grew, collaborators recommended division of functionality into layers of distinct protocols, allowing users direct access to datagram service. Advocates included Bob Metcalfe and Yogen Dalal at Xerox PARC, Danny Cohen, who needed it for his packet voice work and Jonathan Postel of the University of Southern California's Information Sciences Institute, who edited the Request for Comments (RFCs), the technical and strategic document series that has both documented and catalysed Internet development. As Jonathan Postel stated, "We are screwing up in our design of Internet protocols by violating the principle of layering." Encapsulation of different mechanisms was intended to create an environment where the upper layers could access only what was needed from the lower layers. A monolithic design would be inflexible and lead to scalability issues.

In 1975, a two-network IP communications test was performed between Stanford University and University College London. In November 1977, a three-network IP test was conducted between sites in the United States, the United Kingdom and Norway. Several other IP prototypes were developed at multiple research centres between 1978 and 1983. A computer called a router was provided with an interface to each network. It forwards network packets back and forth between them. Originally a router was called gateway, but the term was changed to avoid confusion with other types of gateways.

During development of the protocol the version number of the packet routing layer progressed from version 1 to version 4. It became known as Internet Protocol version 4 (IPv4) as the protocol that is still in use in the Internet, alongside with its current successor, Internet Protocol version 6 (IPv6). In version 4, written in 1978, Jonathan Postel split the Transmission Control Program into two distinct protocols, the Internet Protocol as connectionless layer and the Transmission Control Protocol as a reliable connection-oriented service. Eventually, version 4 of TCP/IP was installed in the ARPANET for production use in January 1983, after the Department of Defense (DoD) declared in March 1982 TCP/IP as the standard for all military computer networking (Figure 11.2). Eventually, when the Internet consisted of a network of networks, Vinton Cerf's innovation would prove invaluable. It remains the basis of the modern Internet https://www.sciencedirect.com/topics/computer-science/arpanet#:~:text=Starting%20with%20four%20nodes%20at,hundreds%20of%20hosts%20in%201983.).

In 1975, ARPANET was transferred to the Defense Communications Agency. In 1979, Robert Kahn became director of the IPTO at DARPA. By this time the US DoD had multiple packet-switched networks, but none of

```
              APPLICATIONS

   FTP | HTTP

      TCP          UDP

              IP

            NETWORK
```

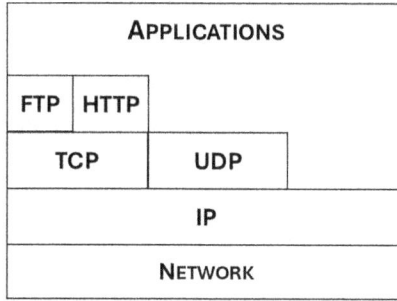

Figure 11.2 Early TCP/IP structure.

**Source: https://www.sciencedirect.com/topics/computer-science/arpa
net#:~:text=Starting%20with%20four%20nodes%20at,hundreds%20
of%20hosts%20in%201983.**

them were compatible with each other. Robert Kahn united the systems by
having the Defense Department adopting TCP/IP, a protocol standard that
Robert Kahn had first imagined in a 1974 paper that he had written with the
prominent software developer Vinton Cerf. Use of TCP/IP spread to other
research laboratories and to the public at large, eventually becoming the
basis for the ultimate packet-switched network: the Internet.

Evolution and growth of ARPANET from 1970 till 1986 can be found
here: https://mercury.lcs.mit.edu/~jnc/tech/arpalog.html

The term "Internet" was adopted in 1983, at about the same time that
TCP/IP came into wide use. In 1983, ARPANET was divided into two parts,
MILNET, to be used by military and defence agencies, and a civilian version
of ARPANET. The word "Internet" was initially coined as an easy way to
refer to the combination of these two networks, to their "internetworking."
The end of ARPANET's days arrived in mid-1982, when its communications protocol, NCP, was turned off for a day, allowing only network sites
that had switched to Vinton Cerf's TCP/IP language to communicate. On
January 1, 1983, NCP was consigned to history, and TCP/IP began its rise
as the universal protocol. The final breakthrough for TCP/IP came in 1985,
when it was built into a version of the UNIX operating system.

From 1986 to 1991, the NSA sponsored the development of security protocols for the Internet under its Secure Data Network Systems (SDNS) program. This brought together various vendors including Motorola who
produced a network encryption device in 1988. The work was openly published from about 1988 by NIST and, of these, Security Protocol at Layer 3
(SP3) would eventually morph into the ISO standard Network Layer
Security Protocol (NLSP). In 1992, the US Naval Research Laboratory
(NRL) was funded by DARPA CSTO to implement IPv6 and to research and
implement IP encryption in 4.4 BSD, supporting both SPARC and x86 CPU
architectures. DARPA made its implementation freely available via MIT.

Under NRL's DARPA-funded research effort, NRL developed the IETF standards-track specifications (RFC 1825 through RFC 1827) for Internet Protocol Security (IPsec). IPsec is a secure network protocol suite that authenticates and encrypts packets of data to provide secure encrypted communication between two computers over an Internet Protocol network. It is used in virtual private networks (VPNs). IPsec includes protocols for establishing mutual authentication between agents at the beginning of a session and negotiation of cryptographic keys to use during the session. IPsec can protect data flows between a pair of hosts (host-to-host), between a pair of security gateways (network-to-network), or between a security gateway and a host (network-to-host). IPsec uses cryptographic security services to protect communications over Internet Protocol (IP) networks. It supports network-level peer authentication, data origin authentication, data integrity, data confidentiality (encryption) and protection from replay attacks. NRL's IPsec implementation was described in their paper in the 1996 USENIX Conference Proceedings. NRL's open-source IPsec implementation was made available online by MIT and became the basis for most initial commercial implementations. The Internet Engineering Task Force (IETF) formed the IP Security Working Group in 1992 to standardise openly specified security extensions to IP, called IPsec. The NRL developed standards were published by the IETF as RFC-1825 through RFC-1827. Secure Sockets Layer (SSL) is IPsec's major rival as a VPN protocol. Though its origins also trace to the 1990s, SSL was a more recent method for implementing VPNs, and it was becoming increasingly popular. The SSL protocol was replaced by a successor technology, Transport Layer Security (TLS), in 2015, but the terms are interchangeable in common parlance and "SSL" is still widely used. The difference between SSL and IPsec VPNs is that SSL/TLS VPNs secure individual web sessions, while IPsec encrypts entire network traffic. However, it is important to remember that neither IPsec, nor later developed Virtual Private Networks did not address underlying insecurities in TCP/IP.

So, what is TCP/IP? The Internet protocol suite, commonly known as TCP/IP, is a framework for organising the set of communication protocols used in the Internet and similar computer networks. The foundational protocols in the suite are the Transmission Control Protocol (TCP), the User Datagram Protocol (UDP) and the Internet Protocol (IP). Early versions of this networking model were known as the DoD model because the research and development were funded by the US DoD through ARPA/DARPA.

The Internet protocol suite provides end-to-end data communication specifying how data should be packetised, addressed, transmitted, routed and received. This functionality is organised into four abstraction layers (and thus is not 100% fully mappable onto the 7-layer reference OSI model from ISO: https://en.wikipedia.org/wiki/OSI_model), which classify all related protocols according to each protocol's scope of networking. An implementation of the layers for a particular application forms a protocol stack. From highest to the lowest, these layers are:

- **Application layer**, providing process-to-process data exchange for applications,
- **Transport layer**, handling host-to-host communication,
- **Internet layer**, providing internetworking between independent networks,
- **Link layer**, containing communication methods for data that remains within a single network segment (link).

The technical standards underlying the Internet protocol suite and its constituent protocols are maintained by the Internet Engineering Task Force (IETF). The Internet protocol suite predates the 7-layerr OSI model, a more comprehensive reference framework for general networking systems.

In 1985, the Internet Advisory Board (later Internet Architecture Board) held a three-day TCP/IP workshop for the computer industry, attended by 250 vendor representatives, promoting the protocol and leading to its increasing commercial use. In 1985, the first Interop conference focused on network interoperability by broader adoption of TCP/IP. The conference was founded by Dan Lynch, an early Internet activist. From the beginning, large corporations, such as IBM and DEC, attended the meeting. IBM, AT&T and DEC were the first major corporations to adopt TCP/IP, despite having competing proprietary protocols. In IBM, from 1984, Barry Appelman's group did TCP/IP development. They navigated the corporate politics to get a stream of TCP/IP products for various IBM systems, including MVS, VM and OS/2. At the same time, several smaller companies, such as FTP Software and the Wollongong Group, began offering TCP/IP stacks for MS DOS MS Windows. The first VM/CMS TCP/IP stack came from the University of Wisconsin.

Some of the early TCP/IP stacks were written single-handedly by a few programmers. Jay Elinsky and Oleg Vishnepolsky of IBM Research wrote TCP/IP stacks for VM/CMS and OS/2, respectively. In 1984, Donald Gillies at MIT wrote a ntcp multi-connection TCP which runs atop the IP/PacketDriver layer maintained by John Romkey at MIT in 1983–1984. John Romkey leveraged this TCP implementation in 1986 when FTP Software was founded. In 1985, Phil Karn created a multi-connection TCP application for ham radio systems (KA9Q TCP).

The spread of TCP/IP was fuelled further in June 1989, when the University of California, Berkeley agreed to place the TCP/IP code developed for BSD UNIX into the public domain. Various corporate vendors, including IBM, included this code in commercial TCP/IP software releases. For MS Windows 3.1, the dominant PC operating system among consumers in the first half of the 1990s, Peter Tattam's release of the Trumpet Winsock TCP/IP stack was key to bringing the Internet to home users. Trumpet Winsock allowed TCP/IP operations over a serial connection (SLIP or PPP). The typical home PC of the time had an external Hayes-compatible modem connected via an RS-232 port with an 8250 or 16550 UART which required this type of stack. Later, Microsoft would release their own TCP/IP add-on stack for MS Windows for Workgroups 3.11 and a native stack in MS Windows 95. These events helped cementing TCP/IP's dominance over other protocols on Microsoft-based networks, which included IBM's Systems Network Architecture (SNA), and on other platforms such as Digital Equipment Corporation's DECnet, Open Systems Interconnection (OSI), and Xerox Network Systems (XNS). Despite rapid proliferation of TCP/IP, between late 1980s and early 1990s, engineers, organisations and nations kept arguing whether the 7-layer OSI model or the TCP/IP suite, would result in the best and most robust computer networks.

An early pair of architectural documents, RFC 1122 and 1123, titled Requirements for Internet Hosts, emphasised architectural principles over layering. They loosely defined a four-layer model, with the layers having names, not numbers, as follows:

- **Application layer** is the scope within which applications, or processes, create user data and communicate this data to other applications on another or the same host. The applications make use of the services provided by the underlying lower layers, especially the transport layer which provides reliable or unreliable pipes to other processes. The communications partners are characterised by the application architecture, such as the client–server model and peer-to-peer networking. This is the layer in which all application protocols, such as SMTP, FTP, SSH, HTTP/HTTPS, operate. Processes are addressed via ports which essentially represent services. The application layer includes the protocols used by most applications for providing user services or exchanging application data over the network connections established by the lower-level protocols. This may include some basic network support services such as routing protocols and host configuration. Examples of application layer protocols include the Hypertext Transfer Protocol (HTTP), the File Transfer Protocol (FTP), the Simple Mail Transfer Protocol (SMTP) and the Dynamic Host Configuration Protocol (DHCP). Data coded according to application layer protocols are encapsulated into transport layer protocol units (such as TCP streams or UDP datagrams), which in turn use lower layer protocols to effect actual data transfer. The TCP/IP model does not consider the specifics of formatting and presenting data and does not define additional layers between the application and transport layers as in the OSI model (presentation and session layers). According to the TCP/IP model, such functions are the realm of libraries and application programming interfaces (APIs). Application layer protocols are often associated with particular client – server applications, and common services have well-known port numbers reserved by the Internet Assigned Numbers Authority (IANA). For example, the HyperText Transfer Protocol uses server port 80 and Telnet uses server port 23. Clients connecting to a service usually use ephemeral ports, that is, port numbers assigned only for the duration of the transaction at random or from a specific range configured in the application. At the application layer, the TCP/IP model distinguishes between user protocols and support protocols. Support protocols provide services to a system of network infrastructure. User protocols are used for actual user applications. For example, FTP is a user protocol and DNS is a support protocol. Application layer in the TCP/IP model is often compared to a combination of the fifth (session), sixth (presentation) and seventh (application) layers of the 7-layer OSI model.

- **Transport layer** performs host-to-host communications on either the local network or remote networks separated by routers. It provides a channel for the communication needs of applications. UDP is the basic transport layer protocol, providing an unreliable connectionless datagram service. TCP provides flow-control, connection establishment and reliable transmission of data. The transport layer establishes basic data channels that applications use for task-specific data exchange. The layer establishes host-to-host connectivity in the form of end-to-end message transfer services that are independent of the underlying network and independent of the structure of user data and the logistics of exchanging information. Connectivity at the transport layer can be categorised as either connection-oriented, implemented in TCP, or connectionless, implemented in UDP. The protocols in this layer may provide error control, segmentation, flow control, congestion control and application addressing (port numbers). Because IP provides only a best-effort delivery, some transport-layer protocols offer reliability.

TCP is a connection-oriented protocol that addresses numerous reliability issues in providing a reliable byte stream:

- data arrives in-order
- data has minimal error (i.e., correctness)
- duplicate data is discarded
- lost or discarded packets are resent
- includes traffic congestion control

The newer Stream Control Transmission Protocol (SCTP) is also a reliable, connection-oriented transport mechanism. It is message-stream-oriented, not byte-stream-oriented like TCP, and provides multiple streams multiplexed over a single connection. It also provides multihoming support, in which a connection end can be represented by multiple IP addresses (representing multiple physical interfaces), such that if one fails, the connection is not interrupted. It was developed initially for telephony applications (to transport SS7 over IP). Reliability can also be achieved by running IP over a reliable data-link protocol such as the High-Level Data Link Control (HDLC).

The User Datagram Protocol (UDP) is a connectionless datagram protocol. Like IP, it is a best-effort, unreliable protocol. Reliability is addressed through error detection using a checksum algorithm. UDP is typically used for applications such as streaming media (audio, video, Voice over IP, etc.) where on-time arrival is more important than reliability, or for simple query/response applications like DNS (see Chapter 12) lookups, where the overhead of setting up a reliable connection is disproportionately large. Real-time Transport Protocol (RTP) is a datagram protocol that is used over UDP and is designed for real-time data such as streaming media.

The applications at any given network address are distinguished by their TCP or UDP port. By convention, certain well-known ports are associated with specific applications. The TCP/IP model's transport or host-to-host layer corresponds roughly to the fourth layer in the OSI model, also called the transport layer.

Quick UDP Internet Connections (QUIC) is rapidly emerging as an alternative transport protocol. Whilst it is technically carried via UDP packets it seeks to offer enhanced transport connectivity relative to TCP. HTTP/3 works exclusively via QUIC.

- **Internet layer** exchanges datagrams across network boundaries. It provides a uniform networking interface that hides the actual topology (layout) of the underlying network connections. It is therefore also the layer that establishes internetworking. It defines and establishes the Internet. This layer defines the addressing and routing structures used for the TCP/IP protocol suite. The primary protocol in this scope is the Internet Protocol, which defines IP addresses. Its function in routing is to transport datagrams to the next host, functioning as an IP router, that has the connectivity to a network closer to the final data destination. The Internet layer does not distinguish between the various transport layer protocols. IP carries data for a variety of different upper layer protocols. These protocols are each identified by a unique protocol number: for example, Internet Control Message Protocol (ICMP) and Internet Group Management Protocol (IGMP) are protocols 1 and 2, respectively. The Internet Protocol is the principal component of the Internet layer, and it defines two addressing systems to identify network hosts and to locate them on the network. The original address system of the ARPANET and its successor, the Internet, is Internet Protocol version 4 (IPv4), that uses a 32-bit IP address and is therefore capable of identifying approximately four billion hosts. This limitation was eliminated in 1998 by the standardisation of Internet Protocol version 6 (IPv6) which uses 128-bit addresses. IPv6 production implementations emerged in approximately 2006.

- **Link layer** defines the networking methods within the scope of the local network link on which hosts communicate without intervening routers. This layer includes the protocols used to describe the local network topology and the interfaces needed to affect the transmission of Internet layer datagrams to next neighbour hosts. The Link layer in the TCP/IP model has corresponding functions in Layer 2 of the 7-layer OSI model.

One of the major and long-lasting features of TCP/IP was introduction of the concept of IP address. IP address works like a postal address, allowing data to be routed to the chosen destination. TCP/IP provides standards for assigning addresses to networks, subnetworks, hosts and sockets, and for using special addresses for broadcasts and local loopback. IP addresses are made up of a network address and a computer (or host or local) address. This two-part address allows a sender to specify the network as well as a specific host on the network. A unique, official network address is assigned to each network when it connects to other Internet networks. However, if a local network is not going to connect to other Internet networks, it can be assigned any network address that is convenient for local use. IP addressing scheme consists of Internet Protocol (IP) addresses and two special cases of IP addresses: broadcast addresses and loopback addresses.

IPv4 uses a 32-bit, two-part address field. The 32 bits are divided into four octets as in the following: 01111101 00001101 01001001 00001111. These binary numbers translate into 125 13 73 15. There are two parts of the IP address are the network address portion and the host address portion. This allows a remote host to specify both the remote network and the host on the remote network when sending information. By convention, a host number of 0 (zero) is used to refer to the network itself. TCP/IP supports three classes of Internet addresses: Class A, Class B and Class C. The different classes of Internet addresses are designated by how the 32 bits of the address are allocated. The particular address Class a network is assigned to depends on the size of the network.

Class A address consists of an 8-bit network address and a 24-bit local or host address. The first bit in the network address is dedicated to indicating the network class, leaving 7 bits for the actual network address. Since the highest number that 7 bits can represent in binary is 128, there are 128 possible Class A network addresses. Of the 128 possible network addresses, two are reserved for special cases: the network address 127 is reserved for local loopback addresses, and a network address of all ones indicates a broadcast address. Therefore, there are 126 possible Class A network addresses and 16,777,216 possible local host addresses. In a Class A address, the highest order bit is set to 0. In other words, the first octet of a Class A address is in the range 1 to 126.

Class B address consists of a 16-bit network address and a 16-bit local or host address. The first two bits in the network address are dedicated to indicating the network class, leaving 14 bits for the actual network address. Therefore, there are 16,384 possible network addresses and

65,536 local host addresses. In a Class B address, the highest order bits are set to 1 and 0. In other words, the first octet of a Class B address is in the range 128 to 191.

Class C address consists of a 24-bit network address and an 8-bit local host address. The first two bits in the network address are dedicated to indicating the network class, leaving 22 bits for the actual network address. Therefore, there are 2,097,152 possible network addresses and 256 possible local host addresses. In a Class C address, the highest order bits are set to 1 and 1. In other words, the first octet of a Class C address is in the range 192 to 223.

Class D (1-1-1-0 in the highest order bits) addresses provide for multicast addresses and are supported by UDP/IP under AIX.

Modern Internet addressing scheme consists of Internet Protocol (IP) addresses and two special cases of IP addresses: broadcast addresses and loopback addresses.

- **Internet addresses:** TCP/IP IPv4 uses a 32-bit, two-part address field.
- **Subnet addresses:** Subnet addressing allows an autonomous system made up of multiple networks to share the same Internet address.
- **Broadcast addresses:** TCP/IP can send data to all hosts on a local network or to all hosts on all directly connected networks. Such transmissions are called broadcast messages.
- **Local loopback addresses:** TCP/IP IPv4 defines the special network address, 127.0.0.1, as a local loopback address.

When deciding which network address class to use, one needed to consider how many local hosts there will be on the network and how many subnetworks there will be in the organisation. Class A, B or C TCP/IP networks can be further divided or subnetted. It becomes necessary as one reconciles the logical address scheme of the Internet (the abstract world of IP addresses and subnets) with the physical networks in use by the real world. If the organisation was small and the network will have fewer than 256 hosts, a Class C address was sufficient. If the organisation was large, then a Class B or Class A address were more appropriate.

As computers read addresses in binary code, conventional notation for IP addresses is the dotted decimal, which divides the 32-bit address into four 8-bit fields. The following binary value: 0001010 00000010 00000000 00110100 can be expressed as: 010.002.000.052 or 10.2.0.52, where the value of each field is specified as a decimal number and the fields are separated by periods.

Configuring the TCP/IP protocol typically requires:

- **An IP address:** For a TCP/IP wide area network (WAN) to work efficiently as a collection of networks, the routers that pass packets of data between networks don't know the exact location of a computer or host for which a packet of information is destined. Routers only know what network the computer or host is a member of and use information stored in their route table to determine how to get the packet to the destination host's network. After the packet is delivered to the destination's network, the packet is delivered to the appropriate computer or host. For this process to work, an IP address has two parts. The first part of an IP address is used as a network address, the last part as a host address.

- **A subnet mask**: The subnet mask is used by the TCP/IP protocol to determine whether a computer or host is on the local subnet or on a remote network. In TCP/IP, the parts of the IP address that are used as the network and host addresses aren't fixed. Unless one has more information, the network and host addresses above can't be determined. This information is supplied in another 32-bit number called a subnet mask.

- **A default gateway**: If a computer or host needs to communicate with a computer or host on another network, it will usually communicate through a device called a router. In TCP/IP terms, a router that is specified on a computer or host, which links the computer's or host's subnet to other networks, is called a default gateway. When a computer or host attempts to communicate with another device using TCP/IP, it performs a comparison process using the defined subnet mask and the destination IP address versus the subnet mask and its own IP address. The result of this comparison tells the computer whether the destination is a local host or a remote host. If the result of this process determines the destination to be a local computer or host, then the computer will send the packet on the local subnet. If the result of the comparison determines the destination to be a remote computer or host, then the computer or host will forward the packet to the default gateway defined in its TCP/IP properties. It's then the responsibility of the router to forward the packet to the correct subnet.

To configure TCP/IP correctly, it's necessary to understand how TCP/IP networks are addressed and divided into networks and subnetworks. The success of TCP/IP as the network protocol is largely because of its ability to connect together networks of different sizes and computers of different types. As mentioned above, these networks are arbitrarily defined into three main classes (along with a few others) that have predefined sizes. Each of them can be divided into smaller subnetworks. A subnet mask is used to divide an IP address into two parts. One part identifies the host (computer), the other part identifies the network to which it belongs. To better understand how IP addresses and subnet masks work, look at an IP address and see how it's organised.

One of the biggest problems that we are facing today is that Internet is built on insecure foundation – TCP/IP. It is appropriate to start discussion of

TCP/IP insecurities noting that packet switching networks are inherently less secure than networks that are based on circuit switching.

As we can see, TCP/IP (and Internet) were created out of a fear that a military strike from the Soviet Union could knock out the whole copper wire-based telephone network that was used at that time for military communications. Collaborative effort that produced unprecedented levels of communication, massive leaps in technology and resulted in creation and proliferation of TCP/IP (and subsequently Internet) also resulted in numerous problems. These problems emanate from the architecture that runs the Internet itself. The fundamental flaw in TCP/IP design is that it was invented with the idea of connecting everything. Unfortunately, when one connects everything, one ends up with an invitation of hackers, cybercriminals and nation state actors for international espionage and/or sabotage. Despite its remarkable success, TCP/IP has experienced a patchwork development process. As the protocol was designed in the 1970s, it was not conceived with the current scale and complexity of the Internet in mind. For example, the need for improved congestion control, error recovery and flow control mechanisms led to the development of TCP extensions, such as Selective Acknowledgments (SACK) and Explicit Congestion Notification (ECN).

As mentioned earlier, IPv4 uses 32-bit addresses, capable of uniquely addressing about 4.3 billion devices. By 1992 it became evident that that would not be enough and the world was already running out of IP addresses. The 1994 RFC 1631 describes Network Address Translation (NAT) as a "short-term solution" to the two most compelling problems facing the Internet at that time: IP address depletion and scaling in routing. NAT is a method of mapping an IP address space into another by modifying network address information in the IP header of packets while they are in transit across a traffic routing device. The technique was originally used to bypass the need to assign a new address to every host when a network was moved, or when the upstream Internet service provider was replaced, but could not route the network's address space. It has become a popular and essential tool in conserving global address space in the face of IPv4 address exhaustion. One Internet-routable IP address of a NAT gateway can be used for an entire private network. By 2004, NAT had become widespread and became the new normal. NAT is not going to disappear overnight, as Google reports a slow 5% annual increase in IPv6 traffic, with global adoption likely to surpass 50% by the end of 2024. In service provider networks, the largest use of network address translation (NAT) tends to be at the point of the subscriber Internet edge, but unfortunately this point is also the largest attack surface, carrying the greatest threats, within the service provider network. Service providers operated under the mistaken assumption that NAT can provide both address translation and security at the subscriber's Internet edge. The security community has tried to dispel that myth, but, unfortunately, it persists. The myth that NAT provides any significant security in light of today's sophisticated attacks needs to be put to rest. There are many implementations of NAT that are inherently insecure. Investigation into some of these has shown increased potential for security holes in NAT deployments. Good example of why NAT is not security is given in: https://0day.work/an-example-why-nat-is-not-security/. From a technical viewpoint, NAT provides:

- No security to IPv6 hosts, as NAT is unnecessary for them.
- No security for stateless NAT hosts.
- No security for stateful NAT host outbound attacks.
- Minimal protection for stateful NAT host ingress attacks, since modern attacks assume the presence of a NAT device and readily compromise or circumvent those devices.
- No tools for responding to security attacks that routinely occur.

Some early examples of TCP/IP insecurities that have been used well-known attacks are:

- Sendmail (debug mode), finger (buffer overflow): Morris worm (11/88)
- IP address spoofing, TCP ISN guessing: Mitnick vs. SDSC (12/94)
- TCP SYN denial-of-service attacks: Panix (9/96)

One of the fundamental flaws within TCP/IP is in its inherent openness, which consequently results in a lack of security. This openness is largely a result of the address-based nature of TCP/IP. In simple terms, the security problem arises because TCP/IP uses the address of a connected device to serve the dual purpose of identifying that device as well. This creates a network vulnerability that is very visible and spoofable to users with malicious intent all over the world. With identity being used simultaneously as a device's address, hackers can simply mock a valid IP address to gain access into organisation's network, where they can steal data, disrupt service and wreak large-scale technological havoc. One of the best examples of this was the very first Internet attack – the Morris worm attack that raged in November 1988, crashing thousands of computers and causing millions of dollars in damage. The worm was using the Internet's essential nature – fast, open and frictionless – to deliver malicious code (see Chapter 3) along communication lines designed to carry harmless files or e-mails. Due to reliance on rsh (normally disabled on untrusted networks), fixes to sendmail, finger, the widespread use of network filtering, and improved awareness of the dangers of weak passwords, the Morris worm type attacks should not succeed on a properly configured system today. But these virus and worm outbreaks have demonstrated that networked computers continue to be vulnerable to new attacks, despite the widespread deployment of antivirus software and firewalls.

The TCP/IP protocol suite (RFC 793), developed under the sponsorship of the US DoD, was designed to work in a trusted environment. The model was developed as a flexible, fault-tolerant set of protocols that were robust enough to avoid failure if one or more nodes went down. The focus was on solving the technical challenges of moving information quickly and reliably, not to secure it. The designers of this original network never envisioned the Internet as it exists today. The problem is that weakness is inherent in the design itself and eradicating it is difficult. Many early TCP/IP protocols are now considered insecure and vulnerable to various attacks, ranging from password sniffing to denial of service. As an example, TCP/IP is shipped with Berkley r-utilities. It is a set of tools featuring remote login (rlogin), remote copying (rcp) and remote command execution (rsh). These commands were developed for password-free access to UNIX machines only. Although the r-utilities tools have some advantages, they should be avoided because they can make access extremely insecure. Also, being inherent in the design means that universal cross-platform vulnerabilities are present in the very infrastructure of the Internet and most other networks, like LANs. The fact that TCP over IP is a low-level protocol means that all the higher-level protocols (for example, HTTP, Telnet and SMTP) are vulnerable by inheritance (for example, hijacking a Telnet connection).

Despite the numerous improvements made to TCP/IP, security remains a significant concern. One of the most critical issues is the lack of encryption in certain applications and protocols, such as Simple Mail Transfer Protocol (SMTP). This exposes email communications to eavesdropping, interception and tampering by malicious actors. Additionally, core design of TCP/IP lacks inherent security features, which resulted in development of various security protocols and mechanisms as "add-ons." Examples include the Secure Sockets Layer (SSL) and its successor, Transport Layer Security (TLS), which provide encryption for web traffic. While these security measures have improved the overall safety of Internet communications, the reliance on external solutions exposes the protocol to potential vulnerabilities and necessitates constant vigilance to stay ahead of emerging threats.

Furthermore, TCP/IP is susceptible to various types of attacks that exploit its inherent weaknesses. Examples include SYN flood attacks, which target the connection establishment process of TCP, and IP spoofing attacks, which allow an attacker to impersonate another user or device on the network. Decades later, after hundreds of billions of dollars spent on computer security, the threat posed by the Internet seems to grow worse each year. Where hackers once attacked only computers, the penchant for destruction has now leapt beyond the virtual realm to threaten banks, retailers, government agencies, a Hollywood studio and, experts worry, critical mechanical systems in dams, power plants and aircraft.

The TCP/IP protocol suite is vulnerable to a variety of attacks ranging from password sniffing to denial of service. Software to carry out most of these attacks is freely available on the Internet. These vulnerabilities – unless carefully controlled – can place the use of the Internet or Intranet at considerable risk. In 1999, B. Harris and R. Hunt written an article "TCP/IP security threats and attack methods" published in Computer Communications,

Vol. 22, issue 10, pp. 885-897 that describes various TCP/IP insecurities (https://www.sciencedirect.com/science/article/abs/pii/S0140 36649900064X). This article describes a range of known attack methods focusing in particular on SYN flooding, IP spoofing, TCP sequence number attack, TCP session hijacking, RST and FIN attacks and the Ping O' Death.

More recent coverage of security flaws in TCP can be found in the paper "Security Flaw in TCP/IP and Proposed Measures" published by Springer in Lecture Notes in Networks and Systems 896 "Cyber Security and Digital Forensics, Select Proceedings of the International Conference, ReDCySec 2023" (pp. 93-107).

There are some specific insecurities associated with UDP, TCP, IP and ICMP. The rest of this chapter is devoted to demonstrating these insecurities and some attacks that are possible due to these insecurities and is significantly based on: https://cjs6891.github.io/el7_blog/texts/cisco-ccna-cyber-ops-secfnd-3/.

The IP is a connectionless protocol that is mainly used to route information across the Internet. The role of IP is to provide best-effort services for the delivery of information to its destination. IP depends on upper-level TCP/IP suite layers to provide accountability and reliability. Layers above IP use the source address in an incoming packet to identify the sender. To communicate with the sender, the receiving station sends a reply by using the source address in the datagram. Because IP makes no effort to validate whether the source address in the packet that is generated by a node is actually the source address of the node, you can spoof the source address and the receiver will think the packet is coming from that spoofed address.

Today many programs for generating spoofed IP datagrams are available for free on the Internet (for example, hping lets you prepare spoofed IP datagrams with a simple one-line command), and one can send them to almost anybody in the world. One can also spoof at various other layers, for example, using ARP spoofing to link a MAC address to the IP address of a legitimate host on the network to divert the traffic that is intended for one station to someone else. The SMTP is also a target for spoofing the email source because SMTP does not verify the sender's address, so you can send any email to anybody pretending to be someone else. As a result, no packet can be trusted, and each packet must earn its trust through the network's ability to classify and enforce policy.

The following are some key IP address-based vulnerabilities that threaten network infrastructures:

- **Man-in-the-middle (MITM) attack:** An MITM attack intercepts a communication between two systems. Essentially, the attacker inserts a device into a network that grabs packets that are streaming past. Those packets are then modified and placed back on the network for forwarding to their original destination. An MITM attack can completely defeat sophisticated authentication mechanisms because the attacker waits until after a communication session is established, which means that authentication has been completed, before starting to intercept packets. An MITM attack does not directly threaten network's stability, but it is an exploit that can target a specific destination IP address. A form of MITM is called "eavesdropping." Eavesdropping differs only in that the perpetrator just copies IP packets off the network without modifying them in any way.

- **Session hijacking:** Session hijacking is a twist on the MITM attack. The attacker gains physical access to the network, initiates an MITM attack and then hijacks that session. In this manner, an attacker can illicitly gain full access to a destination computer by assuming the identity of a legitimate user. The legitimate user sees the login as successful but then is cut off. Subsequent attempts to log back in might be met with an error message that indicates that the user ID is already in use.
- **IP address spoofing:** Attackers spoof the source IP address in an IP packet. IP spoofing can be used for several purposes. In some scenarios, an attacker might want to inspect the response from the target victim (nonblind spoofing). In other cases, the attacker might not care (blind spoofing). Blind IP address spoofing is most frequently used in DoS attacks. Some reasons for nonblind spoofing include sequence-number prediction, hijacking an authorised session, and determining the state of a firewall.
- **DoS attack:** In a DoS attack, an attacker attempts to prevent legitimate users from accessing information or services. By targeting one's computer and its network connection, or the computers and network of the sites that one is trying to use, an attacker may be able to prevent one from accessing email, websites, online accounts, or other services that rely on the affected computers. Common types of DoS attacks include packet floods and service buffer overflow attacks. Other types of DoS attacks rely on specific flaws in various applications and operating systems, such as the "teardrop" attack which can crash older operating systems. For example, in a teardrop attack, the attacker sends IP fragmented packets to a target machine. Since the machine receiving such packets cannot reassemble them due to a bug in TCP/IP fragmentation reassembly, the packets overlap one another, crashing the target network device.
- **DDoS attack:** A DDos attack is a DoS attack that features a simultaneous, coordinated attack from multiple source machines. The best-known example of a DDoS attack is the "smurf" attack.
- **Smurf attack:** A smurf attack exploits the IP broadcast addressing to create a DoS. This attack uses the ICMP. One of the utilities that are embedded in ICMP is ping which is commonly used to test the availability of certain destinations. The attacker installs smurf on a hacked computer. The hacked machine starts continuously pinging one or more networks – with all their attached hosts – using IP broadcast

addresses. Every host that receives the broadcast ping message is obliged to respond with its availability. The result is that the hacked machine gets overwhelmed with inbound ping responses.

- **Resource exhaustion attacks:** Resource exhaustion attacks are forms of DoS attacks. These attacks cause the server's or network's resources to be consumed to the point where the service is no longer responding, or the response is significantly reduced. By targeting IP routers, an attacker may adversely affect the integrity and availability of the network infrastructure, including end-to-end IP connectivity. Router resources that are commonly affected by packet flood attacks include the following: CPU, packet memory, route memory, network bandwidth and vty lines.

ICMP is a connectionless protocol that does not use any port number and works in the network layer. ICMP was not designed to transfer data in the same way as TCP and UDP. Rather, ICMP was intended to carry diagnostic messages to ensure that links were active and to report error conditions when routes, hosts and ports are inaccessible. ICMP datagrams are often associated with commands that are used by network administrators, such as ping (ICMP echo request) and traceroute (ICMP TTL expired in transit). Most ICMP traffic is generated by routers, firewalls and endpoints to detect and diagnose network connection issues. ICMP is used to inform the sender that a problem has occurred while delivering the data. For example, if a host is unable to reach another host on the local network, the sender might receive an ICMP Destination Host Unreachable message. If a network link is down, a router may respond to the sender with an ICMP Destination Network Unreachable message. If traffic is blocked by a firewall, the firewall may respond with an ICMP Host Administratively Prohibited message. Every network device must implement ICMP, but some administrators block ICMP to prevent attackers from gathering information about their internal network.

All types of ICMP traffic are interesting to the security analyst. Either the traffic is user-generated (as in the case of ICMP echo requests) or generated by network devices (as in the case of ICMP Destination Network Unreachable). In the first case, it is beneficial to know when someone, especially inside the network, is generating ICMP traffic to scan the network. Traffic that is generated by network devices indicates network issues and outages.

The following are the security issues of ICMP messages that a security analyst needs to understand:

- **Reconnaissance and scanning:** ICMP can be used to launch information gathering attacks. Attackers can use different methods within the ICMP to find out live host, network topology and OS fingerprinting, and determine the state of a firewall.

- **ICMP unreachables**: ICMP unreachables are commonly used by attackers to perform network reconnaissance. In cybersecurity, network reconnaissance refers to the act of scanning the target network to gather information about the target. For example, during a protocol or port scan, if an ICMP Protocol Unreachable is received, the attacker will know that protocol is not in use on the target device.
- **ICMP mask reply**: A feature that malicious insiders or outsiders can use to map your IP network. This feature allows the router to tell a requesting endpoint what the correct subnet mask is for a given network.
- **ICMP redirects**: A router sends an IP redirect to notify the sender of a better route to the destination. The intended purpose of this feature was for a router to send redirects to the hosts on its directly connected networks. However, an attacker can leverage this feature to send an ICMP redirect message to the victim's host, luring the victim's host into sending all traffic through a router that is owned by the attacker. ICMP redirect attack is an example of a MITM attack, where an attacker will act as the middleman for all communication from the source to the destination.
- **ICMP router discovery**: IRDP allows hosts to locate routers that can be used as a gateway to reach IP-based devices on other networks. Because IRDP does not have any form of authentication, it is impossible for end hosts to tell whether the information they receive is valid or not. Therefore, an attacker can perform a MITM attack using IRDP. Attackers can also spoof the IRDP messages to add bad route entries into a victim's routing table, so that the victim's host will forward the packets to the wrong address, and be unable to reach other networks, resulting in a form of a DoS attack.
- **Firewalk**: Firewalking is an active reconnaissance technique that employs traceroute-like techniques to analyse IP packet responses to determine the gateway Access Control List (ACL) filters and map out the networks. The firewalking technique works by sending out TCP or UDP packets with a TTL that is one greater than the targeted gateway. If the gateway allows the traffic, it will forward the packets to the next hop where they will expire and elicit an ICMP Time Exceeded message. If the gateway host does not allow the traffic, it will likely drop the packets and the attacker will see no response.
- **ICMP tunneling**: ICMP tunnel, also known as ICMPTX, establishes a covert connection between two remote computers, using ICMP echo requests and reply packets. ICMP tunneling can be used to bypass firewalls rules through obfuscation of the actual traffic inside the ICMP packets. Without proper deep packet inspection or log review, network administrators will not be able to detect this type of tunneling traffic through their network. A common ICMP tunneling program

is LOKI that uses ICMP as a tunneling protocol for a covert channel. By using LOKI, an attacker can transmit data secretly by hiding their malicious traffic inside ICMP so that the networking devices cannot detect the transmission.

- **ICMP-based OS fingerprinting**: OS fingerprinting is the process of learning which operating system is running on a device. ICMP can be used to perform an active OS fingerprint scan. For example, if the ICMP reply contains a TTL value of 128, it is probably a Windows machine, and if the ICMP reply contains a TTL value of 64, it is probably a Linux-based machine.
- **Denial of service attacks**: DoS attacks that use ICMP include the following:
  - **ICMP flood attack**: The attacker overwhelms the targeted resource with ICMP echo request (ping) packets, large ICMP packets and other ICMP types to significantly saturate and slow down the victim's network infrastructure. The following figure is an example of the ICMP Flood.

TCP segments reside within IP packets. The TCP header appears immediately after the IP header and supplies information specific to the TCP protocol. TCP provides more functionality than UDP, at the cost of higher overhead, but additional fields in the TCP header help provide reliability, flow control and stateful communication. Reliable communication is the largest benefit of TCP. TCP incorporates acknowledgments to guarantee delivery instead of relying on upper-layer protocols to detect and resolve errors. If a timely acknowledgment is not received, the sender retransmits the data. But requiring acknowledgments of received data can cause substantial delays. TCP implements flow control to address this issue. Rather than acknowledge one segment at a time, multiple segments can be acknowledged with a single acknowledgment segment. TCP stateful communication between two parties happens by way of TCP three-way handshake. Before data can be transferred using TCP, a three-way handshake opens the TCP connection. If both sides agree to the TCP connection, data can be sent and received by both parties using TCP. At the conclusion of the TCP session, a four-way handshake generally closes the TCP connection gracefully where a typical teardown requires a pair of FIN and ACK segments from each TCP endpoint. Examples of application-layer protocols that make use of TCP reliability are DNS zone transfers, HTTP, SMB, SSL/TLS, FTP and so on.

Though TCP protocol is a connection-oriented and reliable protocol, there are still vulnerabilities that can be exploited. These vulnerabilities are explained in terms of the following attacks:

- **TCP SYN flooding**: TCP SYN flooding causes a DoS attack. It exploits an implementation characteristic of the TCP that can be used to make server processes incapable of responding to any legitimate client's requests. Any service, such as server applications for email, web and file storage, that binds to and listens on a TCP socket, is potentially vulnerable to TCP SYN flooding attacks. The basis of the SYN flooding attack lies in the design of the three-way handshake that begins a TCP connection. TCP connections that have been initiated but not finished are called half-open connections. A finite-sized data structure in each host is used to store the state of the half-open connections. An attacking host can send an initial SYN packet with a spoofed IP address, and then the victim sends the SYN-ACK packet, and waits for a final ACK to complete the three-way handshake. If the spoofed address does not belong to a host, then this connection stays in the half-open state indefinitely, thus occupying the finite-sized data structure. If there are enough half-open connections to fill up the entire finite-sized data structure, then the host cannot accept further TCP connection requests, thus denying service to the legitimate TCP connections. Setting a time limit for half-open connections, then deleting them after the timeout, can help with the TCP SYN flooding problem, but the attacker may continuously send the TCP SYN flood attack traffic. The attacked host will not have space to accept new incoming legitimate TCP connections, but the TCP connection that was established before the attack will have no effect. In this type of attack, the attacker has no interest in examining the responses from the victim. When the spoofed address does belong to a connected host, that host sends a reset to indicate the end of the handshake. The following are some variations of the TCP SYN flooding attack methods:

- **Direct attack**: When attackers rapidly send SYN segments without spoofing their IP source address, that is a direct attack. This method of attack is very easy to perform because it does not involve directly injecting or spoofing packets below the user level of the attacker's operating system. It can be performed by simply using many TCP connect() calls. To be effective, attackers must prevent their operating system from responding to the SYN-ACKs in any way, because any ACKs, RSTs, or ICMP messages will allow the listener to move the TCP out of SYN-RECEIVED. This scenario can be accomplished through firewall rules that either filter outgoing packets to the listener (allowing only SYNs out), or filter incoming packets so that any SYN-ACKs are discarded before they reach the local TCP processing code. When detected, this type of attack is very easy to defend against – a simple firewall rule to block packets with the attacker's source IP address is all that is needed. This defence behaviour can be automated, and such functions are available in off-the-shelf reactive firewalls.
- **Spoofing-based attack**: Another form of SYN flooding attacks uses IP address spoofing, which might be considered more complex than the method used in a direct attack. Instead of merely manipulating local firewall rules, the attacker also needs to be able to form and

inject raw IP packets with valid IP and TCP headers. Popular libraries exist to aid with raw packet formation and injection, therefore
attacks that are based on spoofing are actually fairly easy. For
spoofing attacks, a primary consideration is address selection. If the
attack is to succeed, the machines at the spoofed source addresses
must not respond in any way to the SYN-ACKs that are sent to
them. A very simple attacker might spoof only a single source
address that it knows will not respond to the SYN-ACKs, either
because no machine physically exists at the address presently, or
because of some other property of the address or network configuration. Another option is to spoof many different source addresses,
assuming that some percentage of the spoofed addresses will be
unrespondent to the SYN-ACKs. This option is accomplished either
by cycling through a list of source addresses that are known to be
desirable for the purpose, or by generating addresses inside a subnet
with similar properties.

- **Distributed attacks**: The real limitation of single-attacker spoofing-
based attacks is that if the packets can somehow be traced back to
their true source, the attacker can be easily shut down. Although
the tracing process typically involves some time and coordination
between ISPs, it is not impossible. A distributed version of the SYN
flooding attack, in which the attacker takes advantage of numerous
botnet machines throughout the Internet, is much more difficult to
stop. To increase the effectiveness even further, each botnet could
use a spoofing attack and multiple spoofed addresses. Currently,
distributed attacks are feasible because there are several "botnets"
or "drone armies" of thousands of compromised machines that
are used by criminals for DoS attacks. Botnet machines are constantly added or removed from the armies and can change their IP
addresses or connectivity, so it is quite challenging to block these
attacks.

- **TCP session hijacking**: TCP hijacking is the oldest type of session
hijacking. Session hijacking is the attempt to overtake an already
active session between two hosts. TCP session hijacking is different
from IP spoofing, in which you spoof an IP address or MAC address
of another host. With IP spoofing, you still need to authenticate to the
target. With TCP session hijacking, the attacker takes over an already-
authenticated host as it communicates with the target. The attacker will
probably spoof the IP address or MAC address of the host, but session
hijacking involves more than address spoofing. If attacker manages
to predict the ISN, they can actually send the last ACK data packet
to the server, spoofing as the original host, and then hijack the TCP
connection. Systems with poor TCP ISN generation are vulnerable to
blind TCP spoofing attacks. Attackers can make a full connection to

those systems and send, but not receive, data while spoofing a differ-ent IP address. The target's logs will show the spoofed IP address, and the attacker can take advantage of any trust relationship between the server and the client. This attack was popular in the mid-90's when people commonly used rlogin, which is an rsh (similar to SSH) that allows users to log in on another host via the network, communicat-ing using TCP port number 513. While the rlogin family is mostly a thing of the past, other types of session hijacking are still actively being used. Session hijacking can also be done at the application level. At the application level, a hijacker can hijack already existing sessions but can also create new sessions from the stolen data, for example, HTTP session hijacking. Hijacking an HTTP session involves obtaining the session ID of the HTTP session, which is the unique identifier of the HTTP session. One way for the attacker to obtain the session ID is by sniffing the HTTP packets. Tools that can be used to perform session hijacking attacks include Juggernaut, Hunt, TTY Watcher and T-Sight. Hijacking a TCP session requires an attacker to send a packet with a right seq-number, otherwise they are dropped. The attacker has two options to get the right seq-number:

- **Non blind spoofing:** The attacker can see the traffic that is being sent between the host and the target. Non-blind spoofing is the easiest type of session hijacking to perform, but it requires attacker to capture packets as they are passing between the two machines. Spoofing-based attacks were discussed earlier in TCP SYN flooding attack methods.
- **Blind spoofing:** The attacker cannot see the traffic that is being sent between the host and the target. Blind spoofing is the most difficult type of session hijacking because it is nearly impossible to correctly guess TCP sequence numbers. TCP sequence prediction is a type of blind hijacking because an attacker needs to make an educated guess on the sequence numbers between the host and target. In TCP-based applications, sequence numbers inform the receiving machine which order to put the packets in if they are received out of order. Sequence numbers are a 32-bit field in the TCP header. Therefore, they range from 1 to 4,294,967,295. Every byte is sequenced, but only the sequence number of the first byte in the segment is put in the TCP header. To effectively hijack a TCP session, you must accu-rately predict the sequence numbers that are being used between the target and host.
- **TCP reset attack:** The TCP reset attack, also known as "forged TCP reset" or "spoofed TCP reset packet," is a technique of maliciously killing TCP communications between two hosts. A TCP connection is terminated by using the FIN bit in the TCP flags or by using the RST bit. The regular way that a TCP connection is torn down is by using the FIN bit in the TCP flags. One side of the connection sends a packet

with the FIN bit set. The other side of the connection responds with two packets, an ACK, and a FIN of its own. This last FIN is acknowledged by the original station, indicating that the connection has been closed on both sides.

Closing a connection can also be done by using the RST bit in the TCP flags field. In most packets, the RST bit is set to 0 and has no effect. If the RST bit is set to 1, it indicates to the receiving computer that the computer should immediately stop using the TCP connection. A reset indicates that this connection is considered closed, and there is no need to send additional packets. A reset is an abrupt way to tear down the TCP connection. Resets are commonly seen when TCP data packets are sent to a server where no connection has been established, or when SYNs are sent to a port that the server is not listening on. The server should reply with a reset, showing that the connection is closed or unavailable. Resets can also be sent by applications when a user is suddenly kicked out of an application. When the RST bit is used as designed, it can be a useful tool. But it is possible for an attacker to monitor the TCP packets on the connection and then send a spoofed packet containing a TCP reset to one or both endpoints. The headers in the spoofed packet must indicate, falsely, that the RST packet came from the victim host and not from the attacker. Every field in the IP and TCP headers must be set to a convincing spoofed value for the fake RST packet to trick the victim host into closing the TCP connection. Properly formatted spoofed TCP resets can be a very effective way to disrupt any TCP connection that the attacker can monitor.

UDP is a connectionless transport-layer protocol that provides an interface between IP and upper-layer processes. UDP protocol ports distinguish multiple applications running on a single device from one another. Unlike the TCP, UDP adds no reliability, flow-control, or error-recovery functions to IP. Because of UDP's simplicity, UDP headers contain fewer bytes and consume less network overhead than TCP. The UDP segment's header contains only source and destination port numbers, a UDP checksum and the segment length. UDP is useful in situations where the reliability mechanisms of TCP are not necessary, such when a higher-layer protocol might provide error and flow control. UDP is the transport protocol for several well-known application-layer protocols, including NFS, SNMP, DNS, TFTP and real-time services, such as online games, streaming media and VoIP. UDP is vulnerable because the checksum, which is an optional field that is used to detect transmission errors, is easy to recompute for attackers who want to alter application data. UDP has no algorithm for verifying the sending packet source. An attacker can eavesdrop on UDP packets and make up false UDP packets, pretending that the UDP packet is sent from another source (spoofing). The receiver of the packet has no guarantee that the source IP address in the receiving packet is the real source of the packet. For example,

SNMPv1 and DNS (see Chapter 12) messages use UDP as transport protocol, and are vulnerable to eavesdropping. It is easy for an attacker to eavesdrop on and make up false messages using UDP, as long as the attacker knows the format of the messages that are sent and that the messages are not encrypted.

Most attacks involving UDP relate to exhaustion of some shared resource (buffers, link capacity and so on), or exploitation of bugs in protocol implementations, causing system crashes or other insecure behaviour. Both fall into the broad category of DoS attacks. For example, in UDP flood attacks, similar to TCP flood attacks, the main goal of the attacker is to cause system resource starvation. A UDP flood attack is triggered by sending many UDP packets to random ports on the victim's system. The victim's system will notice that no application is listening on that port and reply with an ICMP destination port unreachable packet. If many UDP packets are sent, the victim will be forced to send numerous ICMP destination port unreachable packets. Usually, these attacks are accomplished by spoofing the attacker's source IP address. Software, such as Low Orbit Ion Cannon and UDP Unicorn, can be used to perform UDP flooding attacks.

The SQL Slammer worm attack of 2003 is a classic example of a software security vulnerability involving UDP port 1434. Microsoft SQL Server 2000 contained three vulnerabilities that can allow a remote attacker to execute arbitrary code or crash the server. The vulnerabilities lie in the SQL Server 2000 Resolution Service. SQL Server 2000 allows several instances of the SQL server to be used on a single machine. Because multiple instances cannot use the standard SQL server session port, TCP port 1433, other instances listen on assigned ports. The SQL Server Resolution Service, which operates on UDP port 1434, responds to the clients' query, so the clients can connect to the appropriate SQL server instance. Two buffer overflow vulnerabilities can be exploited by sending an especially malformed packet to the SQL Server Resolution Service on UDP port 1434 to cause the heap or the stack memory to be overwritten. If remote attackers successfully exploit the vulnerabilities, they can execute arbitrary code on the system. An unsuccessful attempt is likely to crash the SQL server service. The arbitrary code would be executed in the security context of the SQL server and may be able to perform any database function. Exploit code for the discussed vulnerabilities is publicly available.

A DoS vulnerability can also be exploited through the SQL keep-alive mechanism over UDP port 1434. The SQL server system uses a keep-alive mechanism to determine which instances are active and which are inactive. When an instance receives a keep-alive packet with the value of 0x0A on UDP port 1434, it generates and returns to the sender a keep-alive packet with the same 0x0A value. If the first keep-alive packet has been spoofed to appear to come from another SQL server system's UDP port 1434, both servers will continually send packets with the value of 0x0A to each other, generating a packet storm that continues until one of the servers is brought offline or rebooted.

Attack surface is the total sum of all the vulnerabilities in a given computing device or network that are accessible to the attackers. Attack surface may be categorised into different areas, such as software attack surfaces (open ports on a server), physical attack surfaces (USB ports on a laptop), network attack surfaces (console ports on a router) and human/social engineering attack surfaces (employees with access to sensitive information).

Attack vectors are the paths or means by which an attacker gains access to a resource (such as end-user hosts or servers) in order to deliver malicious software or malicious outcome. Attack vectors enable an attacker to exploit system vulnerabilities. Many attack vectors take advantage of the human element in the system, because that is often the weakest link. For example, if the attack vector is a malicious file, then the victim needs to be tricked into opening it for the attack to work.

A smaller attack surface can help make the organisation to become less exploitable, reducing the risk. A greater attack surface makes the organisation more vulnerable to attacks, which increases the risk.

Attack surfaces can be divided into the following four categories:

- The network attack surface comprises all vulnerabilities that are related to ports, protocols, channels, devices (smart phones, laptops, routers and firewalls), services, network applications (SaaS) and even firmware interfaces. For example, some network protocols are inherently more insecure than others as they pass data over the network unencrypted. These protocols include Telnet, FTP, HTTP and SMTP. Many network file systems, such as NFS and SMB, pass information over the network unencrypted. Remote memory dump services, like netdump, also pass the contents of memory over the network unencrypted. Memory dumps can contain passwords or, even worse, database entries and other sensitive information. Other services, such as finger and rwhod, reveal information about users of the system. Network printers are also the target of a wide array of attacks from hackers because the operating system drivers, management tools and the printer's software make them vulnerable. Printers can be attacked via the web-based administrative interface, SMTP, FTP and SNMP.
- The software attack surface is the complete profile of all functions in any code that is running in a given system that is available to an unauthenticated user. An attacker or a piece of malware can use various exploits to gain access and run code on the target machine. The software attack surface is calculated across many different kinds of code, including applications, email services, configurations, compliance policy, databases, executables, DLLs, web pages, mobile apps, device OS and so on. Unpatched software, such as Java, Adobe Reader and Adobe Flash, also provide greater software attack surface because they are widely used. Publicly known cybersecurity vulnerabilities are listed in CVE libraries. Common CVE identifiers make it easier to share data

across separate network security databases and tools, and provide a baseline for evaluating the coverage of an organisation's security tools.

- The physical attack surface is composed of the security vulnerabilities in a given system that are available to an attacker in the same location as the target. The physical attack surface is exploitable through inside threats such as rogue employees, social engineering ploys and intruders who are posing as service workers. External threats include password retrieval from carelessly discarded hardware, passwords on sticky notes and physical break-ins. Also, consider a scenario where an intruder steals or downloads the information from an entire drive and extracts the target data in the future.

- The social engineering attack surface usually takes advantage of human psychology: the desire for something free, the susceptibility to distraction, or the desire to be liked or to be helpful. A few examples of human social engineering attacks are fake calls to IT, where the attacker is posing as an employee to get a password; or media drops where an employee might find a flash drive in the parking lot, and when they use that device, they inadvertently execute automatic running code leading to a data breach. Socially engineered Trojans provide another method of attack. An end user browses to a website that is usually trusted, which prompts the end user to run a Trojan. Most of the time the website is a legitimate one, innocent victim that has been temporarily compromised by hackers. Another very popular method is an advanced persistent threat (APT) attacker sends a very specific phishing campaign, which is known as spear-phishing, to multiple employees' email addresses. The phishing email contains a Trojan attachment, which at least one employee is tricked into running. After the initial execution and first computer takeover, an APT attacker can compromise an entire enterprise in a short time.

An attack vector is a path or route by which an attack was carried out. Examples of attack vectors include malware that is delivered to users who are legitimately browsing mainstream websites, spam emails that appear to be sent by well-known companies but contain links to malicious sites, third-party mobile applications that are laced with malware that are downloaded from popular online marketplaces, and insiders using information access privileges to steal intellectual property from employers.

Common security threats include the following:

- **Reconnaissance:** The attacker attempts to gather information about targeted computers or networks that can be used as a preliminary step toward a further attack seeking to exploit the target system. For example, what operating system is on the target systems? Is there a firewall? Which ports are available? What content management system (CMS) does the system run? There are also sources of information such as Facebook, Twitter and Google that can be used to gather information about organisations or persons that are being targeted.

- **Known vulnerabilities:** The attacker finds weakness in hardware and software and then exploits those vulnerabilities. There are several online resources that publish information about vulnerabilities that have been discovered in different systems. Often, a proof-of-concept attack code will be provided with the vulnerability disclosure. Each platform has its own strengths and weakness. Once the target system is identified, it is simply a matter of trying out the different attacks for the targeted system to see if any of them work.
- **SQL injection:** This attack works by manipulating the SQL database queries that the web application sends. An application can be vulnerable if it does not sanitise user input properly, or uses untrusted parameter values in database queries without validation.
- **Phishing:** The attacker sends out spam email to thousands of recipients. The email contains a link to a malicious site that has been set up to look like, for instance, a regular bank's site. When the user enters their credentials in the login form, it actually is captured by the malicious site and then used to impersonate that user on the real site. Spear phishing is another variation of the phishing attack, in which the attacker usually targets specific persons. The RSA breach in 2011, which resulted in unspecified data that are related to their SecurID product being stolen, started with a spear phishing attack.
- **Malware:** Short for "malicious software," malware may be computer viruses, worms, Trojan horses, dishonest spyware and malicious rootkits.
- **Weak authentication:** These attacks exploit poorly designed and/or implemented authentication mechanisms. Weak authentication usually means one or more of the following: weak, guessable passwords are allowed, no lockout enforcement after a specific number of invalid login attempts, or the password reset methods are not secure.

Other common threats, such as security misconfiguration, cross-site scripting, cross-site request forgery and HTTP header manipulation, have not been included in the list above.

A reconnaissance attack is an attempt to learn more about the intended victim before attempting a more intrusive attack, such as an actual access or DoS. The goal of reconnaissance is to discover the following information about targeted computers or networks:

- IP addresses, sub-domains and related information on a target network
- Accessible UDP and TCP ports on target systems
- The operating system on target systems

There are four main subcategories or methods for gathering network data:

- **Packet sniffers:** Packet sniffing, or packet analysis, is the process of capturing any data that are passed over the local network and looking for any information that may be useful to an attacker. The packet sniffer may be either a software program or a piece of hardware with software installed in it that captures traffic that is sent over the network, which is then decoded and analysed by the sniffer. Tools, such as Wireshark, Ettercap, or NetworkMiner, give anyone the ability to sniff network traffic with little practice or training.
- **Ping sweeps:** A ping sweep is another kind of network probe. In a ping sweep, the attacker sends a set of ICMP echo packets to a network of machines, usually specified as a range of IP addresses, and sees which ones respond. The idea is to determine which machines are alive and which aren't. Once the attacker knows which machines are alive, he can focus on which machines to attack and work from there. The fping command is one of the many tools that can be used to conduct ping sweeps.
- **Port scans:** A port scanner is a software program that surveys a host network for open ports. As ports are associated with applications, the attacker can use the port and application information to determine a way to attack the network. The attacker can then plan an attack on any vulnerable service that they find. Examples of insecure services, protocols, or ports include but are not limited to port 21 (FTP), port 23 (Telnet), port 110 (POP3), 143 (IMAP) and port 161 (SNMPv1 and SNMPv2) because protocols using these ports do not provide authenticity, integrity and confidentiality. NMAP is one of the many tools that can be used for conducting port scans.
- **Information queries:** Information queries can be sent via the Internet to resolve hostnames from IP addresses or vice versa. One of the most commonly used queries is the nslookup command. One can use nslookup by opening a Windows or Linux command prompt window on one's computer and entering the nslookup command, followed by the IP address or hostname that one is attempting to resolve.

Initially, an attacker attempts to gain information about targeted computers or networks that can be used as a preliminary step toward a further attack seeking to exploit the target system. A reconnaissance attack can be active or passive.

Whois: Attackers passively start using standard networking command-line tools such as dig, nslookup and whois to gather public information about a target network from DNS registries. The nslookup and whois tools are available on both Windows, UNIX and Linux platforms, and dig (domain information groper) is available on UNIX and Linux systems.

Shodan Search Engine: Another innocuous tool is the Shodan search engine with metadata filter capabilities that can help an attacker identify a specific device, such as a computer, router and server. For example, an attacker can search for a specific system, such as a Cisco 3945 router, running a certain version of the software, and then explore further vulnerabilities.

Robots.txt File: The Robots.txt file is another example where attacker can gather a lot of valuable information from a target's website. The Robots.txt file is publicly available and found on websites that gives instructions to web robots (also known as search engine spiders), about what is and is not visible using the robots exclusion protocol. An attacker can find the Robots.txt file in the root directory of a target website.

After the passive reconnaissances, the attacker can start using active reconnaissance tools such as ping sweeps, traceroutes, port scans, or operating system fingerprinting to actually send packets to discover the target systems. One of the tools that can be used is traceroute to find out the IP addresses of routers and firewalls that protect victim hosts. Ping sweeps of the addresses can present a picture of the live hosts in a particular environment. After a list of live hosts is generated, the attacker can probe further by running port scans on the live hosts. The attacker can use this information to determine the easiest way to exploit a vulnerability.

NMAP Port Scan: Port scanning tools like Network Mapper can cycle through all well-known ports to provide a complete list of all services that are running on the hosts. NMAP is an open-source tool that is specialised in network exploration and security auditing. Design and operation of port scans uses two components: a host address and a port number that is used by host services. An attacker can attempt to connect to a device on a specified array of ports, such as 21 (FTP), 23 (Telnet) and 25 (SMTP). With the information received from these scans, an attacker can find open ports that could allow access to a network and launch more sophisticated attacks.

Vulnerability Scanners: An authorised security administrator can use vulnerability scanners, such as Nessus and OpenVAS, to locate vulnerabilities in their own networks and patch them before they can be exploited. Of course, these tools can also be used by attackers to locate vulnerabilities before an organisation even knows that they exist. After getting a foothold in a network, an attacker can use these same tools to pivot sideways and scan machines on the network to extend their positions.

An access attack is an attempt to access another user account or network device through improper, unauthorised means. Access attacks exploit known vulnerabilities in authentication services, FTP services and web services to gain entry to web accounts, confidential databases and other sensitive information. After gaining access to your network with a valid account, an attacker can obtain lists of valid user and computer names and network information, modify server and network configurations, including access controls and routing tables, and modify, reroute or delete one's data. There are many attacks which can lead to a system being compromised, and allowing the attacker to gain unauthorised access to the system. The following are some prominent types of attacks:

- **Password attack** is typically used to obtain system access. When access is obtained, the attacker is able to read, modify or delete data, and add, modify or remove network resources. For example, tools like "John the ripper," and "Cain and Abel" are password cracker tools.
- **Spoofing/masquerading attack** is a situation in which one person or program successfully masquerades as another by falsifying data and gaining illegitimate access.
- **Session hijacking** is an attack in which the session established by the client to the server is taken over by a malicious person or process.
- **Malware** is used to infect the victim's system with malicious software.

MITM attacks, sometimes referred to as eavesdropping attacks or connection hijacking attacks, exploit inherent vulnerabilities of TCP/IP protocol at various layers. The attack is a derivative of packet sniffing and spoofing techniques and if carried out properly, it can be completely invisible to the victims, making it difficult to detect and stop. Generally, in MITM attacks, a system that has the ability to view the communication between two systems imposes itself in the communication path between those other systems. The main objective is to steal the information being transmitted between two parties. TCP/IP works on a handshake (SYN, SYN-ACK, ACK). This three-way handshake establishes a connection between two different network interface cards, which then use packet sequencing and data acknowledgements to send or receive data. The data flows from the physical layer all the way up to the application layer.

- **ARP poisoning**: An ARP-based MITM attack is achieved when an attacker poisons the address resolution protocol (ARP) cache of two devices with the MAC address of the attacker's NIC. Once the ARP caches have been successfully poisoned, each victim device sends all its packets to the attacker when communicating to the other device and puts the attacker in the middle of the communications path between the two victim devices. It allows an attacker to easily monitor all communication between victim devices. The intent is to intercept and view the information being passed between the two victim devices and potentially introduce sessions and traffic between the two victim devices. A MITM attack can be passive or active. In passive attacks, attackers steal confidential information. In active attacks, attackers modify data in transit or inject data of their own. ARP cache poisoning attacks often target a host and the host's default gateway. ARP cache poisoning puts the attacker as a MITM between the host and all other systems outside of the local subnet.

- **ICMP-based MITM attack:** An ICMP MITM attack is accomplished by spoofing an ICMP redirect message to any router that is in the path between the victim client and server. An ICMP redirect message is typically used to notify routers of a better route; however, it can be abused to effectively route the victim's traffic through an attacker-controlled router. The threat of this attack is mitigated by routers that have static routes and routers that do not accept/process ICMP redirect packets.
- **DNS-based MITM attack:** DNS (see Chapter 12) spoofing is an MITM technique that is used to supply false DNS information to a host so that when they attempt to browse, for example, https://www.xyzbank.com at the IP address XXX.XX.XX.XX, the host is actually sent to an imposter https://www.xyzbank.com that is residing at IP address YYY.YY.YY.YY, which an attacker has created in order to steal online banking credentials and account information from unsuspecting users.
- **DHCP-based MITM attack:** Similar to the DNS attack, DHCP server queries and responses are intercepted. This interception helps the attacker gain complete knowledge of the network, such as host names, MAC addresses, IP addresses and the DNS servers. This information is further used to plant advanced attacks to steal the information. An attacker can initiate a DoS attack on a real DHCP server to keep it busy, and in the meanwhile spoof and respond to the DHCP host queries by itself.

DoS and DDoS attacks attempt to consume all the critical computer or network resources in order to make them unavailable for valid use. DoS attacks are considered a major risk, because they can easily disrupt the operations of a business and they are relatively simple to conduct.

A reflection attack is a type of DoS attack in which the attacker sends a flood of protocol request packets to various IP hosts. The attacker spoofs the source IP address of the packets such that each packet has as its source address the IP address of the intended target rather than the IP address of the attacker. The IP hosts that receive these packets become "reflectors." The reflectors respond by sending response packets to the spoofed address (the target), thus flooding the unsuspecting target.

If the request packets that are sent by the attacker solicit a larger response, the attack is also an amplification attack. In an amplification attack, a small forged packet elicits a large reply from the reflectors. For example, some small DNS queries elicit large replies. Amplification attacks enable an attacker to use a small amount of bandwidth to create a massive attack on a victim by hosts around the Internet. It is important to note that reflection and amplification are two separate elements of an attack. An attacker can use amplification with a single reflector or multiple reflectors. Reflection and amplification attacks are very hard to trace because the actual source of the attack is hidden.

A classic example of reflection and amplification attacks is the smurf attack, which were common during the late 1990s. Although the smurf attack no longer poses much of a threat (because mitigation techniques became standard practice some time ago), it provides a good example of amplification. In a smurf attack, the attacker sends numerous ICMP echo-request packets to the broadcast address of a large network. These packets contain the victim's address as the source IP address. Every host that belongs to the large network responds by sending ICMP echo-reply packets to the victim. The victim is flooded with unsolicited ICMP echo-reply packets.

In a TCP/IP-based network, every device must have a unique unicast IP address to access the network and its resources. Without DHCP, the IP address for each client (a host that is requesting initialisation parameters from a DHCP server) must be configured manually and IP addresses for computers that are removed from the network must be manually reclaimed. With DHCP, the IP address allocation process is automated and managed centrally. The DHCP server maintains a pool of IP addresses and leases an address to any DHCP-enabled client when it starts up on the network. Because the IP addresses are dynamic (leased) rather than static (permanently assigned), addresses that are no longer in use are automatically returned to the pool for reallocation. DHCP was based on BOOTP when the Internet was relatively small. Not only does DHCP run over IP and UDP, which are inherently insecure, the DHCP protocol itself has no security provisions, which causes a serious vulnerability in networks because DHCP deals with critical configuration information. Two classes of potential security problems are related to DHCP:

- **DHCP server spoofing:** The attacker runs DHCP server software and replies to DHCP requests from legitimate clients. As a rogue DHCP server, the attacker can cause a DoS by providing invalid IP information. The attacker can also perform confidentiality or integrity breaches via a MITM attack. The attacker can assign itself as the default gateway or DNS server in the DHCP replies, later intercepting IP communications from the configured hosts to the rest of the network. The following is the DHCP server spoofing attack process:
  - An attacker activates a malicious DHCP server on the attacker's port
  - The client broadcasts a DHCP configuration request.

- The DHCP server of the attacker responds before the legitimate DHCP server can respond, assigning attacker-defined IP configuration information.
- Host packets are redirected to the attacker address because it emulates the default gateway that it provided to the client.
- **DHCP starvation:** A DHCP starvation attack works by the broadcasting of DHCP requests with spoofed MAC addresses. If enough requests are sent, the network attacker can exhaust the address space available to the DHCP servers in a time period. The network attacker can then set up a rogue DHCP server. However, the exhaustion of all the DHCP addresses is not required to introduce a rogue DHCP server.

Those who want to better understand insecurities of the Internet should read the article by Stu Sjouwerman "How The NSA Killed Internet Security in 1978" (https://blog.knowbe4.com/how-the-nsa-killed-internet-security-in-1978).

Does the word "complexity" come to your mind after reading this chapter? It definitely should.

Chapter 12

DNS and BGP

Some readers may find this chapter too technical, they may skip the technical bits and focus on observations, examples and conclusions (Figure 12.1).

Every device connected to Internet has its own IP address (see Chapter 11), which is used by other devices to locate the device. But it is important to remember that IP also enables id enticing and communication. Today IP addresses are being considered personally identifiable information (PII) that is defined as any information connected to a specific individual that can be used to uncover that individual's identity, such as their social security number, full name, email address or phone number.

Domain Name System (DNS) turns domain names into IP addresses, which browsers use to load Internet pages (https://www.fortinet.com/resources/cyberglossary/what-is-dns#:~:text=DNS%20Definition,devices%20to%20locate%20the%20device.). DNS servers make it possible for people to input normal words into their browsers, such as microsoft.com, without having to keep track of the IP address for every website. DNS was designed in the 1980s when Internet access was restricted to government agencies, scientists and the military. Its architects were concerned about reliability and functionality, not security. As a result, DNS has always been vulnerable to a broad spectrum of attacks that will be discussed later in this chapter.

DNS is the phonebook of the Internet (https://www.cloudflare.com/learning/dns/what-is-dns/). Humans access information online through domain names, like google.com or microsoft.com. Web browsers interact through Internet Protocol (IP) addresses. DNS translates domain names to IP addresses (see Chapter 12) so browsers can load Internet resources. Each device connected to the Internet has a unique IP address which other machines use to find the device. DNS servers eliminate the need for humans to memorise IP addresses such as 192.168.1.1 (in IPv4), or more complex newer alphanumeric IP addresses such as 2400:cb00:2048:1::c629:d7a2 (in IPv6).

Remember ARPANET (see Chapter 11)? It did not have DNS. The beginnings were small-scale and did not require any sophisticated or automated tools to remember and manage addresses of computers connected. So,

DOI: 10.1201/9781032672601-12

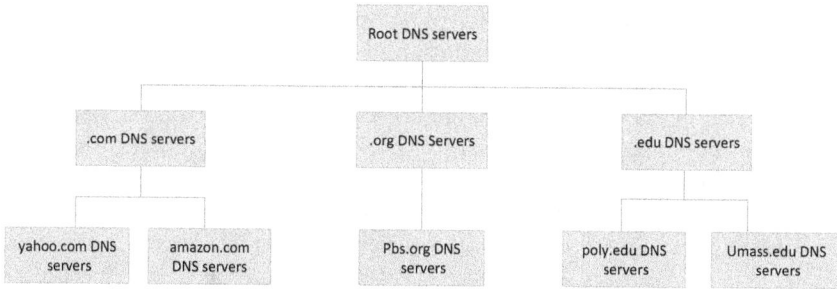

Figure 12.1 Domain name space.

Source: https://commons.wikimedia.org/w/index.php?curid=75965575.

Stanford Research Institute (now SRI International) maintained a text file named HOSTS.TXT that mapped computer names to the numerical addresses of computers on the ARPANET. The first ARPANET directory was developed and maintained by Elizabeth Feinler. Maintenance of numerical addresses, called the Assigned Numbers List, was handled by Jon Postel at the University of Southern California's Information Sciences Institute (ISI), whose team worked closely with SRI. Addresses were assigned manually and computers, including their hostnames and addresses, were added to the primary file by contacting the SRI Network Information Centre (NIC), directed by Feinler, via telephone during business hours. Later, Feinler set up a WHOIS directory on a server in the NIC for retrieval of information about resources, contacts and entities. Feinler and her team managed the Host Naming Registry from 1972 to 1989.

Then she and her team developed the concept of domains. Feinler suggested that domains should be based on the type of organisation computer belongs to. For example, computers at educational institutions have the domain .edu, while commercial organisations have the domain .com, non-profit organisations have the domain .org and organisations with global presence have the domain .net. These were what is called today top-level domains or generic top-level domains (gTLD). By the early 1980s, maintaining a single, centralised host table had become slow and unwieldy and the emerging network required an automated naming system to address technical and personnel issues. Postel has given the task of forging a compromise between five competing proposals of solutions to Paul Mockapetris. Instead, in 1983 Paul Mockapetris, who at the time was at the University of Southern California, created the DNS. In 2012, Paul Mockapetris, who was inducted into the Internet Hall of Fame, admitted that he got the job because "nobody else wanted to do it."

In 1984, four UC Berkeley students, Douglas Terry, Mark Painter, David Riggle and Songnian Zhou, wrote the first Unix name server implementation for the Berkeley Internet Name Domain, commonly referred to as

BIND. In 1985, Kevin Dunlap of DEC substantially revised the DNS implementation. Mike Karels, Phil Almquist and Paul Vixie then took over BIND maintenance. Internet Systems Consortium was founded in 1994 by Rick Adams, Paul Vixie and Carl Malamud, expressly to provide a home for BIND development and maintenance. BIND versions from 4.9.3 onwards were developed and maintained by Internet Software Consortium (ISC), with support provided by ISC's sponsors. As co-architects/programmers, Bob Halley and Paul Vixie released the first production-ready version of BIND version 8 in May 1997. Since 2000, over 43 different core developers have worked on BIND. In November 1987, RFC 1034 and RFC 1035 superseded the 1983 DNS specifications. Several additional Request for Comments have proposed extensions to the core DNS protocols.

DNS vulnerabilities are buried in its original design. Let's have a look at one of DNS design aspects. A Canonical Name (CNAME) record (specified in RFC 1034 and clarified in Section 10 of RFC 2181) is a type of resource record in DNS that maps one domain name (an alias) to another (the canonical name). This can be convenient when running multiple services (like an FTP server and a web server, each running on different ports) from a single IP address. One can, for example, use CNAME records to point ftp.example.com and www.example.com to the DNS entry for example.com, which in turn has an A record which points to the IP address. Then, if the IP address ever changes, one only has to record the change in one place within the network: in the DNS A record for example.com. CNAME records must always point to another domain name, never directly to an IP address. So, CNAME records are a type of DNS record that allows one domain name to be an alias for another domain name. While CNAME records offer flexibility and convenience in managing domain names, they also come with certain disadvantages in terms of cybersecurity. DNS handles CNAME records and there are several restrictions on their use.

One of the main disadvantages of using DNS CNAME records is the potential for DNS hijacking or DNS spoofing attacks (https://www.imperva.com/learn/application-security/dns-hijacking-redirection/). DNS hijacking occurs when an attacker gains unauthorised access to a DNS server and redirects legitimate traffic to malicious websites. By creating a CNAME record pointing to a malicious domain, an attacker can effectively redirect users to a fake website that resembles the original, tricking them into providing sensitive information such as login credentials or financial details. This can lead to identity theft, financial loss or other forms of cybercrime.

Another disadvantage of CNAME records is the impact on DNS resolution time. When a DNS resolver receives a query for a domain with a CNAME record, it needs to perform an additional lookup to resolve the final domain name. This can introduce latency and increase the time it takes to resolve the DNS query. In scenarios where performance is critical, such as high-traffic websites or real-time applications, the additional lookup caused by CNAME records can have a noticeable impact on user experience.

Furthermore, the use of CNAME records can complicate troubleshooting and debugging processes. When multiple CNAME records are chained together, it can be challenging to identify the root cause of DNS resolution issues. Each CNAME record adds an extra layer of complexity, making it harder to pinpoint the exact source of the problem. This can result in longer resolution times and increased frustration for network administrators or system operators trying to diagnose and resolve DNS-related issues.

Additionally, CNAME records can create dependencies and potential points of failure. If a CNAME record points to a domain that is temporarily or permanently unavailable, it can disrupt the resolution of the original domain name. This can lead to service disruptions, broken links or other accessibility issues for users trying to access resources associated with the original domain. It is essential to regularly monitor and maintain CNAME records to ensure their continued availability and prevent potential disruptions.

While DNS CNAME records offer flexibility and convenience in managing domain names, they also introduce certain cybersecurity risks and performance considerations (https://et.eitca.org/cybersecurity/eitc-is-cnf-computer-networking-fundamentals/domain-name-system/introduction-to-dns/the-disadvantage-of-the-dns-cname-records/). DNS hijacking, increased resolution time, troubleshooting complexities and potential points of failure are among the disadvantages associated with the use of CNAME records. It is important for network administrators and system operators to carefully evaluate the trade-offs and implement appropriate security measures to mitigate the risks associated with CNAME records.

Another important part of DNS is an MX record, or mail exchange record. MX records route emails to specified mail servers. It indicates how email messages should be routed in accordance with the Simple Mail Transfer Protocol (SMTP, the standard protocol for all email). MX records essentially point to the IP addresses of a mail server's domain. If a domain doesn't have an MX record or has an invalid MX record, email messages will bounce back or will be rejected, and one's reputation as a sender may suffer. One of the way DNS attack can be performed is an MX record hijacking(https://elie.net/blog/security/how-email-in-transit-can-be-intercepted-using-dns-hijacking).

Computers and various devices that connected to the Internet rely on IP addresses to send a request to the website they are attempting to reach. Without DNS, one would have to keep track of the IP addresses of all the websites one visits (like it was the case with ARPANET before DNS introduction), similar to carrying around a phone book of websites all the time. The DNS server allows one to type in the name of the website. It then goes out and gets the right IP address for this website.

The domain name space consists of a tree data structure. Each node or leaf in the tree has a label and zero or more resource records (RR), which hold information associated with the domain name. The domain name itself consists of the label, concatenated with the name of its parent node on the right, separated by a dot. The tree sub-divides into zones beginning at the root zone. A DNS zone may consist of as many domains and subdomains as the zone manager chooses. DNS can also be partitioned according to class where the separate classes can be thought of as an array of parallel namespace trees. Administrative responsibility for any zone may be divided by creating additional zones. Authority over the new zone is said to be delegated to a designated name server. The parent zone ceases to be authoritative for the new zone. The hierarchical DNS for class Internet, organised into zones, each served by a name server. The Internet Engineering Task Force (IETF) published the original DNS specifications in RFC 882 and RFC 883 in November 1983. These were later updated in RFC 973 in January 1986.

A domain name consists of one or more parts, technically called labels, which are conventionally concatenated and delimited by dots, such as example.com. The right-most label conveys the top-level domain. For example, the domain name www.example.com belongs to the top-level domain com. The hierarchy of domains descends from right to left. Each label to the left specifies a subdivision or subdomain of the domain to the right. This tree of subdivisions may have up to 127 levels. A label may contain zero to 63 characters. The null label of length zero is reserved for the root zone. The full domain name may not exceed the length of 253 characters in its textual representation. In the internal binary representation of the DNS, the maximum length requires 255 octets of storage, as it also stores the length of the name. The definitive descriptions of the rules for forming domain names appear in RFC 1035, RFC 1123, RFC 2181 and RFC 5892.

Hostnames use a preferred format and character set. The characters allowed in labels are a subset of the ASCII character set, consisting of characters a through z, A through Z, digits 0 through 9 and hyphen. This rule is known as the letter, digits, hyphen (LDH) rule. Domain names are interpreted in a case-independent manner. Labels may not start or end with a hyphen. An additional rule requires that top-level domain names should not be all-numeric. The limited set of ASCII characters permitted in the DNS prevented the representation of names and words of many languages in their native alphabets or scripts. To make this possible, ICANN approved the Internationalising Domain Names in Applications (IDNA) system, by which user applications, such as web browsers, map Unicode strings into the valid DNS character set using Punycode. In 2009, ICANN approved the installation of internationalised domain name country code top-level domains (ccTLDs). In addition, many registries of the existing top-level domain names (TLDs) have adopted the IDNA system, guided by RFC 5890, RFC 5891, RFC 5892 and RFC 5893.

DNS protocol uses two types of DNS messages, queries and responses. Both have the same format. Each message consists of a header and four sections: question, answer, authority and an additional space. A header field (flags) controls the content of these four sections.

DNS is maintained by a distributed database system, which is based on a the client-server model. The nodes of this database are the name servers. Each domain has at least one authoritative DNS server that publishes information about that domain and the name servers of any domains subordinate to it. The top of the hierarchy is served by the root name servers, the servers to query when looking up (resolving) a TLD. An authoritative name server is a name server that only gives answers to DNS queries from data that have been configured by an original source, for example, the domain administrator or by dynamic DNS methods, in contrast to answers obtained via a query to another name server that only maintains a cache of data. Every DNS zone must be assigned a set of authoritative name servers. This set of servers is stored in the parent domain zone with name server (NS) records.

A DNS server is a computer with a database containing the public IP addresses associated with the names of the websites. DNS acts like a phonebook for the Internet. Whenever people type domain names like google.com or microsoft.com into the address bar of web browsers, the DNS finds the right IP address. The site's IP address is what directs the device to go to the correct place to access the site's data. Once the DNS server finds the correct IP address, browsers take the address and use it to send data to content delivery network (CDN) edge servers or origin servers. Once this is done, the information on the website can be accessed by the user. The DNS server starts the process by finding the corresponding IP address for a website's uniform resource locator (URL).

In a usual DNS query, the URL typed in by the user has to go through four servers for the IP address to be provided. The four servers work with each other to get the correct IP address to the client, and they include:

- **DNS recursor**: The DNS recursor, which is also referred to as a DNS resolver, receives the query from the DNS client. Then it communicates with other DNS servers to find the right IP address. After the resolver retrieves the request from the client, the resolver acts like a client itself. As it does this, it makes queries that get sent to the other three DNS servers: root nameservers, TLD nameservers and authoritative nameservers.
- **Root nameserver**: The root nameserver is designated for the Internet's DNS root zone. Its job is to answer requests sent to it for records in the root zone. It answers requests by sending back a list of the authoritative nameservers that go with the correct TLD.
- **TLD nameserver**: A TLD nameserver keeps the IP address of the second-level domain contained within the TLD name. It then releases the website's IP address and sends the query to the domain's nameserver.

- **Authoritative nameserver**: An authoritative nameserver is what gives you the real answer to your DNS query. There are two types of authoritative nameservers: a master server or primary nameserver and a slave server or secondary nameserver. The master server keeps the original copies of the zone records, while the slave server is an exact copy of the master server. It shares the DNS server load and acts as a backup if the master server fails.

To use the phone book analogy, one can think of the IP address as the phone number and the person's name as the website's URL. Authoritative DNS servers have a copy of the "phone book" that connects these IP addresses with their corresponding domain names. They provide answers to the queries sent by recursive DNS nameservers, providing information on where to find specific websites. The answers provided have the IP addresses of the domains involved in the query. Authoritative DNS servers are responsible for specific regions, such as a country, an organisation or a local area. Regardless of which region is covered, an authoritative DNS server does two important jobs. First, the server keeps lists of domain names and the IP addresses that go with them. Next, the server responds to requests from the recursive DNS server regarding the IP address that corresponds with a domain name.

Authoritative nameservers keep information of the DNS records. A recursive server acts as a middleman, positioned between the authoritative server and the end-user. To reach the nameserver, the recursive server has to "recurse" through the DNS tree to access the domain's records. After one types in a URL in their web browser, this URL is given to the recursive DNS server. The recursive DNS server then examines its cache memory to see whether the IP address for the URL is already stored. If the IP address information already exists, the recursive DNS server will send the IP address to the browser. The user is then able to see the website for which they typed in the URL.

On the other hand, if the recursive DNS server does not find the IP address when it searches its memory, it will proceed through the process of getting the IP address for the user. The recursive DNS server's next step is to store the IP address for a specific amount of time. This period of time is defined by the person who owns the domain using a setting referred to as time to live (TTL). Once the recursive DNS server gets the answer, it sends that information back to the computer that requested it. The computer then uses that information to connect to the IP address, and the user gets to see the website.

An authoritative server indicates its status of supplying definitive answers, deemed authoritative, by setting a protocol flag, called the "Authoritative Answer" (AA) bit in its responses. This flag is usually reproduced prominently in the output of DNS administration query tools, such as dig, to indicate that the responding name server is an authority for the domain name in question.

When a name server is designated as the authoritative server for a domain name for which it does not have authoritative data, it presents a type of error called a "lame delegation" or "lame response." An authoritative name server can either be a primary server or a secondary server. Historically the terms master/slave and primary/secondary were sometimes used interchangeably but the current practice is to use the latter form. A primary server is a server that stores the original copies of all zone records. A secondary server uses a special automatic updating mechanism in the DNS protocol in communication with its primary to maintain an identical copy of the primary records.

DNS was designed in the early 1980s when the Internet was much smaller, and security was not a primary consideration in its design. As a result, when a recursive resolver sends a query to an authoritative name server, the resolver has no way to verify the authenticity of the response. The resolver can only check that a response appears to come from the same IP address where the resolver sent the original query. But relying on the source IP address of a response is not a strong authentication mechanism, since the source IP address of a DNS response packet can be easily forged or spoofed. As DNS was originally designed, a resolver cannot easily detect a forged response to one of its queries. An attacker can easily masquerade as the authoritative server that a resolver originally queried by spoofing a response that appears to come from that authoritative server. In other words, an attacker can redirect a user to a potentially malicious site without the user realising it.

Recursive resolvers cache the DNS data they receive from authoritative name servers to speed up the resolution process. If a stub resolver asks for DNS data that the recursive resolver has in its cache, the recursive resolver can answer immediately without the delay introduced by first querying one or more authoritative servers. This reliance on caching has a downside, however: if an attacker sends a forged DNS response that is accepted by a recursive resolver, the attacker has poisoned *the* cache of the recursive resolver. The resolver will then proceed to return the fraudulent DNS data to other devices that query for it.

As an example of the threat posed by a cache-poisoning attack, consider what happens when a user visits their bank's website. The user's device queries its configured recursive name server for the bank website's IP address. But an attacker could have poisoned the resolver with an IP address that points not to the legitimate site but to a website created by the attacker. This fraudulent website impersonates the bank website and looks just the same. The unknowing user would enter their name and password, as usual. Unfortunately, the user has inadvertently provided its banking credentials to the attacker, who could then log in as that user at the legitimate bank website to transfer funds or take other unauthorised actions.

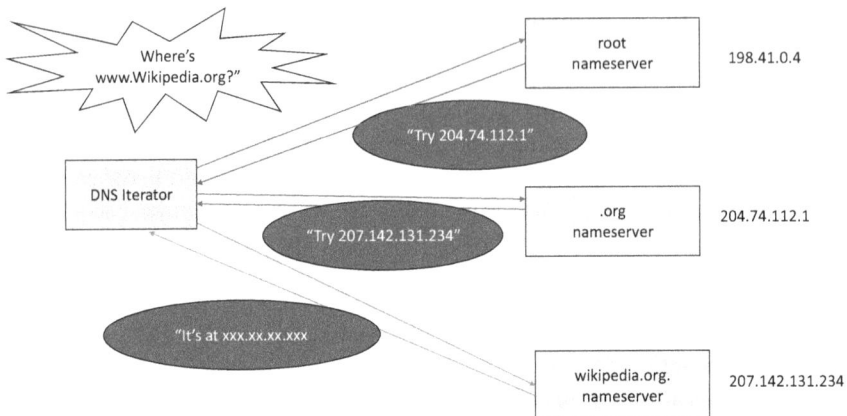

Figure 12.2 Example of DNS iterative resolver.

Source: https://commons.wikimedia.org/w/index.php?curid=75965574.

Domain name resolvers determine the domain name servers responsible for the domain name in question by a sequence of queries starting with the right-most (top-level) domain label. A DNS resolver that implements the iterative approach mandated by RFC 1034 (Figure 12.2). In this case, the resolver consults three name servers to resolve the fully qualified domain name www.wikipedia.org.

For proper operation of its domain name resolver, a network host is configured with an initial cache (hints) of the known addresses of the root name servers. The hints are updated periodically by an administrator by retrieving a dataset from a reliable source. Assuming the resolver has no cached records to accelerate the process, the resolution process starts with a query to one of the root servers. In typical operation, the root servers do not answer directly, but respond with a referral to more authoritative servers, for example, a query for "www.wikipedia.org" is referred to the ".org" servers(s). The resolver now queries the servers referred to and iteratively repeats this process until it receives an authoritative answer. The diagram illustrates this process for the host that is named by the fully qualified domain name "www.wikipedia.org."

In theory, authoritative name servers are sufficient for the operation of the Internet. However, with only authoritative name servers operating, every DNS query must start with recursive queries at the root zone of the DNS and each user system would have to implement resolver software capable of recursive operation. However, this mechanism would place a large traffic burden on the root servers, if every resolution on the Internet required starting at the root. In practice, caching is used in DNS servers to off-load the root servers, and as a result, root name servers actually are involved in only a relatively small fraction of all requests.

To improve efficiency, reduce DNS traffic across the Internet and increase performance in end-user applications, the DNS supports DNS cache servers which store DNS query results for a period of time determined in the configuration (time-to-live) of the domain name record in question. Typically, such caching DNS servers also implement the recursive algorithm necessary to resolve a given name starting with the DNS root through to the authoritative name servers of the queried domain. With this function implemented in the name server, user applications gain efficiency in design and operation. The combination of DNS caching and recursive functions in a name server is not mandatory. These functions can be implemented independently in servers for special purposes. Internet service providers typically provide recursive and caching name servers for their customers. In addition, many home networking routers implement DNS caches and recursion to improve efficiency in the local network. This common approach also reduces the burden on DNS servers as they cache the results of name resolution locally or on intermediary resolver hosts. Each DNS query result comes with a time to live (TTL), which indicates how long the information remains valid before it needs to be discarded or refreshed. This TTL is determined by the administrator of the authoritative DNS server and can range from a few seconds to several days or even weeks. As a result of this distributed caching architecture changes to DNS records do not propagate throughout the network immediately, but require all caches to expire and to be refreshed after the TTL. RFC 1912 conveys basic rules for determining appropriate TTL values. Some resolvers may override TTL values, as the protocol supports caching for up to 68 years or no caching at all. Negative caching, that is, the caching of the fact of non-existence of a record, is determined by name servers authoritative for a zone which must include the Start of Authority (SOA) record when reporting no data of the requested type exists. The value of the minimum field of the SOA record and the TTL of the SOA itself is used to establish the TTL for the negative answer.

The client side of the DNS is called a DNS resolver. A resolver is responsible for initiating and sequencing the queries that ultimately lead to a full resolution (translation) of the resource sought, for example, translation of a domain name into an IP address. DNS resolvers are classified by a variety of query methods, such as recursive, non-recursive and iterative, and resolution process may use a combination of these methods:

- **Non-recursive query:** a DNS resolver queries a DNS server that provides a record either for which the server is authoritative or it provides a partial result without querying other servers. In case of a caching DNS resolver, the non-recursive query of its local DNS cache delivers a result and reduces the load on upstream DNS servers by caching DNS resource records for a period of time after an initial response from upstream DNS servers.

- **Recursive query:** a DNS resolver queries a single DNS server, which may in turn query other DNS servers on behalf of the requester. For example, a simple stub resolver running on a home router typically makes a recursive query to the DNS server run by the user's ISP. A recursive query is one for which the DNS server answers the query completely by querying other name servers as needed. In typical operation, a client issues a recursive query to a caching recursive DNS server, which subsequently issues non-recursive queries to determine the answer and send a single answer back to the client. The resolver, or another DNS server acting recursively on behalf of the resolver, negotiates use of recursive service using bits in the query headers. DNS servers are not required to support recursive queries.
- **Iterative query:** a DNS resolver queries a chain of one or more DNS servers. Each server refers the client to the next server in the chain, until the current server can fully resolve the request. For example, a possible resolution of www.example.com would query a global root server, then a ".com" server and finally an "example.com" server.

Name servers in delegations are identified by name, rather than by IP address. This means that a resolving name server must issue another DNS request to find out the IP address of the server to which it has been referred. If the name given in the delegation is a subdomain of the domain for which the delegation is being provided, there is a circular dependency. In this case, the name server providing the delegation must also provide one or more IP addresses for the authoritative name server mentioned in the delegation. This information is called glue. The delegating name server provides this glue in the form of records in the additional section of the DNS response and provides the delegation in the authority section of the response. A glue record is a combination of the name server and IP address.

For example, if the authoritative name server for example.org is ns1.example.org, a computer trying to resolve www.example.org first resolves ns1.example.org. As ns1 is contained in example.org, this requires resolving example.org first, which presents a circular dependency. To break the dependency, the name server for the TLD ".org" includes glue along with the delegation for example.org. The glue records are address records that provide IP addresses for ns1.example.org. The resolver uses one or more of these IP addresses to query one of the domain's authoritative servers, which allows it to complete the DNS query.

A reverse DNS lookup is a query of the DNS for domain names when the IP address is known. Multiple domain names may be associated with an IP address. The DNS stores IP addresses in the form of domain names as specially formatted names in pointer (PTR) records within the infrastructure top-level domain arpa. For IPv4, the domain is in-addr.arpa. For IPv6, the reverse lookup domain is ip6.arpa. The IP address is represented as a name in reverse-ordered octet representation for IPv4 and reverse-ordered nibble representation for IPv6. When performing a reverse lookup, the DNS client converts the address into these formats before querying the name for a PTR record following the delegation chain as for any DNS query. For example, assuming the IPv4 address 208.80.152.2 is assigned to Wikimedia, it is represented as a DNS name in reverse order: 2.152.80.208.in-addr.arpa. When the DNS resolver gets a PTR request, it begins by querying the root servers, which point to the servers of American Registry for Internet Numbers (ARIN) for the 208.in-addr.arpa zone. ARIN's servers delegate 152.80.208. in-addr.arpa to Wikimedia to which the resolver sends another query for 2.152.80.208.in-addr.arpa, which results in an authoritative response.

Users generally do not communicate directly with a DNS resolver. Instead DNS resolution takes place transparently in applications such as web browsers, e-mail clients and other Internet applications. When an application makes a request that requires a domain name lookup, such programs send a resolution request to the DNS resolver in the local operating system, which in turn handles the communications required. The DNS resolver almost always has a cache containing recent lookups. If the cache can provide the answer to the request, the resolver will return the value in the cache to the program that made the request. If the cache does not contain the answer, the resolver will send the request to one or more designated DNS servers.

DNS resolvers come in various forms, including authoritative DNS servers, recursive resolvers and open resolvers. While authoritative DNS servers are responsible for hosting and providing authoritative answers for specific domain zones, recursive resolvers play a crucial role in navigating the DNS hierarchy to resolve queries by recursively querying other DNS servers until they obtain the final answer. Open resolvers, however, serve a unique function in the DNS ecosystem. Unlike authoritative and recursive resolvers, open resolvers accept and respond to DNS queries from any source, regardless of whether they are within the resolver's administrative domain. Open resolvers operate as public DNS servers, providing resolution services to any requester on the Internet.

Open resolvers are distributed across the Internet, operated by various entities, including ISPs, universities, public DNS service providers and other organisations. This decentralisation contributes to the resilience and redundancy of the DNS infrastructure but also complicates efforts to secure and monitor open resolver operations effectively. While open resolvers offer convenience and accessibility, they also pose inherent security risks. Open resolvers may be misconfigured or lack important DNS updates that leave them susceptible to security threats. Managing and securing DNS authoritative infrastructure becomes challenging when it must be open to open resolvers, which may be misconfigured, unpatched or abused to launch reflection amplification DNS DDoS attacks. Attackers can exploit open resolvers to amplify and reflect DNS traffic as part of DDoS attacks, manipulate DNS responses to redirect users to malicious websites or conduct reconnaissance activities to gather information about target networks. The unrestricted accessibility of open resolvers makes them susceptible to exploitation in DNS amplification and reflection attacks. Attackers leverage open resolvers to amplify DNS traffic directed at a target by sending spoofed queries, leading to volumetric DDoS attacks with magnified impact.

For the most of home users, the Internet service provider (to which their computer is connected) will usually supply this DNS server: such a user will either have configured that server's address manually or allowed DHCP to set it. Some large ISPs have configured their DNS servers to violate rules, such as by disobeying TTLs, or by indicating that a domain name does not exist just because one of its name servers does not respond. However, where systems administrators have configured systems to use their own DNS servers, their DNS resolvers point to separately maintained name servers of the organisation. In any event, the name server thus queried will follow the process outlined above, until it either successfully finds a result or does not. It then returns its results to the DNS resolver; assuming it has found a result, the resolver duly caches that result for future use, and hands the result back to the software which initiated the request.

Some applications like, for example, web browsers may maintain their own internal DNS cache to avoid repeated lookups via the network. This practice can add extra difficulty when debugging DNS issues as it obscures the history of such data. These caches typically use very short caching times in the order of one minute.

DNS includes several other functions and features, as also DNS serves other purposes in addition to translating names to IP addresses. For instance, mail transfer agents use DNS to find the best mail server to deliver e-mail: An MX record provides a mapping between a domain and a mail exchanger. This can provide an additional layer of fault tolerance and load distribution. DNS is also used for efficient storage and distribution of IP addresses of blacklisted email hosts. A common method is to place the IP address of the subject host into the sub-domain of a higher-level domain name, and to resolve that name to a record that indicates a positive or a negative indication. E-mail servers can query "blacklist.example" to find out if a specific host connecting to them is in the blacklist. Many of such blacklists, either subscription-based or free of charge, are available for use by email administrators and anti-spam software.

Hostnames and IP addresses are not required to match in a one-to-one relationship. Multiple hostnames may correspond to a single IP address, which is useful in virtual hosting, in which many websites are served from a single host. Alternatively, a single hostname may resolve to many IP addresses to facilitate fault tolerance and load distribution to multiple server instances across an enterprise or Internet.

To ensure resilience in the event of computer or network failure, multiple DNS servers are usually provided for coverage of each domain. At the top level of global DNS, thirteen groups of root name servers exist, with additional "copies" of them distributed worldwide via anycast addressing.

Dynamic DNS (DDNS) updates a DNS server with a client IP address on-the-fly, for example, when moving between ISPs or mobile hot spots, or when the IP address changes administratively.

As we can see, Internet can't survive without DNS. However, DNS creates numerous insecurities. Probably one of the best descriptions of DNS insecurities was given in https://www.f5.com/labs/articles/threat-intelligence/dns-is-still-the-achilles-heel-of-the-internet-25613:

> Imagine proposing a new application project to your boss. It's a distributed network database that runs across millions of nodes on the Internet. Everyone would own and run their own server but would need to coordinate data storage, retrieval, and update with all the others. This cooperation would be based on a published document describing the relationship - but that's all! There would be no organisation and no master control server in charge, just some simple hierarchies and some registration authorities that keep track of who holds what records. Anyone could query this database anonymously, and the whole distributed system would work out the answer and return it to the requestor. Oh, and the whole thing would run over the fire-and-forget, unreliable User Datagram Protocol (UDP), which can be easily spoofed. Your boss would probably laugh you out of the room for proposing such an unworkable system. Yet, in 1983, the Internet Engineering Task Force proposed a solution and the following year, the first Domain Name System (DNS) server was coded at UC Berkley. This was back in the days when everyone on the net (called ARPANET back then) trusted each other completely and none of the participants were motivated to cause problems. Somehow, good old DNS survived this sheltered childhood and thrives today in our modern swamp of vipers and leeches that is the Internet. It hasn't been without some scars, as DNS still bears some fundamental weaknesses that are still exploited today.

One may ask: "How come that DNS has survived?." The answer is very simple – because it is extremely useful. Firstly, it's the strong inertia of being the first such system with deep legacy and dependence sunk into to the Internet's infrastructure combined with enormous potential cost and complexity of its replacement with something else. Secondly, it's cheap and easy to run and query, with many different services available in both commercial and open-source implementations. And, of course, it's something that is proven and works well on a global scale of interconnected disparate networks because it is distributed and no one controls it. It's obvious that DNS is a critical piece of Internet infrastructure. As security guru Dan Geer said, "Risk is a consequence of dependence." We are stuck with DNS and future of the Internet depends on it.

So, the problem is that DNS is too important to do without, but it's difficult to defend. In fact, DNS services are an excellent target for an attack. Taking out an organisation's DNS service renders it unreachable to the rest of the world except by IP address. If "xyz.com" failed to be published online, every single Internet site and service it runs would be invisible. This means web servers, VPNs, mail services, file transfer sites – everything. Even worse, if hackers could change the DNS records, then they could redirect everyone to sites they controlled. Imagine going to "www.xyz.com" and landing on a page full of banner ads. Since DNS is built upon cooperation between millions of servers and clients over insecure and unreliable protocols, it is uniquely vulnerable to disruption, subversion and hijacking.

In his excellent article (https://www.linkedin.com/posts/andy-jenkinson-96210727_in-1999-when-many-of-todays-security-professionals-activity-7248316498883047424-bVnJ?utm_source=share&utm_medium=member_ios), Andy Jenkinson highlighted that the first warning about DNS came in 1999, when Daniel J. Bernstein (DJB) warned the world about the exposure and vulnerabilities of DNS servers. His concerns went largely ignored.

Then, nearly a decade later, in 2008, Dan Kaminsky demonstrated how DNS security flaws left the entire Internet vulnerable to attacks. Microsoft and others responded with temporary fixes, but these were mere patches and failed to address the underlying issues.

In 2013, Edward Snowden exposed how the NSA, along with their allies, had long exploited the Internet's weaknesses, including DNS and Public Key Infrastructure (PKI), to carry out mass surveillance. This became the MO for Cyberwars and Cyber Crime. Rather than spurring real action to address and improve security, Snowden's and Kaminsky's revelations armed adversaries with knowledge of these vulnerabilities, as the NSA sought to cover-up and keep their methods secret. Despite Kaminsky's warnings and Snowden's disclosures, the majority still ignore or dismiss these critical vulnerabilities. Adversaries continue to exploit these weaknesses, and the Internet remains dangerously exposed. As Dan Kaminsky concluded: "DNS should not have been capable of this much damage – it was – but why?"

DNS attacks picked up and in 2018 large numbers of US Federal Agencies suffered DNS attacks. This served as the catalyst for CISA to reluctantly issue their first Emergency Directive – M-19-01 on DNS Tampering and Abuse.

The DNS system is vulnerable to numerous cyberthreats due to its design limitations and lack of security measures. Such hazards include spoofing, amplification, DoS and the interception of private information. Moreover, DNS attacks are often used as a distraction tactic with other cyberattacks, making it harder for a security team to focus on a potentially more significant threat. With these vulnerabilities, it is essential to have strong DNS security to prevent DNS attacks and protect business continuity.

Let's have a quick look at some known types of major DNS attacks.

Denial of Service (DoS): DoS attacks are a common type of cyber-attack that can be used to disrupt the normal functioning of a DNS. In a DoS attack, a threat actor floods a DNS server with a large traffic volume, causing it to become overwhelmed and stop responding to legitimate DNS requests. DoS attacks are not limited to DNS, but taking out DNS decapitates an organisation. Why bother flooding thousands of websites when killing a single service does it all for the malicious agent?

Distributed Denial of Service (DDoS) attack: In a DDoS attack, multiple systems are used to flood a DNS server with traffic, making it much more difficult to mitigate the attack. Whereas a DoS attack is an attack from one spot, not multiple. DDoS attacks can be particularly challenging to defend against, as they can involve many compromised systems and come from many different sources.

A DoS and DDoS attacks can have negative consequences, including website downtime, lost revenue and reputational damage. One of the most famous examples of this was 2016 attack on Dyn, Inc. On October 21, 2016, three consecutive distributed denial-of-service attacks were launched against the DNS provider Dyn. The attack caused major Internet platforms and services to be unavailable to large swathes of users in Europe and North America. The groups Anonymous and New World Hackers claimed responsibility for the attack, but scant evidence was provided. DDoS attacks which exceeded 40 gigabytes of noise blared at their DNS services. Dyn was running DNS services for many major organisations, so when they were drowned by a flood of illegitimate packets, so were companies like Amazon, CNN, Netflix, Twitter (now X), Reddit, FiveThirtyEight, Visa5 and many others. Dyn's chief strategist said in an interview that the assaults on the company's servers were very complex and unlike everyday DDoS attacks. Dyn said that it was orchestrated using a weapon called the Mirai botnet as the "primary source of malicious attack"

There are many ways to knock out DNS service, the simplest being a stream of garbage from thousands of compromised computers or hosts (bots) in a DDoS attack. Instead of clogging up the pipe, attackers can also overwork the server with DNS Query Flood attacks from thousands of bots.

DNS can also be subverted for use as a denial-of-service weapon against other sites by way of DNS Amplification/Reflection. This works because DNS almost always returns a larger set of data than what was queried. A simple DNS query asking for XYZ.com only amounts to a few hundred bytes at most, while the response will be several orders of magnitude larger. This way an attacker can amplify network traffic through DNS servers, building up a tsunami from a ripple. As DNS runs over UDP, it's a simple matter for attackers to craft fake packets spoofing a query source, so if they can fake thousands of queries from the victim's IP address, that tsunami of responses will return to overwhelm the victim. A bonus for the attacker is that, to the victim, it will appear as if a huge number of DNS servers are attacking it, while, the attacker stays safely hidden.

Pseudo-random subdomain (PRSD) attack: PSRD attack, also known as a random subdomain attack or a water torture attack, is a type of DDoS attack. Directed at a specific web domain, PRSD attacks flood DNS nameservers with thousands of apparently legitimate but malicious DNS requests. As a result, both recursive DNS servers and authoritative DNS nameservers and its infrastructure environment become overloaded with requests and will slow down or crash, preventing the servers from responding to legitimate traffic and causing the domain to become unavailable. These types of attacks are highly effective because rather than simply flooding DNS servers with illegitimate packets, PRSD attacks send DNS requests for subdomains that appear to be legitimate and are harder to recognise as malicious. These attacks are therefore quite powerful, as their legitimate nature means they can bypass many of the DDoS protections and most of the automatic mitigations of most firewalls and·DDoS scrubbers (automatic filters for large attacks or malicious attacks) and therefore overwhelm most nameservers.

TCP SYN Flood attack: TCP SYN Flood is a form of DDoS attack when the attacker floods the target with SYN messages as in TCP state-exhaustion attack.

NXDOMAIN attack: NXDOMAIN flood DDoS attack attempts to overwhelm the DNS server using a large volume of requests for invalid or non-existent records. These attacks are often handled by a DNS proxy server that uses up most (or all) of its resources to query the DNS authoritative server. This causes both the DNS Authoritative server and the DNS proxy server to use up all their time handling bad requests. As a result, the response time for legitimate requests slows down until it eventually stops altogether.

DNS Amplification attack: DNS Amplification is a form of DDoS attack that is used by cybercriminals to overwhelm a website's servers with traffic. This attack is carried out by exploiting the way the DNS works. In a DNS Amplification attack, the attacker spoofs the victim's IP address and sends a query to a DNS server with a spoofed IP address, requesting a significant DNS response. The server then returns a large DNS response to the victim's IP address, much larger than the original query packet. This amplification of the DNS response causes the victim's network to be flooded with traffic, ultimately leading to a DoS attack.

Distributed Reflection Denial of Service (DRDoS) attack: A slightly different type of DDoS attack, in which not the direct queries, but the answers to them will go to the victim. This is the reflection. The cybercriminals will send DNS queries, but the IP of the source will be changed. Servers will respond and will send all that traffic to the target. The traffic can be overwhelming and flood the target, eventually stopping it. A Smurf attack is a popular DNS attack of that type.

tsuNAME DDoS attack: tsuNAME is a flaw in DNS resolver software that enable DDoS attacks against DNS servers. Domains with "cyclic dependencies" can exist, where domain A delegates to domain B and vice versa. Vulnerable DNS resolvers will start looping when presented with domains causing cyclic dependencies. In one case, just two misconfigured domains created a 50% traffic increase for .nz authoritative DNS servers in 2020.

DNS Hijacking attack: DNS hijacking is a cyberattack where an attacker gains access to a user's DNS records and redirects their traffic to a malicious website or server. The attack can result in the theft of sensitive information, installation of malware and financial losses. Who owns what domain name and what DNS servers are designated to answer queries are managed by Domain Registrars. These are commercial services, such as GoDaddy, CrazyDomains, Domain.com, Bluehost, HostGator, HostPapa, Network Solutions Inc., where registered accounts store this information. If attackers can hack these accounts, they can repoint a domain to a DNS server they control. Attacks like this have affected the New York Times, LinkedIn, Dell, Harvard University, Coca Cola and many others. There are several forms of DNS hijacking:

- **Local DNS hijacking:** An attacker installs Trojan software on a computer, then modifies the local DNS settings to reroute the user to harmful websites.
- **DNS hijacking using a router:** Many routers have weak firmware or use the default passwords they were shipped with. Attackers can take advantage of this to hack a router and change its DNS settings, which will affect everyone that uses that router.
- **Man-in-the-middle attacks (MITM):** Attackers use MITM attack techniques to intercept communications between users and a DNS server. They then direct the target to malicious websites.
- **Rogue DNS server:** Hackers can alter DNS records on a DNS server, enabling them to reroute DNS requests to malicious websites. If the site looks legitimate, the user may not even know they are in the wrong place.

DNS rebinding attack: DNS rebinding method allows an attacker to overcome the problem of closed ports on the router. In this case, the attack starts from a web page that executes a malicious client-side script in the browser. This generates an attack on machines elsewhere on the network. Domain name verification is one of the essential building blocks of the same-origin policy enforced by web browsers to exclusively grant the host that created the script access to content. The DNS rebinding attack, however, overcomes this policy by exploiting the system in order to resolve domain names abusively. Simply put, the DNS rebinding attack allows a browser to start communicating with remote servers with which it should not actually exchange data.

Phantom domain attack: Phantom domains are defined as active links to .com domains that have never been registered. Phantom domain attack happens when the attacker sets up "phantom" domains that do not respond to DNS queries which makes it possible for malicious actors to hijack hyperlinks and exploit users' trust in familiar websites. Under normal circumstances, the DNS recursive server contacts authoritative servers to resolve recursive queries. When phantom domain attacks happen, the recursive server continues to query nonresponsive servers, which causes the recursive server to spend valuable resources waiting for responses. When resources are fully consumed, the DNS recursive server may drop legitimate queries, causing serious performance issues.

DNS spoofing or DNS cache poisoning attack: In this type of DNS attack, hackers redirect Internet traffic from legitimate websites to fraudulent ones, which can result in data theft, financial fraud and other malicious activities. DNS spoofing works by altering the DNS cache of a user's computer or the DNS server, replacing the IP address of a legitimate website with that of a fraudulent website. As a result, when users try to access a legitimate website, they are directed instead to a fraudulent one, where their sensitive information may be stolen, or malware can be distributed. With cache poisoning, hackers target caching name servers to manipulate the DNS cache's stored responses. This attack can be carried out in a variety of ways, but it commonly involves flooding the server with forged DNS responses while altering the query ID of each response. Unless Domain Name System Security Extensions (DNSSEC) is implemented, cache poisoning can be difficult to identify and defend against. DNSSEC refers to a collection of extension specifications set up by the Internet Engineering Task Force (IETF) to safeguard data exchanged in the DNS and IP systems. Without DNSSEC, hackers are more likely to execute a successful attack and impact thousands of users who access a nameserver with compromised responses.

The original design of the Domain Name System did not include any security features. It was conceived only as a scalable distributed system. The Domain Name System Security Extensions (DNSSEC) attempt to add security, while maintaining backwards compatibility. RFC 3833 of 2004 documents some of the known threats to the DNS, and their solutions in DNSSEC. DNSSEC was designed to protect applications using DNS from accepting forged or manipulated DNS data, such as that created by DNS cache poisoning. All answers from DNSSEC protected zones are digitally signed. By checking the digital signature, a DNS resolver is able to check if the information is identical (i.e., unmodified and complete) to the information published by the zone owner and served on an authoritative DNS server. While protecting IP addresses is the immediate concern for many users, DNSSEC can protect any data published in the DNS, including text records (TXT) and mail exchange records (MX), and can be used to bootstrap other security systems that publish references to cryptographic certificates stored in the DNS such as Certificate Records (CERT records RFC 4398), SSH fingerprints (SSHFP, RFC 4255), IPSec public keys (IPSECKEY, RFC 4025), TLS Trust Anchors (TLSA, RFC 6698) or Encrypted Client Hello (SVCB/HTTPS records for ECH). However, DNSSEC does not provide confidentiality of data. In particular, all DNSSEC responses are authenticated but not encrypted. DNSSEC does not protect against DoS attacks directly, though it indirectly provides some benefits.

"Forgot Password" cache poisoning attack: "Forgot password" links are common in web applications, but a vulnerability discovered in July 2021 made them vulnerable to DNS cache poisoning attacks. Security researchers discovered that, by performing a cache poisoning attack on 146 vulnerable web applications, they could redirect password reset emails to attacker-controlled servers. This enabled them to click on the link and reset the user's password, providing legitimate access to their account.

Data exposure in managed DNS attack: Research presented at Black Hat USA 2021 demonstrated that bugs in certain managed DNS services could expose corporate DNS traffic containing sensitive information. By registering a domain on Amazon's Route53 DNS service or Google Cloud DNS that had the same name as the DNS name server, the attacker could force all DNS traffic to be sent to their server. This exposed sensitive information and could enable DNS spoofing attacks.

DNS Tunneling attack: DNS tunneling is a technique used by hackers to bypass security measures and steal data. DNS tunneling technique enables threat actors to compromise network connectivity and gain remote access to a targeted computer. It involves encoding data into DNS queries and responses to create a covert communication channel through a DNS server. This technique allows hackers to bypass firewalls and other security measures that may block other types of communication channels. While DNS tunneling can also be used for legitimate purposes, when used maliciously, it poses a significant threat to data security. DNS tunneling has been around for well over 20 years. Both the Morto and Feederbot malware have been used for DNS tunneling. Some well-known DNS tunneling attacks include those from the threat group DarkHydrus, which targeted government entities in the Middle East in 2018, and OilRig, which has been operating since 2016 and is still active. This is how DNS tunnelling attack works:

- The attacker registers a domain, such as badsite.com. The domain's name server points to the attacker's server, where a tunneling malware program is installed.
- The attacker infects a computer, which often sits behind organisation's firewall, with malware. Because DNS requests are always allowed to move in and out of the firewall, the infected computer is allowed to send a query to the DNS resolver. The DNS resolver is a server that relays requests for IP addresses to root and top-level domain servers.
- The DNS resolver routes the query to the attacker's command-and-control (C2) server, where the tunneling program is installed. A connection is now established between the victim and the attacker through the DNS resolver. This tunnel can be used to exfiltrate data or for other malicious purposes. Because there is no direct connection between the attacker and victim, it is more difficult to trace the attacker's computer.

Fast Flux DNS: Fast-flux DNS is a technique cybercriminals use to strengthen their botnet networks. This is accomplished by rapidly changing the DNS entries present on a domain name and setting up multiple subdomains in a rapid and automated manner using DGAs (domain generation algorithms). Fast flux is a technique cybercriminals use to evade detection by rapidly changing the IP addresses associated with a domain. This method is often employed in social engineering campaigns, command-and-control (C2) infrastructures, and even illicit gambling and adult sites.

Smoke Loader C2 Campaign: In a smoke loader C2 campaign, fast flux DNS enables cybercriminals to maintain control over compromised systems by constantly rotating IP addresses. This agility makes it difficult for security researchers and law enforcement to pinpoint and disrupt the malicious infrastructure.

Domain generation algorithm (DGA) attack: Like fast-flux attacks, DGA attacks are used by cybercriminals to rapidly generate subdomain or domain names via an automated method that, combined with the non-logging nature of DNS, makes it a very resilient malware delivery platform.

Domain squatting attack: Domain squatting attacks are often malicious from multiple fronts, wherein a cybercriminal purchases a domain name similar to organisation's domain name, then uses the domain name to set up phishing pages or extort money from organisation in order to sell the domain for a highly inflated amount. For example, if organisation owns "myorganisation.com," the cybercriminal may purchase "myorgaanisation.com" or similarly, "myorganisation.net" and set up phishing pages or extort organisation for money accordingly.

Phishing attack: Phishing attack is a type of social engineering attack when an attacker sends a fraudulent email or message to a victim, which leads the victim to a website that the attacker has crafted to resemble a legitimate website. Then, any information that is input on this fake website is logged by the attacker. Such attacks are intended to steal sensitive information such as login credentials or financial data.

DNS server vulnerabilities: Because DNS services are software, they are likely to contain bugs. It's possible that some of these bugs will create software vulnerabilities that attackers can exploit. That's just the way it is with all software written. Luckily, DNS is old (so we've had time to find most of the bugs) and simple (so bugs are easy to spot), but problems have cropped up. In 2015, there was a rather significant hole found in BIND, an open-source DNS server running much of the Internet. This BIND vulnerability was called CVE-2015-547711 and allowed an attacker to crash a DNS server with a single crafted query. Another software vulnerability in DNS servers is the Recursive DNS spoof cache poisoning technique, which means that an attacker can temporarily change DNS database entries by issuing specifically crafted queries.

DNS spoofing is the resulting threat which mimics legitimate server destinations to redirect a domain's traffic. Unsuspecting victims end up on malicious websites, which is the goal that results from various methods of DNS spoofing attacks. Among the various methods for DNS spoof attacks, these are some of the more common:

- **Man-in-the-middle (MITM) duping:** Where an attacker steps between the web browser and the DNS server. A tool is used for simultaneous cache poisoning on the local device and server poisoning on the DNS server. The result is a redirect to a malicious site hosted on the attacker's own local server.
- **DNS server hijacks:** Where an attacker directly reconfigures the server to direct all requesting users to the malicious website. Once a fraudulent DNS entry is injected onto the DNS server, any IP request for the spoofed domain will result in the fake site.
- **DNS cache poisoning via spam:** DNS cache poisoning is a user-end method of DNS spoofing, in which victim's system logs the fraudulent IP address in victim's local memory cache. This leads the DNS to recall the bad site specifically for the victim, even if the issue gets resolved or never existed on the server-end. The code for DNS cache poisoning is often found in URLs sent via spam emails. These emails attempt to push users into clicking on the supplied URL, which in turn infects their computer. Banner ads and images – both in emails and on untrustworthy websites – can also direct users to this code. Once poisoned, computer will take one to fake websites that are spoofed to look like the real thing. This is where the true threats are introduced to devices.

- **Unauthorised DNS Changes**: Every server has to be managed by someone. This means that organisations are dependent on how strongly they are authenticating the admins to that server as well as ensuring the trustworthiness and competence of those admins. In practice, this vulnerability is often realised by accident when admin's "fat-fingers" a DNS change or incorrectly manages the DNS servers, but ill will can't be excluded either. Because of the nature of DNS records, changes to DNS are cached by query clients, so mistakes can sometimes take long time (hours or even days) to unwind across the Internet.
- **DNS data leakage:** One can't run an unauthenticated Internet database full of important information without the occasional risk of leaking out something important. Attackers will often repeatedly query DNS servers as a prelude to an attack, looking for interesting Internet services that may not be widely known. For example, an organisation may have a site called myservice.example.com which it doesn't advertise to anyone except its employees. If an attacker discovers this site, they've just found a new potential target in an attack. DNS records can also aid phishing expeditions by using known server names in their phony baloney emails. Some organisations run DNS on the inside of the network, advertising local area network (LAN) resources. Some smaller organisations run split-horizon DNS servers that offer up Internet DNS services to the world as well as these LAN-based DNS services on the same box. A wrong configuration on that DNS server can lead to some devastating DNS data leakages as internal names and addresses are shared with attackers. Even giants can be tripped up by this seemingly simple vulnerability.
- **DNS Man-in-the-Middle (MITM)**: Once again, the easily spoofed protocol UDP that DNS uses is the weak link. In this case, an attacker inline between the victim and the DNS server they're querying can intercept and monkey with DNS queries

Another unpleasant side effect of using DNS is DNS tracking/logging. Whenever a domain name is resolved, a DNS server is queried for information. In doing so, information about the user is sent to the ISP in charge of that server, which records user's IP address and thus user's approximate location. TLS/SSL certificates encrypt the communication so that hackers are not able to read the content. But this doesn't hide user's IP address when a user visits a domain name. If someone is able to track an IP address, they can potentially relate it to other stored information like name, address, bank details and much more. Hackers can potentially collect and correlate this information to perpetrate their attacks. In the past, some ISPs have accumulated this information to resell it to third parties, often advertisers, enabling them to implement their strategies in a targeted manner. Users in Europe enjoy greater protection due to the introduction of the GDPR. IP may certainly be associated with other information.

For those interested in DNS abuse it is recommended to look at (and start following him on LinkedIn) multiple Andy Jenkinson's publications on this topic in which he highlights lack of understanding of DNS that led (and continue to lead) to multiple misconfigurations and exposures. As he once said that "options to abuse DNS are endless as DNS touches everything." Quoting Andy Jenkinson again:

> As businesses continue to digitally transform and the interconnected ecosystem on which they depend expands, DNS attacks will only become more frequent and more damaging. Dr Paul Mockapetris and Dr Paul Vixie, another Internet Hall of Fame inductee and DNS expert state, "Over 95% of all Cyberattacks, Malware and Bots rely upon DNS.

It may be also of interest to follow DNS Abuse SIG (https://www.first.org/global/sigs/dns/) and read https://www.first.org/global/sigs/dns/DNS-Abuse-Techniques-Matrix_v1.1.pdf.

Summarising, this is why DNS is inherently vulnerable to cyberattacks:

- **Insufficient Security Measures:** Historically, DNS was designed with a focus on functionality rather than security. While efforts such as DNSSEC aim to address some of these security shortcomings, adoption remains relatively low, leaving many DNS transactions vulnerable to interception and manipulation.
- **Protocol Complexity:** The DNS protocol itself is complex, with various components and interactions between servers (see Chapter 4). This complexity increases the likelihood of implementation errors and vulnerabilities that can be exploited by attackers.
- **Attack Surface:** DNS is a critical component of Internet infrastructure, making it an attractive target for attackers seeking to disrupt services, steal sensitive information or launch large-scale attacks, such as DDoS attacks.
- **Centralisation and Hierarchical Structure:** The hierarchical structure of DNS involves multiple levels of authority, from the root servers down to individual domain name servers. This structure creates multiple potential points of failure and opportunities for attackers to exploit vulnerabilities at various levels.
- **Lack of Authentication:** Traditional DNS lacks built-in authentication mechanisms, making it susceptible to various types of attacks, such as DNS spoofing and cache poisoning. Without cryptographic validation of DNS responses, attackers can manipulate DNS data to redirect users to malicious websites or intercept sensitive information.

- **Weaknesses in Infrastructure**: DNS infrastructure, including DNS servers and resolvers, is often poorly configured or outdated, leaving them vulnerable to exploitation. Additionally, many organisations fail to implement security best practices, such as regular software updates and patch management, further exacerbating the vulnerabilities in DNS infrastructure.
- **Human Factors**: Human error, such as misconfigurations or weak password practices, can also contribute to DNS vulnerabilities. Attackers often exploit these weaknesses through social engineering tactics or by targeting individuals with access to DNS infrastructure.

Overall, the combination of protocol complexity, lack of authentication, weaknesses in infrastructure, and the attractiveness of DNS as a target make it susceptible to cyberattacks and hacking. The result of this is that according to IDC 2022 Global DNS Threat Report on a global scale, 88% of organisations have suffered DNS attacks and 76% of DNS attacks caused application downtime and the average attack took over five and a half hours to mitigate – with organisations encountering an average of seven attacks per year at a cost of $942K per attack. In addition to financial losses, other serious consequences of DNS attacks include data theft, reputation damage, website downtime and malware infections. In 2024, average cost of DNS attack recovery has grown to $1.1 Mln and DNS attacks lead to application outages in 82% of businesses and data theft in 29% of those cases. 80% of organisations consider DNS security is crucial for their protection.

There are around 4.3 billion IPv4 addresses and IPv6 allows 340 trillion trillion trillion IP addresses. One of the common errors of DNS is not realising it's not just the digital cert of the URL that counts but that of the IPv4 server itself. That can allow both secure and not secure version of the same website with all that entails. Equally the same applies for BGP's CNAME and MX. Ultimately and this is the part most don't understand or don't want to, DNS was exploited by the NSA for decades, particularly post 9/11. They never once tried to address and educate, just exploit. Revelations by Dan Kaminsky and later Edward Snowden allowed the bad guys to learn how to exploit DNS for cybercrime.

Cybercriminals and hackers abuse DNS for data exfiltration for secretly stealing information from organisation's network. This technique is explained in detail in Andy Jenkinson's article "How the Domain Name System (DNS) and the Hypertext Transfer Protocol (HTTP) Are Exploited by Cyber Criminals". In another very interesting article ("Domain Name System (DNS) Abuse"), Andy highlights the risks of DNS abuse and points that DNS is possibly the most abused and the most manipulated area of Internet Assets.

And, as they say, better late than never: DNS abuse is any activity that makes use of domain names or the DNS protocol to carry out harmful or illegal activity.

One may wonder why DNS and BGP are lumped together in a single chapter. To begin with, security of BGP is non-existent. But, as in case with DNS, we all rely on Border Gateway Protocol (BGP) every day. For everything. DNS, as well as, BGP both are another insecure layers built on top of already inherent insecurities of TCP/IP.

And after some 30+ years of running BGP, it would be nice to believe that we've learned from this rich set of accumulated experience, and we now understand how to manage operation BGP to keep it secure, stable and accurate. But no, that is not where we are today. Despite its crucial function in routing wholesale amounts of data across the globe in real time, BGP still largely relies on the Internet equivalent of word of mouth for organisations to track which IP address rightfully belong to which Autonomous System Number (ASN).

In September 2022, AWS lost control of its cloud-based IP address pool for more than three hours, which allowed cyber criminals to steal $235,000 in cryptocurrency from users of one of AWS's customers. Using BGP hijacking (form of attack that exploits known weaknesses in this core Internet protocol), hackers gained control over a pool of 256 IP addresses. So, as some people say in this type of situations: "It's always DNS (unless it's BGP)." Early in 2023, Microsoft experienced a three-hour outage of its core M365 offerings due to Azure network issues, wiping out some of its most popular services. I was obvious that Microsoft impacted the internal network with a configuration change. The change didn't immediately cause problems, but issues slowly rippled across the infrastructure. This had all the hallmarks of a dodgy DNS config or a broken BGP update. So, as some people say in this type of situations: "It's always DNS (unless it's BGP)."

What is BGP? BGP is a standardised exterior gateway protocol designed to exchange routing and reachability information among Autonomous Systems (AS) on the Internet. AS is a very large network or group of networks with a single routing policy. In practice, one can think of AS as a collection of routers controlled by a single organisation that uses one or more interior gateway routing protocols and common metrics to route packets among themselves. An interior gateway protocol (IGP) or interior routing protocol is a type of routing protocol used for exchanging routing table information between gateways (commonly routers) within an autonomous system (e.g., a system of corporate local area networks). If an AS uses multiple IGPs or metrics, the AS must be consistent with external ASs in the routing policy.

AS is a collection of connected Internet Protocol (IP) routing prefixes under the control of one or more network operators on behalf of a single administrative entity or domain, that presents a common and clearly defined routing policy to the Internet. Each AS is assigned a unique ASN, which is a number that identifies the AS for use in BGP routing. ASNs are assigned to Local Internet Registries (LIRs) and end-user organisations by their respective Regional Internet Registries (RIRs), which in turn receive blocks of

ASNs for reassignment from the Internet Assigned Numbers Authority (IANA). The IANA also maintains a registry of ASNs which are reserved for private use (and should therefore not be announced to the Internet).

Originally, the definition required control by a single entity, typically an Internet service provider (ISP) or a very large organisation with independent connections to multiple networks, which adhered to a single and clearly defined routing policy. In March 1996, the newer definition came into use because multiple organisations can run BGP using private AS numbers to an ISP that connects all those organisations to the Internet. Even though there may be multiple autonomous systems supported by the ISP, the Internet only sees the routing policy of the ISP. That ISP must have an officially registered ASN. Until 2007, AS numbers were defined as 16-bit integers, which allowed for a maximum of 65,536 assignments. Since then, the IANA has begun to also assign 32-bit AS numbers to regional Internet registries (RIRs).

Origins of BGP can be traced back to 1989 when Kirk Lougheed, Len Bosack and Yakov Rekhter were sharing a meal at an IETF conference. They famously sketched the outline of their new routing protocol on the back of napkins, hence often referenced to as the "Two Napkin Protocol." It was first described in 1989 in RFC 1105 and has been in use on the Internet since 1994. IPv6 BGP was first defined in RFC 1654 in 1994, and it was improved to RFC 2283 in 1998. The current version of BGP is version 4 (BGP4), which was first published as RFC 1654 in 1994, subsequently updated by RFC 1771 in 1995 and RFC 4271 in 2006. RFC 4271 corrected errors, clarified ambiguities and updated the specification with common industry practices. BGP used for routing within an AS is called Interior Border Gateway Protocol (iBGP). In contrast, the Internet application of the protocol is called Exterior Border Gateway Protocol (eBGP). As later admitted by Yakov Rekhter (one of BGP Fathers), security at that time "wasn't even on the table."

The main difference between iBGP and eBGP peering is in the way routes that were received from one peer are typically propagated by default to other peers:

- New routes learned from an eBGP peer are re-advertised to all iBGP and eBGP peers.
- New routes learned from an iBGP peer are re-advertised to all eBGP peers only.

These route-propagation rules effectively require that all iBGP peers inside an AS are interconnected in a full mesh with iBGP sessions. Route distribution occurs by learning routes from a neighbour and advertising to other neighbours.

BGP uses TCP, which is capable of crossing network boundaries (that is, multi-hop capable). BGP neighbours, called peers, are established by manual configuration among routers to create a TCP session on port 179. A BGP speaker sends 19-byte keep-alive messages every 30 seconds (protocol default value, tuneable) to maintain the connection. Among routing protocols, BGP is unique in using TCP as its transport protocol.

Similar to DNS, BGP is not secure and allows to be abused. The challenge with BGP is that the protocol does not directly include security mechanisms and is based largely on trust between network operators that they will secure their systems correctly and will not send incorrect data. Mistakes happen, though, and problems could arise if malicious attackers were to try to affect the routing tables used by BGP. The task of trying to build a secure BGP system is a bit like trying to stop houses from burning. We could try to enforce behaviours of both the building industry, of our furniture and fittings, and of our behaviours that make it impossible for a house to catch fire. Or we could have a fire brigade to put out the fire as quickly as possible. For many years, we've opted for the latter option as being an acceptable compromise between cost and safety.

There are parallels here with BGP security. It would be an ideal situation where it would be impossible to lie in BGP. Where any attempt to synthesis BGP information could be readily identified and discarded as being bogus. But this is a very high bar to meet. And some 30 years of effort are showing just how hard this task really is.

It's hard because no one is in charge. It's hard because BGP can't be audited, as there is no standard reference data set to compare it with. It's hard because it is impossible to arbitrate between conflicting BGP information because there is no standard reference point. It's hard because there are no credentials that allow a BGP update to be compared against the original route injection, as BGP is a hop-by-hop protocol. And it's hard because BGP is the aggregate outcome of a multiplicity of opaque local decisions. There is also the problem that it is just too easy to be bad in BGP. Accidental misconfiguration in BGP appears to be a consistent problem, and it's impossible to determine the difference between a mishap and a deliberate attempt to inject false information into the routing system.

It is extremely challenging to identify a "correct" routing system, and it is far easier to understand when and where an anomaly arises and react accordingly. This situation could be characterised as: we know what we don't want when we see it, but that does not mean that we can recognise what we actually want even when we may be seeing it! This is partially due to the observation that the absence of a recognisable "bad" does not mean that all is "good"!

BGP security is a very tough problem. The combination of the loosely coupled decentralised nature of the Internet and a hop-by-hop routing protocol that has limited hooks on which to hang credentials relating to the veracity of the routing information being circulated unites to form a space that resists most conventional forms of security. It's a problem that has its consequences, in that all forms of Internet services can be disrupted, and users and their applications can be deceived in various ways where they are totally oblivious of the deception.

Let's throw some stats on BGP incidents in 2024. The number of unique AS that perpetrated BGP Route Leaks in Q2 2024: 3,044 (vs 3,017 in the previous quarter). At the same time, the number of unique AS that perpetrated BGP Hijacks: 13,626 (vs. 15,000 in the previous quarter).

Common types of BGP security risks include:

- **Route hijacks,** when a router advertises routes that are more attractive than the legitimate ones
- **Route leaks,** when a router advertises routes that it should not.
- **Route instability,** when a router changes its routes frequently or withdraws them abruptly.

BGP route hijack occurs when a "hostile" AS decides to advertise a prefix that is not its own, allowing attackers maliciously reroute Internet traffic. Attackers accomplish this by falsely announcing ownership of groups of IP addresses, called IP prefixes, that they do not actually own, control or route to. A BGP hijack is much like if someone were to change out all the signs on a stretch of freeway and re-route automobile traffic onto incorrect exits. Because BGP is built on the assumption that interconnected networks are telling the truth about which IP addresses they own, BGP hijacking is nearly impossible to stop – imagine if no one was watching the freeway signs, and the only way to tell if they had been maliciously changed was by observing that a lot of automobiles were ending up in the wrong neighbourhoods. BGP hijacks can disrupt essential services and connectivity. Rerouted traffic can lead to network instability (which can cause critical services to either become inaccessible or experience degraded performance) or even worse – to wrong websites. This can result in significant financial losses, reputational damage and operational challenges for organisations and service providers. BGP hijacking poses a significant threat to critical national infrastructure, such as power grids, financial systems and government networks. The implications for national security and the economy can be severe. For example, an attacker successfully hijacks traffic intended for a power grid. By disrupting the flow of legitimate traffic, they can force the grid offline, leading to widespread power outages and significant disruptions. However, for a hijack to occur, attackers need to control or compromise a BGP-enabled router that connects one AS to another AS, so not just anyone can carry out a BGP hijack.

BGP route leaks occur when AS incorrectly announces routing information to another AS, resulting in network traffic being directed through unintended paths. In RFC 7908, IETF provides a working definition of a BGP route leak as "the propagation of routing announcement(s) beyond their intended scope. That is, an announcement from AS of a learned BGP route to another AS is in violation of the intended policies of the receiver, the sender, and/or one of the AS along the preceding AS path." Then RFC 7908 continues this with "the result of a route leak can be redirection of traffic through an unintended path that may enable eavesdropping or traffic analysis and may or may not result in an overload or black hole. Route leaks can be accidental or malicious but most often arise from accidental misconfigurations." BGP route leaks can lead to suboptimal routing, increased latency and even complete loss of connectivity in severe cases. Monitoring and mitigating BGP route leaks is essential for maintaining a secure and reliable Internet routing infrastructure. Factors such as BGP insecurity, prefix hijacking and configuration errors contribute to the complexity of managing BGP route leaks.

Route instability is one of the most important and pathological problems of the Internet. This kind of instability can cause loss of service, waste of network resources and service degradation of applications that require QoS. Route instability refers to the rapid change of network reachability and topology information, and results in a large number of routing updates that are passed to the core Internet routers. Since the end of the NSFNet backbone in April of 1995, the Internet has been growing "explosively" in both size and topological complexity. To appreciate the scale of this growth, one needs to understand that routing tables within core Internet routers currently contain upwards of 80,000 routes. Adding to this complexity, routers in the Internet core exchange a total of somewhere between three and six million routing prefixes each day, and a single BGP update typically contains multiple route advertisements and withdrawals.

Other threats related to BGP are:

- **Wrong Peering Setup/Changes**: BGP vulnerability caused by an incorrect peering configuration between AS. It can happen for various reasons, such as hacker attacks, equipment failure or poor maintenance practices, which affects the BGP operation and therefore the entire network.
- **Route Flapping**: network event generated by a high rate of updates of the status of a route (e.g., available and not available). Route flapping produces an unstable state, resulting in a loss of data packets and a decrease in the traffic circulating in the network.
- **BGP Manipulation**: an attack in which a hacker modifies the content of the routing table in order to send data to other destinations without the sender's knowledge.
- **BGP DoS**: attack in which hackers send a large amount of data or requests to a machine or network device in order to reduce the computational resources for processing legitimate BGP traffic. For example, vulnerability in eBGP implementation of Cisco NX-OS Software could allow an unauthenticated, remote attacker to DoS condition on an affected device.

The most significant point of concern in BGP is its lack of effective security measures which makes Internet vulnerable to different forms of attacks.

Many solutions have been proposed to combat BGP security issues but not a single one is deployable in practical scenario. By 2007 several BGP enhancements proposals were developed with the view to improve BGP security: secure-BGP (sBGP), secure-origin BGP (soBGP) and pretty-secure BGP (psBGP). These enhancements come with certain advantages, as well as with some limitations. Both sBGP and soBGP use a single-level PKI for AS number authentication, a decentralised trust model for verifying the propriety of IP prefix origin, and a rating-based stepwise approach for AS_PATH (integrity) verification. Whilst psBGP trades off the strong security guarantees of S-BGP for presumed-simpler operation, for example, using a PKI with a simple structure, with a small number of certificate types and of manageable size to defend against various (non-malicious and malicious) threats from uncoordinated BGP speakers, and to be incrementally deployed with incremental benefits.

One may ask: "Why is securing BGP so hard?" Reasons are very similar to those mentioned in explanation of why DNS is inherently vulnerable to cyberattacks plus several other related to the nature and design of BGP:

- **No one is in charge**: There is no single "authority model" for the Internet's routing environment. There are various bodies that oversee the Internet's domain namespace and IP address space, but the role of a routing authority is still a vacant space. The inter-domain routing space is a decentralised, distributed environment of peers. The characterisation of this routing space implies that there is no objective reference source for what is right in routing, and equally no clear way of objectively understanding what is wrong.
- **Routing is by rumour**: We use a self-learning routing protocol that discovers the network's current inter-AS topology (or part of that topology to be more accurate). The basic algorithm is very simple, in that we tell our immediate eBGP neighbours what we know, and we learn from our immediate BGP neighbours what they know. The assumption in this form of information propagation is that everyone is honest, and everyone is correct in their operation of BGP. But essentially this is a hop-by-hop propagation, and the reachability information is not flooded across the network in the form of an original route reachability advertisement. Instead, each BGP speaker ingests neighbour information, applies local policy constraints, generates a set of advertisements that include locally applied information and, subject to outbound policy constraints, advertises that information to its neighbours. This is in many ways indistinguishable from any other form of rumour propagation. As there is no original information that is necessarily preserved in this protocol it is very challenging to determine if a rumour (or routing update) is correct or not. And impossible to determine which BGP speaker was the true origin of the rumour.
- **Routing is relative, not absolute**: Distance Vector protocols (such as BGP) work by passing their view of the best path to each destination

to their immediate neighbours. They do not pass all their available paths, just the best path. This is a distinct point of difference to the operation of Shortest Path First (SPF) algorithms, which flood link-level reachability information across the entire network, so that each SPF speaker assembles an identical (hopefully) view of the complete topology of the network. What this means is that not only does each BGP speaker only have a partial view of the true topology of the network, it is also the case that each BGP speaker assembles a view that is relative to their location in the network.

One important factor in many aspects of the Internet is the ability to support piecemeal deployment. Indeed, this loosely coupled nature of many aspects of the Internet is now so pervasive that central orchestration of many deployed technologies in the Internet is now practically impossible. The Internet is just too big, too diverse and too loosely coupled to expect any quick change. Any activity that requires some general level of coordination of actions across a diversity of networks and operational environments is a forbidding prospect.

However, recently BGP insecurity has attracted attention of the White House and in early September 2024 it indicated that it hopes sort out the weak security of Internet routing and specifically of BGP. Earlier In June 2024, the US Justice Department (DoJ) and the Defense Department (DoD) wrote to the FCC regarding the comms agency's decision to look into secure Internet routing. Endorsing the need to address BGP risks, the DoJ and DoD pointed to the way that China Telecom Americas (CTA) advertised erroneous traffic routing in 2010, 2015, 2016, 2017, 2018 and 2019 to send American network traffic to China. CTA had its FCC license revoked in 2021.

Due to a DNS setting error (which the security researcher who discovered it said was almost certainly a cut-and-paste problem), Mastercard had a DNS record with a missing character for almost 5 years. That error would have allowed attackers to potentially take over the subdomain, create a bogus site that mimics the legitimate Mastercard site and then trick customers into revealing sensitive details and credentials (https://www.csoonline. com/article/3808152/mastercards-multi-year-dns-cut-and-paste-nightmare. html). What is frightening about this mistake is not how much damage cyberthieves could have done, but how easy it is to make and how difficult it is to discover. CIP CEO Andy Jenkinson, reviewed the Mastercard problem and labelled it "appalling."

The following interesting article by Elias Heftrig, Haya Schulmann, Niklas Vogel, Michael Waidner "The Harder You Try, The Harder You Fail: The KeyTrap Denial-of-Service Algorithmic Complexity Attacks on DNSSEC" (https://arxiv.org/pdf/2406.03133) illustrates DNSSEC design flaw discovered in 2023. This design flaw makes all popular DNS implementations and services vulnerable to the so-called KeyTrap attack and is a result of the flawed design philosophy of DNSSEC. Flaws in the DNSSEC specification are rooted in the interaction of a number of recommendations that when combined can be exploited as a powerful attack vector. As a result, with just a single DNS packet, the KeyTrap attacks lead to a 2.000.000 times spike in CPU instruction count in vulnerable DNS resolvers, stalling some for as long as 16 hours. KeyTrap was disclosed to vendors and operators on November 2, 2023, confidentially reporting the vulnerabilities to a closed group of DNS experts, operators and developers from the industry. This prompted major DNS vendors to refer to KeyTrap as "the worst attack on DNS ever discovered." Exploiting KeyTrap, an attacker could effectively disable Internet access in any system utilising a DNSSEC-validating resolver.

In July 2024, another DNS insecurity has been described in the blog titled "Ducks Now Sitting (DNS): Internet Infrastructure Insecurity" (https://eclypsium.com/blog/ducks-now-sitting-dns-internet-infrastructure-insecurity/). It is called "The Sitting Ducks attack."

So, is DNS and BGP (design, implementation and management) complexity low, medium, or high? In my books it definitely falls into "high".

Chapter 13

Compliance, conformance and security

It is not unusual for people to use the words "compliance" and "conformance" as interchangeable. Common habit of using the two terms interchangeably means that many are unaware of the subtle differences between them. This is a common mistake prone to many of us. At first it may seem that "to conform" and "to comply" essentially means the same thing, notably, to agree to do something or to follow certain rules. However, the strict definitions of these two terms illustrate something different entirely. Let's have a closer look.

Virtually every organisation, regardless of the industry, conducts its activities under set standards, guidelines and regulations. In some cases, these guidelines detail internal procedures that must be adhered to in order to ensure the organisation's products/services keep meeting the standards considered satisfactory to the consumer. In other cases, these guidelines are prescribed by the external regulatory bodies and deviations come with strict penalties. The distinction between internal and external requirements creates the need to differentiate between compliance and conformance.

In general English, the term "conformity" is, often, simply considered to be the harmonisation between person's behaviour and the standards of a particular group. For example, a person conforms when he/she seeks to adopt the same behaviour, beliefs, attitudes and practices of those in the group or the wider society. Conformity for taking a party photograph is that everyone should be smiling and making friendly gestures to the camera. As opposed to a state of compliance, conformity is not prescribed by a legal body. In fact, refusal to conform is viewed as an act of independence or rebellion. If a person does not conform to certain social norms or conventions, they face rejection. And, so is the case with the use of conformity in Management Systems Standards. Conformance pertains to aligning with established standards, guidelines or specifications, often set by international bodies like, for example, ISO. Conformance encompasses meeting the prescribed criteria, whether they are industry standards, organisational policies, customer requirements or other relevant benchmarks. Organisations may conform to the various standards, specifications, industry best practices or

DOI: 10.1201/9781032672601-13

customer-specific requirements to improve their products or services quality and reliability. Effectively, conformance refers to meeting the specifications or criteria set by a standard or test method, which is often voluntary. It implies that a product, service or process has met the requirements and specifications defined by a certain standard, albeit not legally mandated.

The term "compliance" implies a more, formal, serious type of act. It is defined as the act or process of adhering to and fulfilling a given order or command. Compliance recognises a situation, where certain rules or orders have been met. In ISO 37301:2021 – Compliance management systems, compliance is defined in clause 3.26 as "meeting all the organisation's compliance obligations." Compliance typically refers to adhering to external regulations, laws or mandates set forth by governing bodies or authorities. Compliance pertains to meeting mandatory statutory and regulatory requirements imposed by local, state, federal and international authorities. It involves aligning operations with legal mandates to ensure adherence to applicable laws and regulations. In essence, it involves meeting these applicable mandatory requirements to avoid legal repercussions or penalties. Effectively, compliance indicates the adherence to legal and regulatory requirements. It's about fulfilling an external authority's legislative and contractual requirements.

In ISO/IEC Guide 2: Standardisation and related activities – General vocabulary (2004) conformance (also referred to as conformity) is defined as the fulfilment of a product, process or service of specified requirements. These requirements are typically specified in a standard or specification as either part of a conformance clause or in the body of the specification. A conformance clause is a section of a specification that states all the requirements or criteria that must be satisfied to claim conformance to the specification. An example of conformity is when an organisation can demonstrate that they conduct a Management Review. So, conformity is very much linked to the achievement of the requirements within the applicable ISO/IEC Standard.

ISO/IEC compliance is the adherence to international standards and guidelines set by ISO and IEC. These standards are designed to ensure that products, services and processes meet certain requirements and are consistent across different countries and organisations. Intent of ISO/IEC compliance is to ensure that products and services are safe, reliable and of required quality. Demonstration of compliance with ISO/IEC standards is usually achieved through certification and auditing activities, which involve assessment of management systems, testing and verification of products and services to ensure they meet the standards.

Conformance and compliance play distinct roles. While conformance revolves around voluntary adherence to standards, enhancing quality and efficiency, compliance is anchored in meeting legal, statutory and regulatory requirements and obligations that are often based on international or national standards. As such, conformity audits evaluate adherence to standards, while compliance audits assess adherence to legal, statutory and regulatory requirements.

Within ISO and ISO/IEC standards, there is a clear view that:

- Conformity refers to when an organisation seeks to meet the requirements of a standard. Conformance can be seen as formal and/or informal requirements that organisation commits itself to meet (corporate, business specifics, customer specifics, product/process/service specifics, industry guidelines, etc.).
- Compliance relates to a situation in which the organisation fulfils a compliance obligation or legal requirement stipulated by a legal or higher authority. Compliance can be seen as mandatory requirements for an organisation to meet at all times (applicable local, state, national/federal, international laws and regulations).

Unfortunately, ISO and IEC are giving too many different definitions of these two important terms, which are provided in ISO and IEC terminology databases for use in standardisation at the following addresses:

- ISO Online browsing platform: available at https://www.iso.org/obp
- IEC Electropedia: available at https://www.electropedia.org/

So, conformance is what organisation commits itself to (formally or informally) and compliance is what is required from organisation based on local (i.e., state, federal), national and international laws and regulations.

As conformance and compliance are not the same, one needs to understand the difference between certificate of conformance (CoCf) and certificate of compliance (CoC):

- Certificate of conformance is a declaration **issued by an organisation,** confirming that a product has been produced in accordance with specified requirements and standards. It serves as evidence that the product meets the agreed-upon specifications, management system (e.g., quality, information security, etc.) standards and contractual obligations.
- Certificate of compliance is a document **issued by a regulatory body or authorised third party,** verifying that a product meets specific regulatory standards or requirements. It attests that the product complies with applicable laws, regulations or industry standards, ensuring safety, quality, etc.

Certificate of compliance focuses on regulatory compliance, verifying that a product, service or process meet legal or industry-specific requirements. In contrast, certificate of conformance pertains to the product's, service's or system's adherence to agreed-upon specifications, management systems standards and contractual obligations. Certificate of compliance is typically issued by a regulatory authority or authorised third party, while certificate of conformance is issued by the organisation itself. Self-declaration of conformity is formally allowed by ISO/IEC 1750:2004 – Suppliers declaration of conformity as one of the methods of attestation of conformity assessment that relates to first-party or self- declaration of conformity.

Compliance and security are not the same thing. Just because organisation is compliant with a certain standard or regulation does not mean it is fully protected against cyberthreats. Security is the implementation of technical controls, cultural norms and procedures that protect digital assets from threats – security helps to manage risks. Compliance is meeting requirements of a third party for business or legal reasons. One of the biggest myths across the IT world is that being compliant means being secure. Compliance is not security. Compliance does not result in good security but good security often results in compliance. Organisation could be secure without being compliant and vice versa. Security doesn't equal compliance nor vice versa. However, compliance, or not it cannot be easily mitigated and typically ensures large fines. And this drives the behaviour.

Incidentally, similar problems do relate to conformity to management system standards. Certificate of conformity to ISO 9001 does not guarantee that the organisation is producing high-quality products.

Equating compliance with security is a common misconception that can lead to a false sense of security. Believing that compliance alone is sufficient for security can make organisations complacent, underestimating the need for continuous monitoring, constant vigilance and improvement of security practices across organisation. Compliance and security are two pillars upon which organisations base their operational and strategic decisions. However, prioritising one at the expense of the other may lead to vulnerabilities and inefficiencies in the organisation. The main difference is in how compliance and security can be measured. In many cases, compliance is a yes-or-no answer like, for example, does organisation have X policy in place? When it comes to security, there are many shades of grey.

Various cybersecurity specialists commented on this topic. One example is what was said by Gary Hibberd, Professor of Communicating Cyber at Cyberfort Group:

> Being compliant limits your approach to security to the narrow confines of the standard you are using. [it is] like looking through "rose-tinted-glasses", everything will appear okay because that is the lens you are using. But in fact, your approach could be one-dimensional and miss important aspects of cybersecurity. The result is that you may be compliant but not necessarily secure.

To illustrate that compliance is not equal to security let's look at some of the examples of organisations that have suffered breaches despite being compliant:

- **Target:** In 2013, Target suffered a data breach that compromised the credit and debit card information of 40 million customers. Target was compliant with the Payment Card Industry Data Security Standard (PCI DSS) but failed to detect and respond to the breach in a timely manner.
- **Equifax:** In 2017, Equifax, one of the largest credit reporting agencies in the US, suffered a data breach that exposed personal information of 143 million consumers. Equifax was compliant with the Payment Card Industry Data Security Standard (PCI DSS) and other regulations, but the breach occurred due to a vulnerability in a web application.
- **SolarWinds:** In December 2020, it was discovered that SolarWinds, a leading IT management software company, had been hacked. The breach affected over 18,000 customers, including numerous US federal agencies. SolarWinds was compliant with various regulations, but the breach occurred due to a vulnerability in its software supply chain.
- **Microsoft Exchange Server:** In March 2021, it was discovered that multiple vulnerabilities in Microsoft Exchange Server had been exploited by state-sponsored attackers. The breach affected at least 30,000 organisations in the United States and around the world. Microsoft Exchange Server was compliant with various regulations, but the breach occurred due to a vulnerability in its software.
- **Fidelity Investments:** Asset manager Fidelity Investments has revealed that it suffered a breach in August 2024 that resulted in the personal data of over 77,000 customers being exposed. At the same time, Fidelity holds ISO 27001 information security management system certification from NQA, accredited by the ANSI National Accreditation Board (ANAB). NQA's marketing of its ISO 27001 certifications suggests such breaches cannot happen under their watch.

Every organisation wants to be secure in the long term, but compliance might order organisation to focus on implementing certain safeguards within a short period of time. Given this situation, some organisations might elect to focus on compliance now and look at security later. This approach can be a slippery slope, as compliance frameworks and standards are always changing. Subsequently, organisations might need to spend additional budget to align with those new versions each time they become publicly available and this can be a costly exercise.

There are a number of reasons why relying only on compliance can be problematic.

- Compliance frameworks create "rose-tinted-glasses" effect, especially among those without deep understanding of security. Typical thinking in this case is that as frameworks are developed by groups of professionals, they are comprehensive and nothing has been left out. But – there is always but – these frameworks effectively ignore inherent holes like Princeton architecture, TCP/IP, DNS, BGP and focus on prevention and mitigation of the next level risks and thus create false perception of achieving security through being compliant.

- Compliance controls are not always comprehensive or clear. Frequently, controls within compliance frameworks aren't prescriptive and can be interpreted in many different ways, leading to ambiguity. It is important to remember that compliance standards often (if not usually) designed for a broad range of organisations and as a result, they might not fully address the unique security needs and risk profiles of a particular organisation.
- Most compliance requirements are a point-in-time snapshot of organisation's environment. Just because all requirements are being met at that time doesn't mean they always will be.
- Compliance frameworks don't cover all possible vectors of attacks. This allows for significant gaps, especially in a rapidly evolving environment of emerging new threats.
- Compliance frameworks are not always up-to-date. Threat landscape is continually evolving, with attackers developing new techniques and exploiting novel vulnerabilities. While compliance standards are updated periodically, they can't always keep pace with the rapidly evolving threat landscape. As a result, organisations meeting current standards can still be vulnerable to new and more sophisticated attacks. Compliance standards cannot always anticipate or adapt to these changes quickly enough.
- Compliance approach quite often leads to a checkbox mentality, as it is easy to adopt a checklist mentality, focusing on meeting specific compliance requirements without totally understanding the underlying security principles. This approach can lead to serious gaps in security posture, leaving organisation open to cyberattacks. It can also result in a false sense of security. Believing that compliance alone is sufficient for security can make organisation complacent, underestimating the need for continuous monitoring, constant vigilance and improvement of security practices across organisation.

Unfortunately, many organisations practice checkbox compliance. This is where they implement what's necessary in a compliance framework not because they see any value in it but because they are mandated to do so in order to operate. They tick off the required policies and use those compliance efforts to claim that they're secure and protected against a variety of threats. This is problematic for a few reasons. Firstly, no compliance framework is comprehensive or an accurate representation of what organisations are deploying across their entire networks, for that matter. That's because technology and the digital threat landscape are always changing. Secondly, checkbox compliance sends a specific kind of message. Organisations essentially tell regulators that they understand the importance of security but are just unwilling to prioritise it. So, they'll just do certain measures and nothing else. This limits organisations' ability to explain - to regulators and/or customers - what they've implemented and why in the event of a breach.

- Compliance frameworks don't account for the human factor like, for example, insider threats. Compliance frameworks can mandate controls to mitigate insider threats, but they cannot eliminate the risk posed by malicious or negligent insiders. Operational risks are a major contributor to cybersecurity breaches.
- Compliance frameworks don't account for technological limitations and dependencies like, for example reliance on legacy systems or introduction of new "bleeding edge" technologies. Many organisations often rely on legacy systems that may not fully support modern security controls, making full compliance challenging while leaving security gaps. Rapid pace of technological innovation can outstrip the guidelines set by compliance frameworks, leaving many vulnerable to exploitation through new technologies. As Caryll Arcales, a global security specialist said: "Due to the changes in technology, one limitation of compliance is that it does not align with and lags behind the latest trends in cybersecurity."
- Compliance framework measures may not be sufficient to defend against Advanced Persistent Threats (APT), which involve sophisticated stealth attackers targeting specific companies for extended periods of time.

A couple of good real-life examples of why compliance is not equal to security can be found at https://www.linkedin.com/pulse/ compliance-security-our-false-sense-keshri-sekhon

Chapter 14

Standards

When it comes to cybersecurity, the National Institute of Standards and Technology (NIST) and the International Organisation for Standardisation (ISO) dominate the standards scene. It is worth noting that SOC 2 has a place in this conversation too, but the focus will be on NIST's Cybersecurity Framework (CSF) and ISO/IEC 27000 family, as they represent two main approaches to compliance standards and frameworks to ensure protection of the integrity and safety of organisation and customer data.

The NIST CSF is a voluntary framework developed by the NIST in collaboration with private sector, academia and government agencies. It provides a common language and a set of best practices for identifying, protecting, detecting, responding and recovering from cyberthreats. It is designed to be flexible, adaptable and scalable for any organisation, regardless of size, sector or maturity level. The NIST CSF consists of five core functions, 23 categories and 108 subcategories that describe the desired outcomes of effective cybersecurity.

The ISO/IEC 27000 series is a family of international standards developed by the ISO and the International Electrotechnical Commission (IEC) that specify the requirements and guidelines for establishing, implementing, maintaining and improving an Information Security Management System (ISMS). An ISMS is a systematic approach to managing the confidentiality, integrity and availability of information assets. The ISO/IEC 27000 series comprises more than 50 standards, but the most relevant ones for cybersecurity are ISO/IEC 27001 and ISO/IEC 27002. ISO/IEC 27001 defines the requirements for an ISMS, while ISO/IEC 27002 provides a code of practice for information security controls.

While both NIST CSF and ISO/IEC 27000 family of standards support formalised approach to security, they aren't interchangeable. The NIST CSF is designed as a guide, whereas ISO/IEC 27001 is designed as a standard. The difference here is that NIST CSF serves as an instruction manual and ISO/IEC 27001 is more of a test that requires certain measures to pass. There is no certification or audit process in the NIST CSF. It's a guide that organisations can use to establish their cybersecurity. There are no proof-points that

DOI: 10.1201/9781032672601-14

show that organisation is adhering to the NIST CSF; however, organisations can self-report that they've used this framework.

Both NIST CSF and ISO/IEC 27001 have their benefits, and choosing one (or both) comes down to organisation's priorities, needs and compliance requirements. Here are a few things to consider:

- The NIST CSF is best for organisations in the early stages of their cybersecurity journey or those looking for an organised, intentional approach. ISO/IEC 27001 is best for strengthening an existing cyber-security program.
- ISO/IEC 27001 will help organisation by demonstrating trust through a standardised certification. It's common for large organisations to require an ISO/IEC 27001 certification from the vendors they do busi-ness with, while the NIST CSF is rarely a noted requirement from customers.

The NIST CSF guides organisations in building a powerful information security program, while ISO/IEC 27001 ensures that organisation is keep-ing up with the latest best practices and helps organisation to articulate its cybersecurity posture to prospects and partners.

The NIST CSF and the ISO/IEC 27000 series have some similarities that bring multiple benefits to cybersecurity. Both framework and standards are based on risk management principles, are aligned with other international standards and best practices and can be tailored to any organisation regard-less of size, sector or geography. Moreover, they are meant to be used as a continuous improvement cycle, rather than a one-time compliance exercise, as they encourage organisations to review and update their security policies, procedures and practices regularly. This commitment to security is recog-nised and respected by regulators, customers, partners and stakeholders.

The main differences between them are shown in Table 14.1 (https://www.vanta.com/collection/iso-27001/nist-csf-vs-iso-27001).

This chapter will focus on ISO/IEC 27001 that has its origins British Standard BS 7799.

Table 14.1 Comparison between NIST and ISO/IEC

	NIST	ISO/IEC 27001
Purpose	Designed as a guide	Designed as a compliance standard
Compliance process	No certification, serves as a guide	Requires formal audit that results in certification
Maturity	Used in early stages	Used by more mature organisations
Cost	Free to download and implement	Requires buying a standard and hiring an auditor

In the early 1990s, the UK government's Department of Trade and Industry (DTI) asked the Commercial Computer Security Centre (CCSC) to create a set of evaluation criteria for determining security of IT products (this led to the creation of ITSEC.) The CCSC was also asked to create a code of best practices for information security. The result was a document known as DISC PD003. Work on DISC PD003 continued and was split into two major parts: BS 7799-1 and BS 7799-2.

In 1995, The British Standards Institution (BSI) published BS 7799 that consisted of several parts. The standard was significantly based on three principles of confidentiality, integrity and availability (CIA) which was a major step forward:

- **Confidentiality**: All information is confidential and only available to authorised personnel
- **Integrity**: Ensuring that data is securely stored and protected
- **Availability**: Data is available for authorised use at all times

The first part of BS 7799 contained the best practices for information security management and was revised in 1998. In the late 1990s, the BS 7799-1 document was organised into 10 sections, each one outlining a series of controls and control objectives. This document laid the groundwork for the ISO/IEC 27002 standard. After a lengthy discussion with the worldwide standards bodies, it was eventually adopted by ISO as ISO/IEC 17799, "Information Technology - Code of practice for information security management." in 2000. ISO/IEC 17799 was then revised in June 2005 and finally incorporated in the ISO 27000 series of standards as ISO/IEC 27002.

The second part of BS 7799 was first published by BSI in 1998, known as BS 7799 Part 2, titled "Information Security Management Systems - Specification with guidance for use." It created a formal standard for developing an ISMS and eventually evolved into ISO/IEC 27001. BS 7799-2 focused on how to implement an ISMS, referring to the information security management structure and controls identified in BS 7799-2. In 2000, this standard has been adopted in Australia and New Zealand as AS/NZS 7799.2:2000. This later became ISO/IEC 27001:2005. In November 2005, Part 2 of BS 7799 was adopted by ISO and IEC as ISO/IEC 27001.

Part 3 of BS 7799 was published in 2005, covering risk analysis and management. It is aligned with ISO/IEC 27001:2005.

In December 2000, ISO adopted BS 7799-1 as the basis for creating its ISO/IEC 17799 standard.

ISO/IEC held a meeting in Oslo in April 2001 to discuss major revisions to ISO/IEC 17799, and work on a new version of the standard continued from 2001 to 2004. The new version of ISO 17799 was voted on and confirmed in April 2005 in Vienna and published in June 2005. Meanwhile, in October 2005, BS 7799-2 was formally adopted as ISO/IEC 27001.

Fast forward to 2024 and we can talk now about ISO/IEC 27000 family of standards as broad in scope and applicable to organisations of all types and sizes and in all sectors, including public and private companies, government entities and not-for-profit organisations. The common thread regardless of organisation size, type, geography or sector is that the organisation is aiming to demonstrate best practice in its approach to information security management. Best practice can be interpreted differently of course. As technology continually evolves, new standards are being developed to address the changing requirements of information security in different industries and environments. The ISO/IEC 27000 family of standards, also known as the ISMS family of standards or, more simply, ISO27K, covers a broad range of information security standards published by both ISO and IEC.

ISO 27000 recommends best practices – best practices for managing information risks by implementing security controls – within the framework of an overall ISMS. It is very similar to standard management systems such as those for quality assurance and environmental protection. ISO/IEC purposely broadened the scope of the ISO 27000 series so it covers security, privacy and IT issues as well. Organisations of all shapes and sizes can benefit from it. The information security controls should be tailored to the needs of each organisation so that they can treat the risks as they deem appropriate. As of the time of writing this chapter, published ISO27K standards related to "information security, cybersecurity and privacy protection" are:

1. ISO/IEC 27000 – Information security management systems – Overview and vocabulary
2. ISO/IEC 27001 – Information security, cybersecurity and privacy protection – Information security management systems – Requirements: specifies requirements for an information security management system in the same formalised, structured and succinct manner as other ISO standards specify other kinds of management systems
3. ISO/IEC 27002 – Information security, cybersecurity and privacy protection – Information security controls: essentially a detailed catalogue of information security controls that might be managed through the ISMS
4. ISO/IEC 27003 – Information security management system implementation guidance
5. ISO/IEC 27004 – Information security management – Monitoring, measurement, analysis and evaluation
6. ISO/IEC 27005 – Guidance on managing information security risks
7. ISO/IEC 27006 – Requirements for bodies providing audit and certification of information security management systems
8. ISO/IEC 27007 – Guidelines for information security management systems auditing (focused on auditing the management system)
9. ISO/IEC TR 27008 – Guidance for auditors on ISMS controls (focused on auditing the information security controls)

10. ISO/IEC 27009 – Information technology – Security techniques – Sector-specific application of ISO/IEC 27001 – Requirements
11. ISO/IEC 27010 – Information security management for inter-sector and inter-organisational communications
12. ISO/IEC 27011 – Information security management guidelines for telecommunications organisations based on ISO/IEC 27002
13. ISO/IEC 27013 – Guideline on the integrated implementation of ISO/IEC 27001 and ISO/IEC 20000-1
14. ISO/IEC 27014 – Information security governance
15. ISO/IEC TR 27015 – Information security management guidelines for financial services (now withdrawn)
16. ISO/IEC TR 27016 – Information security economics
17. ISO/IEC 27017 – Code of practice for information security controls based on ISO/IEC 27002 for cloud services
18. ISO/IEC 27018 – Code of practice for protection of personally identifiable information (PII) in public clouds acting as PII processors
19. ISO/IEC 27019 – Information security for process control in the energy industry
20. ISO/IEC 27021 – Competence requirements for information security management systems professionals
21. ISO/IEC TS 27022 – Guidance on information security management system processes – under development
22. ISO/IEC TR 27023 – Mapping the revised editions of ISO/IEC 27001 and ISO/IEC 27002
23. ISO/IEC 27028 – Guidance on ISO/IEC 27002 attributes
24. ISO/IEC 27031 – Guidelines for information and communication technology readiness for business continuity
25. ISO/IEC 27032 – Guideline for cybersecurity
26. ISO/IEC 27033-1 – Network security – Part 1: Overview and concepts
27. ISO/IEC 27033-2 – Network security – Part 2: Guidelines for the design and implementation of network security
28. ISO/IEC 27033-3 – Network security – Part 3: Reference networking scenarios – Threats, design techniques and control issues
29. ISO/IEC 27033-4 – Network security – Part 4: Securing communications between networks using security gateways
30. ISO/IEC 27033-5 – Network security – Part 5: Securing communications across networks using Virtual Private Networks (VPNs)
31. ISO/IEC 27033-6 – Network security – Part 6: Securing wireless IP network access
32. ISO/IEC 27033-7 – Network security – Part 7: Guidelines for network virtualisation security
33. ISO/IEC 27034-1 – Application security – Part 1: Guideline for application security
34. ISO/IEC 27034-2 – Application security – Part 2: Organisation normative framework

35. ISO/IEC 27034-3 – Application security – Part 3: Application security management process
36. ISO/IEC 27034-4 – Application security – Part 4: Validation and verification (under development)[15]
37. ISO/IEC 27034-5 – Application security – Part 5: Protocols and application security controls data structure
38. ISO/IEC 27034-5-1 – Application security – Part 5-1: Protocols and application security controls data structure, XML schemas
39. ISO/IEC 27034-6 – Application security – Part 6: Case studies
40. ISO/IEC 27034-7 – Application security – Part 7: Assurance prediction framework
41. ISO/IEC 27035-1 – Information security incident management – Part 1: Principles of incident management
42. ISO/IEC 27035-2 – Information security incident management – Part 2: Guidelines to plan and prepare for incident response
43. ISO/IEC 27035-3 – Information security incident management – Part 3: Guidelines for ICT incident response operations
44. ISO/IEC 27035-4 – Information security incident management – Part 4: Coordination (under development)[16]
45. ISO/IEC 27036-1 – Information security for supplier relationships – Part 1: Overview and concepts
46. ISO/IEC 27036-2 – Information security for supplier relationships – Part 2: Requirements
47. ISO/IEC 27036-3 – Information security for supplier relationships – Part 3: Guidelines for information and communication technology supply chain security
48. ISO/IEC 27036-4 – Information security for supplier relationships – Part 4: Guidelines for security of cloud services
49. ISO/IEC 27037 – Guidelines for identification, collection, acquisition and preservation of digital evidence
50. ISO/IEC 27038 – Specification for Digital redaction on Digital Documents
51. ISO/IEC 27039 – Intrusion prevention
52. ISO/IEC 27040 – Storage security[17]
53. ISO/IEC 27041 – Investigation assurance
54. ISO/IEC 27042 – Analysing digital evidence
55. ISO/IEC 27043 – Incident investigation
56. ISO/IEC 27050-1 – Electronic discovery – Part 1: Overview and concepts
57. ISO/IEC 27050-2 – Electronic discovery – Part 2: Guidance for governance and management of electronic discovery
58. ISO/IEC 27050-3 – Electronic discovery – Part 3: Code of practice for electronic discovery
59. ISO/IEC 27050-4 – Electronic discovery – Part 4: Technical readiness
60. ISO/IEC TS 27110 – Information technology, cybersecurity and privacy protection – Cybersecurity framework development guidelines

61. ISO/IEC 27557 – Information security, cybersecurity and privacy protection – Application of ISO 31000:2018 for organisational privacy risk management
62. ISO/IEC 27701 – Information technology – Security Techniques – Information security management systems – Privacy Information Management System (PIMS).
63. ISO 27799 – Information security management in health using ISO/IEC 27002 (guides health industry organisations on how to protect personal health information using ISO/IEC 27002)

Without questioning importance and quality of each of the listed above documents the sheer size of this list raises questions about ability of any organisation to digest and implement them all, especially - small organisations.

An ISMS provides a structured and systematic approach for managing the information security of an organisation and involves putting policies, procedures and controls into writing to create an official system that instructs, monitors and improves information security. An ISMS will also cover topics such as how to protect sensitive information from being stolen or destroyed, and detail all the mitigations necessary to achieve information security goals. Information security encompasses certain broad policies that control and manage security risk levels across an organisation. It is comprised of a set of policies, processes and procedures for systematically managing organisation's information assets, including, but not limited to, sensitive data. The goal of an ISMS is to minimise risks and ensure business continuity by proactively limiting the impact of a security breach. An ISMS also typically addresses employees' behaviour and processes, as well as data, and technology. It can be targeted towards a particular type of data, such as customer data, or it can be implemented in a comprehensive way that becomes part of the organisation's culture.

Organisations should rely on security guidance and suggestions when appropriate. As information security and risk management are dynamic disciplines, the ISMS concept incorporates continuous feedback and improvements to respond to the changes in threats or vulnerabilities that occurred as a result of incidents. Information security experts suggest that compliance with the ISO 27000 series is the first step towards an information security program that will properly protect your organisation.

The standards, however, are not specific to any industry and this makes them able to be applied in any business, regardless of size and industry.

At the core of ISO/IEC 27000 family is ISO/IEC 27001, the latest version of this key standard was published on October 25, 2022. ISO/IEC 27001 is an Information Security Management System standard and supports effective Information Security Management that helps organisations meet requirements for confidentiality, integrity and availability of information. It is a globally recognised standard. The main body of ISO/IEC 27001:2022 consists of ten sections (i.e., clauses). The first three clauses provide general introductory information, terms and definitions. Clauses four to ten contain mandatory requirements that organisation must follow to become ISO/IEC 27001 compliant. In order to achieve "continuous improvement" within the information security management system, the ISO/IEC 27001:2022 standard specifies that organisation should address seven main areas – also known as "clauses":

- Context of the organisation
- Leadership
- Planning
- Support
- Operation
- Performance evaluation
- Improvement

Immediately after the ten clauses, Annex A contains 93 information security controls (this number has been decreased down from 114 in the previous version of ISO/IEC 27001:2013 to 93 in ISO/IEC 27001:2022, including 11 new controls introduced in this version of the standard) grouped according to themes. Organisation is not expected to implement each of these controls.

Rather, when organisation is performing information security risk treatment process (defined in clause 6), organisation needs to go through Annex A to determine what controls this organisation needs and then verify that no necessary controls have been omitted. So, Annex A in ISO/IEC 27001 is a part of the standard that lists a set of classified security controls that organisations use to demonstrate compliance with ISO/IEC 27001.

The controls are broken down into four numbered sections. These sections correspond with Clauses five to eight of a linked standard, ISO 27002, which provides more detailed guidance on how ISOIEC 27001 controls can be implemented. The four categories are as follows:

- Clause 5: Organisational (37 controls)
- Clause 6: People (8 controls)
- Clause 7: Physical (14 controls)
- Clause 8: Technological (34 controls)

Summary of Annex A controls can be found at: https://www.scribd.com/document/631573670/ISO-27001-controls-2022.

There are no questions about benefits of ISMS implementation. Having policies, processes and procedures documented and followed enables their repeatable implementation, hopefully following the best practices in each of the areas covered. Regular checks that these policies, processes and procedures are followed provides assurance that this is the case indeed. Continuous improvement and regular reviews/updates (typically – annual) enable ongoing refinement, identification and closure of any gaps and enables organisation to rely on best practices as they keep evolving.

Having said this, it is important to note that practicality and usefulness of ISMS are heavily dependent on what is called Statement of Applicability (SoA), as defined in 6.1.3 of the main requirements for ISO/IEC 27001, which is part of the broader 6.1, focused on actions to address risks and opportunities. In ISO/IEC 27001:2022, an SoA is a document that lists the Annex A controls that an organisation will implement to meet the requirements of the standard. It is a mandatory step for anyone planning on pursuing ISO/IEC 27001 certification. Organisation's SoA should contain four main elements:

- A list of all controls that are necessary to satisfy information security risk treatment options, including those contained within Annex A
- A statement that outlines why all of the above controls have been included
- Confirmation of implementation
- The organisation's justification for omitting any of the Annex A controls

The SoA is therefore an integral part of the mandatory ISO/IEC 27001 documentation. Properly and well-defined SoA is a foundation of meaningful and useful ISMS, while wrongly defined SoA may result in a waste of time, money and resources without delivering expected benefits to the organisation. Correct definition of SoA is based on correct identification of information assets and risks to them. As such, failure to properly identify information assets and associated risks results in ill-defined SoA. I the past, the author of this chapter (having been an ISMS auditor for SAI Global) has witnessed first-hand numerous organisations with ill-defined SoAs and as a result – multiple ISMS implementations with questionable value.

The goal of an ISMS is not necessarily to maximise information security, but rather to reach an organisation's desired level of information security. Depending on the specific needs of the industry and organisation, these levels of control may vary. For example, since healthcare is a highly regulated field, a healthcare organisation may develop a system to ensure sensitive patient data is fully protected.

Implementation of ISMS helps organisations meet regulatory compliance and contractual requirements and provides a better grasp on legalities surrounding information systems. Since violation of legal regulations comes with hefty fines, having an ISMS can be especially beneficial for highly regulated industries with critical infrastructure, such as finance or healthcare.

It is important to remember that due to its origins ISO/IEC 27000 family of standards is still significantly based on "pre-Internet" paradigms and though the standards have dramatically evolved, their foundation is still significantly based on the "perimeter security" paradigm.

To summarise, ISMS, like Essential 8 (see Chapter 15) is a useful (but expensive) tool that helps with implementation of and adherence to good practices, but is actually more a compliance tool.

Chapter 15

Essential 8 Myth

Australian Signals Directorate (ASD) is a government agency with a long and rich history in cybersecurity dating back to 1947, when it was established as Defence Signals Bureau. Back then it was primarily responsible for intercepting and decoding foreign signals intelligence. Over the years, the agency has undergone several name changes to reflect its evolving role, including changing the name to Defence Signals Directorate (DSD) in 1977. Over the years, the organisation evolved to meet the growing challenges of the digital age. In 2010, it was restructured and given a wider remit. This change reflected the increasing importance of cybersecurity in national defence and the need for a dedicated agency to combat emerging threats. Since then, it has played a critical role in protecting Australia's critical infrastructure, government agencies and businesses from cyberthreats. It has developed various strategies, guidelines and frameworks to enhance cyber resilience and security posture of Australian organisations.

In 2010, DSD established the Cyber Security Operations Centre (CSOC) to develop a comprehensive understanding of ICT security threats to critical Australian systems and to coordinate a response to those threats across government and industry. In 2013, DSD transformed into ASD and in 2018 ASD became a statutory agency within the Defence portfolio.

CSOC was a Defence-based capability that hosted liaison staff from other government agencies. ASD saw the need for collocation of all contributing agencies' cyber security capabilities and as a result of this in 2014 CSOC evolved into Australian Cyber Security Centre (ACSC), while still being a part of ASD. ACSC is a whole-of-government organisation. Before the establishment of the ACSC, Australian government had a number of different agencies and organisations that were responsible for different aspects of cyber security.

In 2017, ACSC, then a division of ASD, released the Information Security Manual (ISM). This comprehensive guide offers practical advice on safeguarding systems and data. It provides guidance and standards for the protection of information and information systems from unauthorised access,

DOI: 10.1201/9781032672601-15

use, disclosure, disruption, modification or destruction. ISM intended audience includes Chief Information Security Officers (CISOs), Chief Information Officers, cyber security professionals and information technology managers. ISM is based on industry standards and best practices and is intended to be used in conjunction with an organisation's risk management framework. It is organised into four key activities:

- Govern
- Protect
- Detect
- Respond

ISM provides guidance on governance, physical security, personnel security and information and communications technology security topics. It is not required by law, unless specifically mandated by legislation or other lawful authority. If ISM conflicts with legislation or law, the latter takes precedence. ISM does not provide a comprehensive consideration of legislative and legal considerations, and organisations are encouraged to familiarise themselves with relevant legislation, such as the Archives Act 1983, Privacy Act 1988 and Telecommunications (Interception and Access) Act 1979.

Initially introduced in 2010 and last updated in 2017, the ACSC released a set of prioritised strategies to help organisations mitigate and protect against various types of cyberthreats. These strategies are based on the ACSC experience responding to cyber security incidents, conducting vulnerability assessments and performing penetration testing on Australian government organisations. The ACSC strategies are designed to address a range of cyberthreats, including: targeted cyber intrusions (also known as advanced persistent threats), ransomware attacks and other external adversaries (that destroy data and prevent operation of computers/systems/networks), malicious insiders who steal data, malicious insiders who destroy data. ACSC strategies are further classified into five relative security effectiveness ratings:

- Essential
- Excellent
- Very good
- Good
- Limited

ACSC considers the strategies with an "essential" rating to be the minimum baseline for all organisations to follow in order to effectively protect against cyberthreats. ACSC has also released additional guidance on implementing these strategies and on measuring the maturity of their implementation.

Essential 8 is the "essential" minimum baseline security for organisations and is a subset of the Strategies to Mitigate Cyber Security Incidents. It provides practical guidance on how to protect organisations' systems and data from cyberthreats, how to implement mitigation strategies in a phased approach and how to measure the maturity of implementation. Essential 8 strategies are primarily focused on MS Windows-based Internet-connected networks and are designed to complement each other in order to provide coverage against a range of cyberthreats. While the principles behind Essential 8 can be applied to other systems, such as cloud services and enterprise mobility, alternative guidance may be more appropriate for these environments.

Essential 8 is a set of cybersecurity controls developed by ASD, an Australian government intelligence agency. These controls are designed to provide a practical framework for organisations to improve their cyber resilience and security posture. Essential 8 covers a range of key security controls, including application whitelisting, patching operating systems and mitigating techniques against phishing and ransomware attacks.

The Essential 8 consists of eight mitigation strategies, including:

- **Application control:** only allowing approved applications to run on a system.
- **Patching applications:** applying updates and patches to software to fix vulnerabilities.
- **Configuring Microsoft Office macro settings:** applying least privileges to Microsoft Office macros
- **User application hardening:** disabling, removing, restricting and monitoring applications to limit ability for compromise
- **Restricting administrative privileges:** limiting the number of users with administrative privileges on a system.
- **Patching operating systems:** applying updates and patches to the operating system to fix vulnerabilities.
- **Multi-factor authentication:** requiring more than one form of authentication to access systems or data.
- **Regular backups:** regularly backing up important data to protect against data loss.

Initially published in February 2017, Essential 8 was mandated by Australian Federal Government for federal departments, with additional requirements set by the Attorney General's Department's Protective Security Policy Framework (PSPF). Since introduction of Essential 8, there have been several updates and modifications to the strategies. The latest update was published on November 27, 2023 (https://www.cyber.gov.au/resources-business-and-government/essential-cyber-security/essential-eight/essential-eight-maturity-model-changes). It's important to note that Essential 8 is not a static set of mitigation strategies and likely to continue to evolve over time as the cybersecurity landscape changes.

Essential 8 is a subset of the Strategies to Mitigate Cyber Security Incidents and ISM's mandatory security controls. Therefore, it can be considered as a stepping stone towards increasing organisation's security posture for future compliance with ISM. Essential 8 controls can be directly mapped to ISM.

Essential 8 adds upon the Strategies to Mitigate Cyber Security Incidents by defining four maturity levels. Maturity levels are designed and based on the level of adversary tradecraft (tools, tactics, techniques and procedures) and targeting that an organisation is aiming to mitigate.

- **Maturity level 0:** There are weaknesses in an organisation's cyber security posture that could be exploited by adversaries.
- **Maturity level 1:** Focuses on mitigation strategies against adversaries that use widely available tools and techniques to gain access to systems.
- **Maturity level 2:** Focuses on mitigation strategies against adversaries that are willing to invest more time and effort in their attacks and use more advanced tools and techniques to bypass security controls and evade detection.
- **Maturity level 3:** Focuses on mitigation strategies against threat actors with advanced capabilities that are willing to invest significant time, money and effort in their attacks and may use customised tools and techniques to compromise a target.

NSW Government mandated implementation of Essential 8:

> ACSC has developed and published the Essential 8 strategies for mitigating cyber incidents. The Essential 8 are embedded in Mandatory Requirements 3.3 to 3.10. Agencies must implement the Essential 8 to applicable ICT environments with a minimum requirement of Level 1 maturity, as part of the baseline set in the Mandatory Requirements. Mitigation strategies for Level 2 and Level 3 maturity should then be considered alongside other mitigation strategies based on the threats and risks identified by the agency as part of the threat-based requirements (https://www.digital.nsw.gov.au/delivery/cyber-security/policies/essential-eight#:~:text=The%20Essential%20Eight%20are%20embedded,set%20in%20the%20Mandatory%20Requirements.)

It is interesting that technically this mandate has an "out of jail" ticket in the form that it is directed only towards "applicable ICT environments."

Some people may be confused between Essential 8 and ISO/IEC 27001 (see Chapter 14) and ask: "Why do we need both?." Essential 8 focuses on eight key areas, while ISO/IEC 27001 provides a comprehensive set of controls and processes for information security management. ISO/IEC 27001 is more comprehensive and detailed, while Essential 8 is more lightweight and easier to implement. It appears (at least in the opinion of the author of this chapter) that introduction of Essential 8 was an attempt to deal with complexity and cost of implementing and maintaining ISO/IEC 27001-compliant ISMS by simplifying the approach and shifting focus on some foundational hygiene, capabilities and controls.

As ISO/IEC 27001 was mentioned, it would be a remiss not to mention NIST Cybersecurity Framework (NIST CSF). Although there are a lot of similarities between Essential 8 and NIST CSF (see Chapter 14), there are also some differences between them:

- Essential 8 focuses more on the prevention of cyber security threats and post-incident recovery, while NIST CSF focuses on a holistic approach to cyber security, including prevention, detection, response and recovery.
- Essential 8 is a set of eight security controls, while NIST CSF is a framework that includes five core functions and associated components.
- Essential 8 is focused on the implementation of security controls, while NIST CSF is focused on the implementation of a risk management process.
- Essential 8 is tailored for Australian organisations (with focus on MS Windows-based systems), while NIST CSF is designed to be applicable to organisations of any size and industry in any country (using any type of technology).
- Essential 8 is designed to be implemented in a short period of time, while NIST CSF is designed to be implemented over a longer period of time.

Let's have a closer look at Essential 8. The very first warning can be found on the ASD website:

> While no set of mitigation strategies are guaranteed to protect against all cyberthreats, organisations are recommended to implement eight essential mitigation strategies from the Strategies to Mitigate Cyber Security Incidents as a baseline. This baseline, known as the Essential 8, makes it much harder for adversaries to compromise systems. (https://www.cyber.gov.au/resources-business-and-government/essential-cyber-security/essential-eight).

As always, the devil is in the detail – how "much harder" is "much harder?" Let's explore each of the eight areas of Essential 8. In the first place, we should remember, that, as mentioned earlier, Essential 8 is primarily focused on MS Windows-based Internet-connected networks (what about other operating systems or SaaS solutions?). Except for multi-factor authentication, seven out of eight Essential 8 focus areas are just good IT Service Management (ITSM) practices that should be practiced by any good IT department. Over 30 years ago author of this chapter was managing a medium sized software development environment and practiced all of them that already were relevant those days (https://cybertheory.io/essential-eight-is-this-really-an-answer/). Regular back-ups deserve a separate discussion, as a lot of organisations do not understand this aspect.

So, what is so special about backups?

Firstly, regular backups without regular restores are useless, as there is no level of confidence that in case of necessity successful restoration will happen. Author of this chapter saw only one organisation that was doing quarterly "switch" exercise: full restoration on the disaster recovery site and operation from this site for the next quarter. And then "switch" back. About 10 years ago author of this chapter was managing a major incident (instead of test database somebody by mistake executed SQL script on production database and noticed the mistake several hours later, when the database was already buggered) that impacted moto-registry offices. The system was supposed to be fault tolerant and operated in a hot–hot mode and thus both databases (primary and secondary) got corrupted at the same time. There was a backup and it failed to restore. There was a bit older backup and it failed to restore too. Organisation ended up manually undoing SQL scripts statement by statement resulting in 2 or 3 days of moto-registry closure and significant financial losses. Author of this chapter did not see any test backup restora-tions across a dozen of organisations he has dealt with over the last 15 years.

Secondly, there is a need for an offline back up and off-site backup storage as in case of cyberattack one of the first things that attacker does is poison-ing online backups. It is getting expensive with tens, sometime hundreds, of terabytes to be backed up, but it is necessary. During the days author of this chapter was managing that software development he used to keep on site a copy of the last 6 months full backups and two copies of off-site backups for 12 months yearly full and monthly full backups stored at two different loca-tions. Some organisations treat cloud replication as a form of backup. Wrong! Moreover, numerous organisations just back up data, not servers themselves. Wrong too!

Vagueness in backup requirements specified in Essential 8 for each of the maturity levels (Appendices A, B and C in https://www.cyber.gov.au/resources-business-and-government/essential-cyber-security/essential-eight/essential-eight-maturity-model) opens the door for multiple interpre-tations. For example, does "Restoration of data, applications and settings from backups to a common point in time is tested as part of disaster recov-ery exercises" mean that this needs to happen on a monthly basis, or quar-terly basis, or once a year - when disaster recovery is typically exercised?

Nobody is going to question importance of multi-factor authentication (MFA). Of all access security recommendations, MFA is arguably the most consistent. And there is a good reason why many best practice recommenda-tions and compliance frameworks now place MFA at the top of the list of security configurations needed to help protect against compromise. MFA can be the crucial layer preventing a breach, as passwords alone are often easy work for hackers. However, MFA isn't infallible – and a weak or breached password is still almost always a key factor when a user is breached. The bottom line is that MFA is not as bulletproof as many think and there are multiple ways MFA can be compromised:

- **MFA fatigue attack (also known as MFA bombing or MFA spamming):** It is a social engineering attack where attackers repeatedly push second-factor authentication requests to the target victim's email, phone or registered devices. The goal is to coerce the victim into confirming their identity via notification, thus authenticating the attackers attempt at entering their account or device. It is based on a feature of some modern authentication apps, as they provide a push notification that prompts the user to either accept or deny the login request. While this is convenient for the end user, attackers can use it to their advantage. If they've already compromised a password, they can attempt to log in and generate an MFA prompt to the legitimate user's device. Then attackers hope that the user either thinks it's a legitimate prompt and accepts it or gets tired of the continuous prompts and accepts it to stop their phone notifications.
- **Service desk social engineering:** Attackers can use social engineering to trick helpdesks into bypassing MFA altogether by pretending they've forgotten their password and gaining access via a phone call. If service desk agents don't enforce verification at this stage (which they more often than not don't, especially in larger organisations with thousands of employees, or in case of outsourced call centres with high levels of personnel turnover), they might unwittingly give a hacker an initial foothold in their organisation's environment. This exact scenario played out in the attack on MGM Resorts in September 2023. After gaining initial access by fraudulently calling the service desk for a password reset, the attack group (Scattered Spider) were able to use their foothold in the environment to launch a ransomware attack.
- **Adversary-in-the-middle (AITM) attack:** AITM is a MITM attack and essentially tricks a user into thinking they're logging into a legitimate network, application or website, when in fact they're putting their details into a fraudulent lookalike. This means hackers can intercept passwords and manipulate MFA prompts and other types of security. For example, a spear phishing email might land in employee's inbox impersonating a known source. The link they click on will take them to a fake site where hackers will harvest their credentials for reuse. In theory, MFA would stop this by requiring a second form of authentication. However, attackers will use a tactic called a "2FA pass-on" where as soon as the victim has entered their credentials into the fake site, the attacker enters those same details into the legitimate site. This will trigger an MFA request, which the victim is expecting and will likely accept, giving the attacker full access.

- **Session hijacking:** Session hijacking is an MFA breach attack similar to an AITM attack, as it involves an attacker positioning themselves in the middle of a legitimate process and exploiting it. When a user authenticates using their password and MFA, many applications use a cookie or session token to remember the user is authenticated and grant access to protected resources. The cookie or token prevents the user from having to authenticate multiple times. But if an attacker uses a tool such as Evilginx to steal the session token or cookie, they can masquerade as an authenticated user, effectively bypassing the multi-factor authentication configured on the account.
- **Sim swap:** Attackers know MFA often relies on cell phones as the "thing you possess" to complete an authentication process. A SIM swap attack is where cybercriminals trick service providers into switching services to a SIM card they control, effectively hijacking the victim's cell service and phone number. This allows attackers to receive the MFA prompts to the hijacked service and grant themselves access. In Australia this technique was used extensively and a number of people lost their money as a result of it.
- **Exporting generated tokens:** Another tactic attackers can use is compromising the back-end system that generates and validates multi-factor authentication. In a bold attack in 2011, attackers were able to steal the "seeds" possessed by RSA for generating SecurID tokens (code-generating key fobs used for multi-factor authentication). Once the seed values were compromised, attackers were able to clone the SecurID tokens and even create their own. Sometimes, attackers will seek the help of malicious insiders, who are paid to provide session tokens for MFA approval. Threat group LAPSUS$'s Telegram channel has confirmed that in the past they have indeed bought accesses from a company's employee, and are actively looking for other insiders to work as providers. Microsoft has also reported that LAPSUS$ were able to obtain passwords and session tokens with the use of RedLine stealer. These credentials and session tokens are then sold on underground forums.
- **Endpoint compromise:** One way to avoid MFA completely is to compromise an endpoint with malware. Installing malware on a device lets hackers create shadow sessions following successful logins, steal and use session cookies or access additional resources. If system allows users to remain logged on after an initial authentication (by generating a cookie or session token), hackers could keep their access for a significant period. Hackers may also look to exploit recovery settings and backup procedures that could be less safe than MFA processes. People often forget passwords and regularly need new or modified accesses. For example, a common recovery method is sending an email link to a secondary email address (or an SMS with a link). If this backup address or phone is compromised, hackers gain full access to their target.

- **Exploiting single sign-on (SSO):** SSO is a double-edged sword. It is convenient for users as they only need to authenticate once. However, it can be exploited by hackers who use it to log into a site requiring just a compromised password, then use SSO to gain access to other sites and applications that would normally require MFA. A sophisticated form of this technique was used in the 2020 SolarWinds hack, where hackers exploited SAML (a method for exchanging authentication between multiple parties in SSO). The hackers gained an initial foothold, then got access to the certificates used to sign SAML objects. With these, they were able to impersonate any user they wanted to, with full access to all SSO resources.
- **Finding technical deficiencies:** Like any software, MFA technology has bugs and weaknesses that can be exploited. Most MFA solutions have had exploits published which temporarily exposed opportunities for hacking. For example, 0ktapus leveraged CVE-2021-35464 to exploit a ForgeRock OpenAM application server, which front-ends web applications and remote access solutions in many organisations.

Let's explore achievability of various maturity levels of Essential 8. The very first thing that one should consider is whether achieving certain level of maturity is a sustainable state, or it is just measured at a certain point in time. The hypothesis the author of this chapter has is that using sustainable state approach without exceptions is unachievable for any organisation, especially for larger organisations. How risky are these exceptions? It is not clear from the approach taken by Essential 8, but if one takes a "black and white" approach exceptions mean failure to achieve certain level of maturity.

Now, let's look at some of the examples supporting this hypothesis.

Let's take, for example, Maturity Level1. It requires, for example, disabling or removing MS Internet Explorer, but every organisation author of this chapter dealt with lately has had at least one, sometimes more, legacy systems that will not work with any other browser. As such, MS Internet Explorer must stay operational at least for certain group(s) of employees. Another requirement stipulated for Maturity Level 1 is that MS Office macros in files originating from Internet are blocked. As always, the devil is in the detail – shall a file that came from another organisation be classified as "originating from Internet?" Blocking MS Office macros files received form service providers or other government agencies is a pretty unworkable arrangement. It is a typical example of a clash between usability in security.

Now, let's extend our example to Maturity Level 2. It requires that an automated method of asset discovery is used at least fortnightly. From the first glance, not very difficult to achieve. However, to achieve the intent of this requirement, organisation must have in place a solid IT asset management (ITAM) which is the process of ensuring that organisation's assets are accounted for, deployed, maintained, upgraded and disposed of when the time comes. Author of this chapter is yet to see large multi-site organisation that has an up-to-date asset management database that accounts for all IT assets. Patching is always on a collision course with business needs, especially when it must be done within 48 hours, which, by the way, doesn't not give enough time for testing. Again, pretty much all organisations author of this chapter dealt with since 2010 had periods of significant duration when no patching was allowed – be it go-live of a new system, financial year end, construction activities on rail track or elections.

Now, let's look at Maturity Level 3, that, for example, requires removal of applications that are no longer supported by vendors. Easier said, than done as migration can be very expensive and can take years...

Reality is that organisations cannot and do not comply despite their best efforts and major expenditure. The dynamic of continuously evolving threat landscape means that organisations are in a forever loop of maintaining and updating software, mitigating risks and addressing vulnerabilities like a dog chasing its tail but never succeeding in doing so.

While Essential 8 provides a very pragmatic foundation for cybersecurity, there are challenges and considerations that organisations must be aware of during implementation. These include:

Resource Constraints: Implementing the Essential 8 requires resources, including skilled personnel, technology, and time. Organisations with limited resources may struggle to achieve higher maturity levels without external support.

Rapidly Evolving Threats: Cyberthreats are constantly evolving, and the Essential 8 must be continuously updated and adapted to address new risks. Organisations need to stay informed about the latest threats and adjust their cybersecurity strategies accordingly.

Compliance and Regulations: In some industries, compliance with specific regulations or standards may be required in addition to implementing the Essential 8. Organisations must ensure that their cybersecurity practices align with both the Essential 8 and any applicable regulatory requirements.

Cultural and Organisational Change: Effective cybersecurity requires more than just technical solutions; it also involves cultural and organisational change. Employees need to be educated about cybersecurity best practices, and leadership must prioritise and support cybersecurity initiatives.

Essential 8 meant to mitigate 85 percent of common types of cyberattacks or cyberthreats. Don't you believe it. The number of threats has increased greatly and keeps increasing on a daily basis. Not only is it difficult or almost impossible to quantify or accurately measure the effectiveness of cyber security controls, but it is disingenuous to make such a statement.

Because Essential 8 controls are targeted only at a basic level of types of threats. Six to be precise.

Essential 8 is by no means a "silver bullet." Because of many reasons. But mainly because it does not address any of the underlying insecurities – be it Princeton computer architecture or insecurities of TCP/IP, DNS and BGP. It is not addressing growing complexities. It stays silent on new systems deployment and pitfalls of Agile approach that is widely used now to build these new systems.

So, the question is: will washing your hands before and after one goes into an infectious disease ward without any other protection prevent them from catching an infectious disease? No, there is no guarantee that it will. But it will definitely decrease the chances of getting one. Decrease by how much? This is a very good question.

We must also remember that compliance is backward looking. The best analogy is to think about how you would drive your car by only looking in the rear-view mirror. The end result would be a major crash. You wouldn't get far.

In a dynamic world where information can travel around the world in a fraction of a second, where technological advancement cycles are measured in months or even weeks, compliance constitutes the bare minimum and is always behind what's happening in the world. Sometimes quite far behind. Years behind. At least 1–2 years.

Chapter 16

OT and IoT

Operational Technology (OT) refers to hardware and software that interact with physical plant and equipment for the purpose of monitoring, controlling and/or operating such equipment in an automated fashion. Internet of Things (IoT) describes devices with sensors, processing ability, software and other technologies that connect and exchange data with other devices and systems over the Internet or other communication networks. IoT encompasses electronics, communication and computer science engineering.

OT is a critical aspect of modern industry because it automates the operation of machinery and equipment by connecting or integrating physical and digital worlds. In other words, OT is a category of real-time control computing and communications systems used to manage, monitor and control or operate industrial plant and equipment or devices.

Unlike Information Technology (IT), which focuses on data management and processing, OT is concerned with the operational processes of industries such as manufacturing, utilities (electricity, gas and water), transport and logistics (telematics), resources and mining among others. OT systems include:

1. **Industrial Control Systems (ICS):**
   - Programmable Logic Controllers (PLCs): These are used to automate machinery and processes, providing real-time control and monitoring.
   - Distributed Control Systems (DCS): Used in large-scale industrial processes, DCS allow for centralised control of multiple processes across a facility.
2. **Supervisory Control and Data Acquisition (SCADA) Systems:**
   - SCADA systems monitor and control industrial processes across various locations, collecting data from sensors and devices to enable remote management.

DOI: 10.1201/9781032672601-16

3. **Critical Infrastructure Systems (CIS):**
   - Denotes the systems, applications and platforms or assets that are crucial for the functioning of our societies and related economies. This technology is considered necessary or essential because any disruption would impact general health and safety, destroy economic stability and possibly cause chaos, anarchy and significant loss of life and property.
4. **Human-Machine Interfaces (HMIs):**
   - HMIs provide visual representations of machinery and processes, allowing operators to interact with and control equipment.
5. **Safety Instrumented Systems (SIS):**
   - These systems are designed to monitor safety conditions and initiate safety measures in case of hazardous events, ensuring operational safety.
6. **Remote Terminal Units (RTUs):**
   - RTUs collect data from field devices and send it to the SCADA system, often used in utilities and remote monitoring applications.
7. **Field Devices:**
   - Sensors: Devices that detect physical properties (temperature, pressure, etc.) and convert them into signals.
   - Actuators: Devices that perform actions based on control signals (e.g., opening or closing valves).
8. **Industrial Internet of Things (IIoT):**
   - IIoT connects devices and sensors to the Internet, enabling data collection and analysis to optimise processes and enhance decision-making. IIoT operates within industrial environments. IoT on the other hand is concerned with mostly consumer-level devices and apps such as wearable gadgets and smart homes.
9. **Asset Management Systems:**
   - These systems track and manage physical assets throughout their lifecycle, helping organisations optimise maintenance and operational efficiency.
10. **Process Automation Software:**
    - Software solutions that automate workflows, data collection and reporting within industrial processes.
11. **Control Room Systems:**
    - Integrated environments where operators monitor and control processes, often featuring advanced visualisation and alert systems.

The roots of OT can be traced back to the early 20th century with the advent of industrial automation. Initially, manufacturing processes were controlled manually, but as industries grew, the need for more efficient and reliable operations led to the introduction of mechanical devices and later, electronic systems.

We can classify the emergence of OT into the following four major broad phases of development.

- **Early Automation (1900s–1960s)** – involves the introduction of mechanical controllers, such as relays and pneumatic systems that allowed basic automation of processes.
- **Digital Revolution (1970s)** – resulted from the development of digital computers. This paved the way for more sophisticated control systems. PLCs began to replace relay-based systems, allowing for more flexible and complex automation. Automation is hugely beneficial because it ensures completely repeatable results given a set of conditions resulting in higher precision, consistency, productivity and quality compared to manual or human or manual production or control.
- **Networking and Integration (1980s–1990s)** – was due to the rise of networking technologies which enabled the integration of different OT systems. The introduction of SCADA systems allowed for centralised monitoring and control of large-scale industrial processes.
- **Convergence of IT and OT (2000s–Present)** – was based on the emergence of the Internet and advances in connectivity, meaning the lines between IT and OT have blurred. Organisations are increasingly utilising data analytics, Cloud computing and IoT to enhance operational efficiencies.

OT security has evolved significantly over the years, driven by technological advancements, changing threats and the increasing convergence of IT and OT environments. In the context of the four major stages of OT development described above, we can also trace the progression of OT security into the following categories although not necessarily aligned to the same timeframes.

- **Phase 1. Isolated Systems or Fully "air-gapped" OT Systems:** This technique is still used in the defence technology arena and is highly effective. We might call this technique "denial of access" or more accurately a restriction of unauthorised access. Access is available only to approved staff or personnel who are physically present and fully identified.

  In the early days of industrial automation, the primary focus was on functionality and reliability. OT systems were often designed to operate in isolated environments with limited (often – no) connectivity to external networks. Security was not a significant concern because these systems were not seen as vulnerable targets to external threats.

  During this period, the security emphasis was largely on physical safeguards. Access controls, surveillance and physical barriers were implemented to protect critical infrastructure. Up until early 2000s the notion of cybersecurity for OT was virtually non-existent, as many believed that isolation inherently ensured safety.

- **Phase 2. Network-Centric Security:** This phase relates to OT systems partially connected to each other and was caused by the emergence of networking. As industries began to embrace digital technologies in the late 1990s and early 2000s, introduction of networking in OT environments changed the landscape dramatically. SCADA systems and other OT tools started to connect to corporate networks, increasing efficiency but also introducing new attack surfaces and vulnerabilities.

  With the growing connectivity came an increase in cyberthreats. High-profile incidents, such as Stuxnet attack on Iran's nuclear facilities, highlighted the vulnerability of OT systems, even those that are air-gapped. This event marked a turning point, highlighting the need for dedicated cybersecurity measures in OT environments.

  It is important to note that air-gapping[1] did not help in the case of Stuxnet attack as specifically designed malware was brought in via a USB device. This is very well described in the great book "Stuxnet to Sunburst - 20 Years of Digital Exploitation and Cyber Warfare" by Andrew Jenkinson, CRC Press, 2022. The challenge is that although many vendors claim their platforms are air-gapped, in reality they are not. The term "air-gapped" is so misused or abused that it has become almost meaningless as a claim. Hence it needs to be re-defined.

- **Phase 3. Convergence of IT and OT:** By the mid-2010s, organisations began to recognise that OT security could no longer be an afterthought. The convergence of IT and OT created a need for integrated security strategies. Many industries adopted IT security practices, such as firewalls, intrusion detection systems, the Purdue Five-Layer Model, DMZ's[2] and unidirectional gateways and/or data diodes[3] and adapted them for OT environments.

  During this period, various standards and frameworks emerged to guide OT security practices. The National Institute of Standards and Technology (NIST) and the International Society of Automation (ISA) released guidelines for securing industrial control systems. These frameworks provided organisations with actionable steps to bolster their OT security posture.

- **Phase 4. Asset-Centric Security:** This phase has perhaps only just begun in the last couple of years and has a long way to go. To survive the latest sophisticated cyberthreats newly designed or newly engineered superior physical OT systems where cybersecurity is a primary driver. This phase relies on dynamic segmentation, new technology architectures, and more stringent standards and controls as well as advanced technologies such as machine learning/AI[4], anomaly detection and behavioural analytics.

  The potential consequences of successful breaches – ranging from operational disruption to safety hazards – have heightened the urgency of OT security. The concept of Zero Trust, which operates under the principle that no user or device should be trusted by default, is gaining traction in OT security. By implementing strict access controls and continuous monitoring, organisations aim to minimise the risk of breaches and unauthorised access.

Arrival of Internet, world-wide-web, mobile devices (smart phones and tablet computers), and the cloud "as-a-service"[5] model (see Chapters 7 and 8) revolutionised software development, destroyed organisational technology boundaries or perimeters and shifted integrated organisations into a global technology environment. Over time the benefits of connectivity resulting from the automation of asset maintenance, works management, telematics, the use of drones for example for vegetation management, among many others – combined with the fact that threats or risks are mostly invisible – meant that the momentum towards convergence became unstoppable.

The convergence of IT and OT is now a pivotal trend in modern industry, driven by the need for greater efficiency and data analytics, as organisations seek to harness the full potential of their digital assets and automation, understanding this convergence is crucial for future success.

Primarily the convergence momentum is driven by economics – more specifically enhanced efficiency and productivity. As industries increasingly rely on data-driven decision-making, integrating IT and OT systems allows organisations to streamline operations. Real-time data from OT devices can be analysed through IT platforms to optimise processes, reduce downtime and enhance productivity.

With integration of IT and OT organisations can leverage advanced analytics and machine learning tools. This enables deeper insights into operational performance, predictive maintenance and ability to anticipate and respond to potential issues before they escalate.

Moreover, bringing IT and OT teams together fosters collaboration and knowledge sharing. Cross-functional teams can develop holistic strategies that leverage both domains' strengths, leading to innovative solutions and improved operational outcomes.

And finally, rapid pace of change in technology requires organisations to be flexible to address changes in market conditions and take advantage of the latest technological developments. Integrated IT and OT systems facilitate quicker responses to market demands, regulatory changes and technological advancements. The downside is significant cost increase, greater complexity and greater vulnerability. These are to a large extent hidden or unexpected consequences, although they shouldn't be.

In closing, it is important to note that the differences[6] between OT and IT may make full convergence an impossibility. The implication is that although OT and IT are interoperable or can be managed and operated side by side there are functions and components that must be kept separate because they require different tools and capabilities. In simple words, OT covers the security and safety of industrial components such as manufacturing plants, electricity grids, mining logistics and telematics[7].

As OT continues to evolve, organisations face several challenges:

Cybersecurity Risks: With increased connectivity comes heightened vulnerability. OT systems are often targeted by cyberthreats due to their critical nature. Ensuring robust cybersecurity measures while maintaining operational efficiency is a significant challenge for all organisations.

- **Cultural Differences** – IT and OT have historically operated in silos, leading to cultural differences in communication, priorities and approaches. Bridging these gaps requires a concerted effort to foster collaboration and mutual understanding.
- **Legacy Systems** – many industries still rely on legacy OT systems that may not support modern technological advancements. Upgrading or replacing these systems can be costly and disruptive and may foster resistance to change. Also, legacy systems usually don't easily integrate with modern IT platforms.
- **Skill Shortages** – there is a growing demand for professionals skilled in both IT and OT. Lack of qualified personnel can hinder organisations from fully realising the potential of their operational technologies.
- **Integration Issues** – convergence of IT and OT presents integration challenges. Many OT systems were designed to operate in isolated environments, making them difficult to integrate with modern IT infrastructures. This can lead to data silos and inefficiencies. Moreover, integration is mostly achieved via APIs[8] which increase cybersecurity vulnerability or attack surface resulting in greater opportunities for threat actors.
- **Regulatory and Compliance Challenges** – industries like energy and pharmaceuticals are heavily regulated. Ensuring compliance with these regulations while adopting new technologies can be complex, costly and time-consuming.

Given OT is a vital component of modern industries and is playing a crucial role in optimising processes and ensuring safety, these challenges must be successfully overcome for organisations to prosper in an increasingly complex and interconnected world.

CRITICAL NATIONAL INFRASTRUCTURE (CNI)

Protection of critical national infrastructure (CNI) has become a major concern for most governments worldwide due to rising geopolitical conflicts, escalating cyberthreats and rising complexity of technological environments due to technological advancement. Australian government introduced the Security of Critical Infrastructure Act 2018 (SOCI Act) which represents significant legislative effort to bolster the security and resilience of essential services in Australia.

We have chosen to cover SOCI legislation in this chapter in order to render a wholistic overview of critical infrastructure. For all other legal aspects of cybersecurity please see Chapter 20.

The SOCI Act was designed to enhance security framework surrounding Australia's critical infrastructure. It recognises that certain assets and services are essential for the nation's security, economy and community well-being. The SOCI Act aims to mitigate risks posed by both physical threats and cyberattacks, ensuring that essential services remain operational even in the face of challenges.

The SOCI Act applies to the following categories of assets or industry sectors:

1. Communications – telecommunications meaning all networks used to supply carriage services (network service providers), broadcasting and Domain Name Systems.
2. Financial services and markets – including banking, financial markets, clearing and settlement, payments systems, insurance, superannuation, health insurance, life insurance, derivative trade repositories and financial analysts/benchmarks.
3. Data storage or processing – any provider of data storage or processing services to the government or an end user that is a responsible entity for a critical infrastructure asset.
4. Water and sewerage systems – water supply and wastewater services.
5. Defence industry – any asset or service that enables a critical defence capability.
6. Higher education and research – encompass universities.
7. Energy – any asset "critical to ensuring the security and reliability of an energy market" used in connection with the operation of the energy market system. Covers electricity, Energy Market Operators, gas and liquid fuels.
8. Food and grocery – all major food and grocery retailers.
9. Healthcare and medical – hospitals primarily.
10. Transport – covers aviation, freight infrastructure, freight services, roads, railways, airports, ports and maritime facilities services and public transport.

Each of these sectors is interconnected, meaning a disruption in one can have cascading effects on others. As such, SOCI Act recognises the importance of safeguarding these sectors to ensure national stability.

However, there are some key aspects that SOCI Act doesn't seem to cover. For example, federal and state electoral commissions or delivery of democracy. According to consultation papers exclusion of electoral commissions from SOCI Act primarily stems from the nature of their operations and the existing legal frameworks that already govern them. Electoral commissions are generally considered to be part of the democratic process, which is regulated by separate legislation focused on electoral integrity and administration rather than infrastructure security, but they should have been considered as a sub-class that should be included into SOCI Act.

Other potential gaps in SOCI Act relate to emerging technologies such as machine learning or AI, blockchain and digital platforms that are integrated into or becoming integral to critical infrastructure but are not fully encompassed. In this context, perhaps integrity of electoral processes and systems perhaps should be covered by SOCI Act.

There are further extensions of SOCI Act that may be considered in the future. For example, a broader definition of public health infrastructure expanding beyond hospitals. And education infrastructure broader than universities.

The main objectives of SOCI Act are as follows:

- **Identification and Regulation** – SOCI Act mandates that entities responsible for critical infrastructure must register with the government. This allows for better monitoring and management of these essential services.
- **Security Obligations** – SOCI Act imposes security obligations on registered entities, requiring them to implement risk management plans and report incidents that could impact their operations.
- **Government Intervention** – in instances where a registered entity faces significant threats, government is empowered to intervene, ensuring the protection of critical services and infrastructure.
- **Collaboration and Information Sharing** – SOCI Act encourages collaboration between government and private sector, facilitating information sharing to enhance overall resilience against potential threats.

Australia is not alone in legislating to protect its CNI. The Unites States, United Kingdom, Canada, Japan and the European Union have also legislated to protect CNI.

UNITED STATES

In the United States, the protection of CNI is managed through a combination of federal, state and local initiatives. The Department of Homeland Security (DHS) plays a central role in coordinating efforts to enhance security and resilience of critical infrastructure sectors.

The Homeland Security Act of 2002 established the framework for CNI protection in the United States. Additionally, Cybersecurity and

Infrastructure Security Agency (CISA) was formed to address cybersecurity threats specifically. CISA works closely with private sector stakeholders to develop guidelines, share information and conduct threat assessments.

The United States emphasises collaboration between government and private sector, given that a significant portion of critical infrastructure is privately owned. Initiatives such as Critical Infrastructure Partnership Advisory Council (CIPAC) facilitate dialogue and coordination between federal and private sector representatives. These partnerships are vital for developing comprehensive risk management strategies and improving incident response capabilities.

Recognising the growing threat of cyberattacks, US government has launched programs like Cybersecurity Framework, which provides guidelines for organisations to enhance their cybersecurity practices. National Institute of Standards and Technology (NIST) also develops standards and best practices for CNI protection, particularly in the context of cybersecurity.

UNITED KINGDOM

The United Kingdom has developed a comprehensive strategy for CNI protection, focusing on both physical and cybersecurity measures. National Cyber Security Centre (NCSC), part of the Government Communications Headquarters (GCHQ), plays a pivotal role in safeguarding UK's critical infrastructure.

Civil Contingencies Act 2004 provides a legal framework for emergency planning and response, ensuring that local authorities and other organisations are prepared for disruptions. UK's approach is heavily based on risk assessment and management, based on government issuing guidance and regulations to critical sectors.

UK conducts regular national risk assessments to identify vulnerabilities in critical infrastructure and prioritise resources accordingly. Crisis Management Framework outlines procedures for managing incidents, including clear communication channels between stakeholders.

In response to the increasing cyberthreat landscape, UK launched the Cybersecurity Strategy 2016–2021, focusing on enhancing security of critical systems and networks. NCSC provides support to businesses and public sector organisations through training, incident response and threat intelligence sharing.

EUROPEAN UNION

The European Union (EU) has recognised the importance of protecting critical infrastructure across its member states. EU's approach emphasises cooperation, regulation and shared standards to enhance resilience.

Directive on Security of Network and Information Systems (NIS Directive), adopted in 2016, aims to enhance cybersecurity across EU. It establishes security requirements for operators of essential services and digital service providers, ensuring they implement appropriate security measures and report incidents.

European Program for Critical Infrastructure Protection (EPCIP) focuses on improving resilience of critical infrastructure across member states. EU facilitates information sharing, risk assessments and joint exercises among member countries to enhance collective security.

EU has also developed Cybersecurity Act, which strengthens mandate of the EU Agency for Cybersecurity (ENISA) and establishes a framework for cybersecurity certification. This ensures that products and services meet established security standards, further protecting critical infrastructure.

CANADA

In Canada, protection of critical infrastructure is a shared responsibility among federal, provincial and territorial governments, as well as private sector. Public Safety Canada department plays a key role in coordinating efforts.

Emergency Management Act and National Strategy for Critical Infrastructure provide the framework for assessing and managing risks to critical infrastructure. Canada emphasises whole-of-society approach, involving all levels of government and private entities in security initiatives.

Similar to the United States and the United Kingdom, Canada relies heavily on public-private partnerships. Critical Infrastructure Partnership Advisory Committee (CIPAC) fosters collaboration between various stakeholders, facilitating information sharing and joint risk management strategies.

Canada's Cyber Security Strategy aims to protect critical infrastructure from cyberthreats. Canadian Centre for Cyber Security provides guidance, incident response and threat intelligence to organisations operating critical infrastructure.

JAPAN

Japan faces unique challenges in protecting its critical infrastructure, particularly due to its vulnerability to natural disasters such as earthquakes and tsunamis. The country has developed a comprehensive approach to infrastructure protection that emphasises resilience and recovery.

GLOBAL TRENDS

Australian government's consultation paper "2023–2030 Australian Cyber Security Strategy: Legislative Reforms" recognises that the global environment is evolving by moving towards regulated standards that accelerate adoption of security-by-design principles. This is happening in United Kingdom, the EU and Singapore as well as the US in terms of mandatory requirements for government procurement.

Consequently, it is desirable for Australia to synchronise its regulatory environment with international developments to ensure consistent consumer protections and minimise regulatory impact on global vendors supplying products and services to multiple global markets.

OT experiences similar kinds of threats to IT but also suffers specific to OT threats. The key difference for OT is that the consequences can be more severe given the nature of OT. Moreover, OT technology architectures tend to be more complex compared to IT. Therefore, although some of the same tools may be used to protect OT, generally speaking OT requires dedicated cybersecurity tools and techniques.

OT specific threats relate to safety (especially potential loss of human life), uptime – where outages may have major consequences – for example electricity disruption including brownouts and blackouts, reduction of critical asset lifespans through damage, increased risk of fire, contamination of water supply, derailment, crashes and theft of plant and equipment that is difficult and costly to replace, among many others.

The following OT threats are the most critical.

THE NATION STATE ACTORS THREAT OR CYBER WAR

We are talking about the prelude to war in cyberspace as an additional theatre of war.

Historically, one of the earliest documented instances of cyber warfare was the 1982 sabotage of a Soviet gas pipeline, attributed to the CIA. This event foreshadowed the potential for cyber tools to disrupt national security.

The 1990s saw the emergence of the Internet as a global phenomenon, leading to growing concerns about cybersecurity. In 1996, the US Department of Defense (DoD) recognised the need for cybersecurity policies by creating "DoD Information Assurance Strategy." This marked one of the first formal acknowledgments of importance of securing digital assets.

Turn of the millennium brought an escalation in cyberthreats. Infamous "I LOVE YOU" worm in 2000 and the 2007 cyberattacks on Estonia highlighted the destructive potential of cyber operations. In response, governments began to recognise the need for a comprehensive approach to cybersecurity.

The pivotal moment came in 2011, when the DoD released its "Strategy for Operating in Cyberspace." This document formally categorised cyberspace as a distinct domain of warfare, alongside land, sea, air and space. It articulated that cyber operations could be used to achieve national objectives, thereby placing cyber activities within the realm of military strategy. This marked the first time cybersecurity was explicitly declared a theatre of war.

This document outlined how cyber operations could be integrated into military strategy and emphasised that cyberthreats could be considered acts of war. The concept has evolved since then, with increasing recognition of cyberspace as a critical aspect of war.

Following the 2011 declaration, the recognition of cybersecurity as a theatre of war gained traction globally. High-profile incidents, such as the 2015 cyberattack on the Ukrainian power grid and the 2016 US presidential election interference, underscored the real-world implications of cyber conflicts. These events demonstrated that cyberattacks could have profound effects on national sovereignty and democratic processes.

In response, NATO and other international bodies began to adopt similar stances. NATO's 2016 Warsaw Summit declared that a cyberattack could invoke Article 5 of the NATO treaty, which states that an attack on one member is an attack on all. This emphasised the collective defence aspect of cybersecurity in the modern era.

Today, the battlefield extends beyond traditional borders, with cyber operations playing a crucial role in international relations and conflicts. Nations are increasingly investing in offensive and defensive cyber capabilities, leading to an arms race in cyberspace. Establishment of cyber commands in many military organisations reflects strategic importance of cybersecurity.

As technology continues to evolve, so will the tactics and strategies employed in cyber warfare. The recognition of cybersecurity as a theatre of war has transformed how nations approach conflict, signalling the new era where digital capabilities can determine the outcomes of global engagements.

US DoD has established US Cyber Command as a subcommand of US Strategic Command. The UK has established National Cyber Force (NCF) to consolidate offensive cyber activity by enabling an offensive capability to combat national state security threats or hostile states, extremism, hackers, terrorism, disinformation and election interference.

The EU established Cyber Defence Policy Framework (CDPF) that supports development of cyber defence capabilities of EU Member States as well as strengthening of cyber protection of the EU security and defence infrastructure, without prejudice to national legislation of Member States and EU legislation. Cyberspace is the fifth domain of operations, alongside the domains of land, sea, air and space: successful implementation of EU missions and operations is increasingly dependent on uninterrupted access to secure cyberspace, and thus requires robust and resilient cyber operational capabilities[9].

Critical infrastructure and by association OT are therefore a key target in a cyber war. The objectives are simple. Disruption is aimed at causing chaos and disruption in daily life. Such attacks can impact essential services, leading to public panic and loss of trust in authorities. Disruption can be achieved through sabotage, where state actors may attempt to disrupt essential services such as electricity, water supply and transportation networks. For example, cyberattacks on power grids can lead to widespread blackouts, impacting both civilian life and military operations.

Economic impact is the result of disrupting normal operations. Targeting industries essential to nation's economy, such as manufacturing or energy, can have cascading effects that weaken nation's economic stability and security. This can have dire financial consequences, affecting not only targeted entities but also broader economic systems and society at large.

In a country like Australia completely dependent economically on importing and exporting goods, and where shipping by sea and air are the only ways to do so – supply chain security is crucial. State actors may target suppliers of OT systems to introduce vulnerabilities. Compromised software updates or hardware components can create backdoors into critical infrastructure, allowing for undetected access and control. Nations increasingly rely on foreign technology providers for OT systems. This dependency can expose critical infrastructure to state-sponsored threats, particularly if the supplier is linked to hostile governments.

Then there are political objectives. Cyberattacks can be used as tools of coercion or to signal discontent, serving as a form of modern warfare to achieve political goals without traditional military engagement. State threats often involve hybrid warfare tactics that blend cyberattacks with physical sabotage. For example, cyber intrusion could be used to manipulate OT systems while simultaneously carrying out physical attacks on infrastructure. Threat of cyberattacks can also serve as a psychological tool to create fear and uncertainty among the populace, undermining trust in government and institutions.

Finally, espionage – compromising infrastructure can provide attackers with sensitive information, facilitating further strategic advantages. For example, data theft where state-sponsored actors target OT systems to steal sensitive information, including intellectual property and proprietary data. This espionage can provide foreign nations with technological advantages in various industries. Obtaining strategic intelligence is also a key objective. Gaining access to OT systems can help adversaries understand details of nation's critical infrastructure capabilities and vulnerabilities, enabling them to plan future operations.

The following high-profile incidents or events highlight the vulnerabilities of critical infrastructure and OT:

- **Stuxnet (2009–2010):** This sophisticated malware that specifically targeted Iran's nuclear enrichment facilities, demonstrating how cyber warfare can be used to sabotage critical operations. It marked a turning point in recognising cyber capabilities as a viable tool for state actors.
- **Ukraine Power Grid Attack (2015):** Russian hackers infiltrated Ukraine's electrical grid, causing widespread blackouts and exposing vulnerabilities in the country's energy infrastructure. This incident underscored the real-world implications of cyber-attacks on critical services.
- **Colonial Pipeline Ransomware Attack (2021):** A ransomware attack on the Colonial Pipeline led to fuel shortages across the eastern United States. This attack highlighted susceptibility of essential services to cyberthreats and prompted discussions about cybersecurity policies.

Most if not all organisations – even the largest global conglomerates – are not in a position to respond and defend against a nation state threat on their own. Coordination with relevant government departments such as defence, and assistance from them is vital. In Australia, there is Australian Signals Directorate (ASD), ACSC and Department of Home Affairs. In USA, there is Homeland Security, Cybersecurity & Infrastructure Security Agency (CISA), National Security Cyber Assistance Program, and National Security Agency (NSA). In Europe, there is the European Network and Information Security Agency (ENISA). In the United Kingdom, there is National Cyber Security Centre (NCSC) and in Canada - Canada Centre for Cyber Security.

Key strategies for protection of critical infrastructure and OT are now a national security priority and include:

- **Robust Cyber Hygiene:** Implementing regular updates, patch management and training for personnel to recognise threats.
- **OT Specific Incident Response Planning:** Developing comprehensive response plans to quickly mitigate damage in the event of a cyber-attack on critical infrastructure.
- **Enhanced OT Cybersecurity Frameworks:** Governments must establish more robust cybersecurity frameworks that specifically address OT systems. This includes developing regulations, standards and best practices to secure critical infrastructure.
- **Collaboration - Public-Private Partnerships:** Fostering partnerships between government agencies, private sector organisations and cybersecurity firms to share threat intelligence and best practices.
- **International Cooperation:** Given the global nature of cyberthreats, international cooperation is essential. Sharing intelligence, resources

and best practices among nations can create a more unified front against state-sponsored attacks.
- **Investment in OT Specific Cyber Defence Technologies:** This includes anomaly detection and intrusion prevention systems. Also includes allocating resources to improve cybersecurity posture of critical infrastructure systems, including adopting advanced technologies for threat detection.
- **Training and Workforce Development:** Building skilled workforce capable of addressing OT security challenges is vital. Investment in training programs and educational initiatives can prepare professionals to protect critical infrastructure from state threats.

HACKTIVISM – THE INTERSECTION OF TECHNOLOGY, HACKING AND ACTIVISM

Hacktivism, a portmanteau of "hacking" and "activism," refers to use of technology and hacking techniques to promote personal, social and/or political causes. This phenomenon has grown in prominence with the rise of the Internet, allowing individuals and groups to challenge authority, expose corruption and advocate for change in innovative ways. As technology advancement continues unabated, the role of hacktivism in shaping public discourse and influencing political movements is also evolving.

The concept of hacktivism emerged in the 1990s, drawing from the traditions of both hacking and activism. Early pioneers, like the group Cult of the Dead Cow, used their skills to expose security vulnerabilities and advocated for Internet freedom. The rise of Internet provided a platform for activists to organise, mobilise and disseminate information rapidly, making hacktivism an attractive option for those seeking to challenge the status quo.

Several groups have become synonymous with hacktivism, each with distinct motivations and methods:

- **Anonymous:** Perhaps the most well-known hacktivist collective, Anonymous has targeted organisations ranging from the Church of Scientology to various government entities. Their operations often involve DDoS (distributed denial of service) attacks, website defacements and release of sensitive information.
- **LulzSec:** Operating for a brief but impactful period in 2011, LulzSec aimed to expose vulnerabilities of major corporations and government organisations for "the lulz" (or laughs), often underlining issues of privacy and security.
- **WikiLeaks (2010–2016):** Founded by Julian Assange, WikiLeaks played a crucial role in dissemination of classified information, revealing government misconduct and sparking global debates about transparency, privacy and whistleblowing.

- **Occupy Wall Street Hackers (2011)**: During the Occupy movement, various hacktivist groups worked to support protests by targeting financial institutions, aiming to draw attention to issues of economic inequality and corporate greed.
- **Operation Payback (2010)**: Anonymous targeted organisations like PayPal and Mastercard in retaliation for their refusal to process donations for WikiLeaks, effectively disrupting their services to protest censorship.
- **Operation BART (2011)**: In response to Bay Area Rapid Transit (BART) police's decision to shut down cell service during the protest, Anonymous hacked BART's website, leaking user data and emphasising issues of free speech.
- **Iranian Cyber Activism (2009)**: During the Green Movement, hacktivists targeted government websites and provided tools for circumventing censorship to support protesters against Iranian government.
- **#OpIsrael (2013)**: Anonymous launched a series of attacks against Israeli government and corporate websites in response to Israeli-Palestinian conflict, aiming to raise awareness about the situation.
- **The GrayZone's Cyber Attacks (2019)**: In response to the treatment of refugees and migrants, various hacktivist groups attacked websites linked to border enforcement agencies in Europe, aiming to raise awareness about human rights abuses.

Hacktivists employ a variety of methods to achieve their goals, including DDoS attacks where overwhelming a website with traffic renders it unusable. This method is often used to protest against organisations perceived as oppressive.

Other methods include Website Defacement or altering the appearance of a website to display a political message or expose wrongdoing, Data Leaks which entails gaining unauthorised access to sensitive information and releasing it to the public. This tactic aims to hold powerful entities accountable.

Finally, we have Social Media Campaigns utilising platforms like X (formerly known as Twitter) and Facebook to spread awareness, organise protests and mobilise support.

Ethical implications of hacktivism are hotly debated. Proponents argue that hacktivism serves as a necessary tool for social change, especially in contexts where traditional methods of protest may be ineffective or dangerous. They contend that exposing corruption and injustice is a moral imperative.

Critics, however, caution against potential for collateral damage. DDoS attacks and data breaches can inadvertently harm innocent parties or disrupt essential services. Moreover, legality of hacktivism often remains murky, with participants facing potential criminal charges.

As technology continues to advance, landscape of hacktivism will likely evolve. With the rise of artificial intelligence, blockchain and other emerging technologies, hacktivists may develop new strategies to address contemporary issues such as surveillance, censorship and data privacy.

Furthermore, as more people engage with digital platforms, hacktivism may become increasingly mainstream, blending with traditional activism to create a more multifaceted approach to social change. However, this also raises questions about effectiveness and sustainability of such tactics in a world that is constantly adapting to counter them.

Hacktivism challenges the boundaries of what is considered acceptable in the pursuit of social justice. As it continues to evolve, it poses important questions about ethics, legality and role of technology in modern activism. While hacktivism may add a voice to the ongoing conversation about power, responsibility and the pursuit of justice ultimately the risk it poses to innocent third parties is unacceptable.

These are sophisticated threat actors posing significant risks to individuals, and business and government organisations today.

Phishing is a cyber-attack that uses social engineering and typically involves sending fraudulent and deceptive emails that appear to come from a legitimate source. The goal is to trick individuals into providing sensitive information, such as usernames, passwords or credit card details and steal sensitive information and credentials.

Phishing emails often contain urgent messages, such as alerts about suspicious activity on an account, prompting the recipient to click on a link. This link may direct them to a counterfeit website designed to look legitimate. Once users enter their information, the attackers collect it for malicious use.

Five years ago, one of the clients experienced a major data breach as a result of phishing. This organisation was by no means small. Their IT budget was less than A$10 million, but less than 10% was spent on security. Needless to say, they were completely unprepared for a phishing threat. The receiver of phishing email was a sophisticated progressive business executive determined to leverage advanced technology. The phishing email was brilliant in its simplicity.

This organisation's entire commercial database was stolen. The cost of dealing with this breach was more than A$2 million. Not including reputational loss, and loss of business.

There are many examples of email phishing.

A very effective phishing campaign a couple of years ago focused on electricity retail customers offering better deals on electricity and gas and targeting major retailers such as Origin Energy, Energy Australia and AGL. Except the phishing emails came from Turkey. And Somalia.

There are phishing emails purporting to come from the Australian Tax Office, toll road companies, police, telcos, banks and airlines or their frequent flyer schemes. The possible combinations and permutations are endless. There are fake invoice scams, changes to payment details, requests for personal information, claims of payment failure or copy infringement, impersonation (Business Email Compromise – BEC) and many more besides.

Moreover, there are many types of phishing. For example, spear-phishing focuses on specific individuals and groups using direct messaging and involves detailed profiling of the target and the organisation they work for. Whaling targets senior executives, CEO's and their assistants relying on information from social media and other business sources.

Voice phishing or vishing is voice call deception and involves scammers calling potential victims directly. New technology enables threat actors to spoof caller IDs and pretend to be from a trusted source. Typically, callers claim to be from a Telco and tell their targets that there is a problem with the Internet connection. Or that a family member needs emergency funds or other monetary assistance. Ot that they are from the tax office and need one's tax file number or social security number to identify them. Or that a warrant has been issued for the one's arrest, or that one's bank account has been compromised.

HTTPS[10] phishing attempts to get the target to click on an unsafe link embedded in the email. Clone phishing intercepts a valid email, makes an almost identical copy, embeds a malicious link and forwards it on to the potential victim. SMS phishing uses SMS text messages rather than emails. Pop-Up phishing involves browser notifications, urgent pop-up messages.

Social media has now become a popular attack vector for phishing attacks. This is achieved using special offers, discounts, surveys, contests, friend requests, fake videos, comments and posts. Angler phishing involves attackers posing as customer support.

Evil Twin phishing involves setting up an unsecured Wi-Fi hotspot. Apparently, this happens a lot in our universities. Website spoofing involves setting up a fake website. Email spoofing involves creating an entirely fake email domain.

And DNS[11] spoofing – also known as DNS server poisoning or pharming attacks – are a more technical process which involves cybercriminals hacking a Domain Name Server (DNS). This is a server that translates domain names into IP addresses. When this happens, the attacker can automatically redirect a URL entry to a malicious website under an alternate IP address.

Image based phishing involves an embedded image link to infected websites. Search engine phishing creates fake content as a lure and directs targets to fake pages containing promises of free products, free travel, discounts, job offers, investment opportunities or claims the user's computer is infected by a virus.

Watering hole phishing is a tactic that targets a particular organisation or a group by infecting a third-party website they visit on a frequent basis. Finally, Man-in-the-Middle phishing involves an attacker intercepting and altering a communications chain, controlling the communication flow and using deception to steal information from both parties.

The easiest way is to only deal with known and trusted parties. That is not foolproof and may not always be possible. Always check where the email is coming from. Often there is a negligible one letter spelling difference between the real and fake sources or websites. Be wary and sceptical. Check everything. If unsure delete.

When it comes to major financial transactions such as buying a home or a motor vehicle, insist on a commercial clause refusing to exchange any details via email, thus banning email as means of communication or exchange. Check and double check details especially payment details which should be included in the contract.

It is critical to understand that banks, the tax office and most major organisations do not use emails in case of emergency or to contact their customers. They will do it through their own apps.

It is imperative we all protect ourselves using end point protection including anti-virus and firewalls, keeping all software up-to-date and patched, using multi-factor authentication and ultimately not sharing any personal information. Whenever someone is using urgency to pressure you to act – think. Are they really who they say they are? If unsure, refuse and hang-up. You can always call the organisation directly to verify.

Ultimately it is easy to make a mistake. Personal data is vulnerable even if one doesn't actually do anything wrong. For example, social media site may have been breached and leaked some details to the Darknet. This happened to LinkedIn.

Credit agencies such as Equifax, and software providers such as Norton now offer services scanning the Darknet for personal details. "Have I Been Pwned": Check if your email has been compromised in a data breach enables users to check if their email addresses have been compromised. There are many tools available such as Email Header Analyzer – WhatIsMyIP.com®, Virus Total, website scanners and remote connectivity analysers, among many others.

KEY VULNERABILITY – SCADA USE OF TCP/IP

SCADA systems play a vital role in managing and monitoring industrial processes using real-time monitoring, control and data acquisition for critical infrastructure including electricity grids, water and gas networks, resources and mining, transport and manufacturing.

Use of IP networks for SCADA systems communication leverages existing established and extensive IP communications infrastructure. This enables SCADA systems to communicate across enormous distances, access data remotely, and due to interoperability of IP networks connect to a variety of devices, technologies and platforms resulting in enhanced SCADA functionality and reach.

Conventionally, SCADA systems used proprietary communication protocols and networks. While TCP/IP offers versatility and interconnectivity, it was not originally designed with the specific needs and constraints of SCADA environments in mind. Despite that SCADA has increasingly adopted IP-based networks. IP, or Internet Protocol, is now the main standard communication protocol for transmitting data across networks including the Internet. This has created unforeseen risks for SCADA systems.

The first consequence is an increased attack surface. This is one of the most critical vulnerabilities of using TCP/IP in SCADA systems. By connecting to wider networks, SCADA systems can be accessed remotely, making them susceptible to cyberattacks. Hackers can exploit availability of remote access, allowing them to manipulate system controls, alter data or disrupt operations.

Another consequence relates to legacy systems compatibility. Many SCADA systems are built on legacy hardware and software that were not designed with modern security protocols in mind. These older systems often lack necessary protections to guard against TCP/IP vulnerabilities, making them easy targets for exploitation. As updates and patches become scarce for outdated systems, they remain exposed to evolving threats.

Third major risk relates to data integrity. TCP/IP does not inherently provide robust mechanisms for ensuring data integrity. In SCADA systems, where accurate data is crucial for decision-making and operational safety, the lack of security measures can lead to unauthorised data manipulation. Attackers can intercept and alter data packets, leading to faulty readings and potentially hazardous decisions.

Fourth risk is greater exposure to DoS attacks. SCADA systems are vital for maintaining functionality of critical infrastructure. DoS attack can overwhelm SCADA networks, rendering them inoperable. Such attacks can disrupt services, lead to financial losses and compromise public safety. Reliance on TCP/IP increases likelihood of these types of attacks, as attackers can flood the system with traffic from multiple sources.

Fifth area of exposure is lack of encryption. While encryption is a standard security measure for many network communications, it is not always implemented in SCADA systems utilising TCP/IP. Without encryption, sensitive data transmitted between devices can be intercepted and read by malicious actors. This lack of protection can expose critical operational details, creating opportunities for further exploitation.

Sixth we have misconfiguration risk. The complexity of TCP/IP networks can lead to misconfigurations that introduce vulnerabilities. SCADA systems often require specialised settings that differ from standard network configurations. Inadequate knowledge of these requirements can result in poorly secured networks, leaving them vulnerable to attacks. Regular audits and expertise in configuration management are essential but often overlooked.

Seventh we have Social Engineering threats. As SCADA systems become more connected, human element in cybersecurity becomes increasingly critical. Employees may be targeted through social engineering tactics to gain access to SCADA networks. Phishing attacks, impersonation and other deceptive practices can lead to unauthorised access, further compromising the system's security.

Finally in eighth place higher complexity associated with regulatory compliance. Many industries that rely on SCADA systems are subject to regulatory compliance standards. Use of TCP/IP can complicate compliance efforts, especially if security measures are insufficient. Non-compliance can result in severe penalties and loss of trust among stakeholders, highlighting importance of robust security frameworks.

To address these vulnerabilities, we must reconsider SCADA reliance on TCP/IP for communications. Implementing dedicated communication protocols specifically designed for industrial control systems that can enhance security and reliability. Protocols such as Modbus[12], DNP3[13] or proprietary systems can reduce exposure to external threats while providing necessary functionality for SCADA operations.

Cultural Differences between OT Engineering and IT – Cultural divide between OT and IT is characterised by several key differences as follows.

Firstly, we have different priorities and focus. OT professionals prioritise reliability, safety and performance in physical processes. Their focus is often on minimising downtime and ensuring continuous operation. In contrast, IT professionals prioritise data integrity, cybersecurity and system functionality. This fundamental difference in priorities can lead to conflicting objectives.

Secondly, there are different levels of risk appetite (or, in other words, risk tolerance). OT environments often have very low risk tolerance due to potential physical consequences of failures (e.g., safety hazards, equipment damage). IT, while is also concerned about risks, often operates in a more dynamic environment where risks can be mitigated through various means.

Thirdly, IT and OT have different communication and working styles. Communication within OT tends to be more practical and focused on specific operational outcomes, while IT communication can be more technical and abstract. This difference can lead to misunderstandings and misalignment when teams collaborate.

Fourthly, there are different approaches to change. OT environments often involve legacy systems that are critical to operations and may resist change due to concerns about disruption. IT, however, is typically more agile and open to adopting new technologies and methodologies.

OT SECURITY – CLEAR AND PRESENT DANGER

In a world with escalating global conflicts or war and greater geopolitical risk, OT security for critical infrastructure and essential services becomes even more pivotal to the survival of countries and continued successful daily life within our communities. The latest developments are indeed very scary.

We now have critical communications infrastructure such as pagers and walkie-talkies being used as a means of attack in war. Not long-ago exploding pagers would have simply been viewed as unlikely far-fetched possibility in reality and more of a dream-gadget for a James Bond movie. Now it has become a reality. The degree of difficulty and sophistication achieved to deploy such technology as a weapon is astonishing.

Hacks of water systems around the United States have been rising and, in some cases, have been tied to geopolitical rivals of the United States, such as Russia, Iran and China.

In March 2024, the US government warned state governors that foreign hackers are carrying out disruptive cyberattacks against water and sewage systems throughout the country. In their letter, National Security Advisor Jake Sullivan and Environmental Protection Agency Administrator Michael Regan warned that "disabling cyberattacks are striking water and wastewater systems throughout the United States." The letter singled out alleged Iranian and Chinese cyber saboteurs. The letter cited a recent case in which hackers accused of acting in concert with Iran's Revolutionary Guards had disabled a controller that opens new tab at a water facility in Pennsylvania. They also called out a Chinese hacking group dubbed "Volt Typhoon" which they said had "compromised information technology of multiple critical infrastructure systems, including drinking water, in the United States and its territories." They said in the letter: "These attacks have the potential to disrupt the critical lifeline of clean and safe drinking water, as well as impose significant costs on affected communities." In early October 2024 American Water, the largest water utility in the United States, revealed that it had been targeted in a cyberattack and had to shut down some systems including billing. American Water provides drinking water and wastewater services to more than 14 million people with regulated operations in 14 states and on 18 military installations.

Where do we go from here? Sadly, it will get much worse. Despite the pin-point accuracy of such techniques when used legitimately, how long will it be before this technology falls into the wrong hands and will be used with devastating effect by less scrupulous countries, organisations or even individuals? The flood-gates are open. Expect more stunning innovation in military and defence technology in the next few years. Hopefully the high cost, high degree of skill and restricted availability of components will contain the spread of such techniques.

ECONOMIC REALITY FOR OT CYBERSECURITY

Although there is now a greater focus on OT security, reality is replacing legacy OT systems is costly and possibly economically unviable or impossible in the short term. Many organisations have billions of dollars of infrastructure, plant and equipment that is old. Consequently, it will take time to achieve this progressively over the next decade or more.

Moreover, achieving the highest level of security probably difficult if not impossible for the following reasons:

- The cost is prohibitive.
- There are unacceptable usability compromises if the highest level of security is implemented.
- Secure technology is still evolving or improving or not yet available.

CYBERSECURITY STANDARDS AND FRAMEWORKS FOR OT

The Purdue Model – A framework for industrial control systems

Purdue Model, also known as the Purdue Enterprise Reference Architecture (PERA), is a widely recognised framework designed to improve organisation and integration of industrial control systems (ICS). Developed in the 1990s by researchers at Purdue University, this model provides a structured approach to manage complexities of manufacturing and production environments, facilitating better communication, security and operational efficiency.

At its core, Purdue Model categorises industrial systems into distinct layers, each with specific functions and responsibilities. These layers are designed to facilitate the flow of information from the physical process to higher-level business applications. The model is typically represented in a five-layer architecture:

Level 0: Physical Process: This layer includes actual physical equipment and processes, such as sensors, actuators and machinery. It is where data is generated and where physical actions are taken.

Level 1: Control: At this level, control systems such as PLCs and DCS manage the physical processes. They monitor inputs from sensors and execute control actions based on predefined algorithms.

Level 2: Supervision: This layer includes systems that provide supervisory control and monitoring, often through Human–Machine Interfaces (HMIs). It allows operators to visualise the process, adjust set points and respond to alarms or alerts.

Level 3: Operations Management: The focus here is on operations management systems, such as Manufacturing Execution Systems (MES). These systems collect data from lower levels to manage production schedules, inventory and overall operations.

Level 4: Business Planning and Logistics: This top layer encompasses enterprise resource planning (ERP) systems and other business applications that manage overall company strategy, financials and supply chain logistics.

Purdue Model offers several key benefits for organisations operating in industrial environments. First benefit is that of a clear structure. By defining distinct layers, the model provides a clear structure for understanding how various components of the industrial system interact. This clarity helps streamline operations and facilitates troubleshooting.

Next, we have an improved communication. The separation of functions across layers encourages better communication between teams. For example, IT department can focus on Level 4 applications, while operations staff manages Levels 0 through 3.

Purdue Model supports a defence-in-depth approach to security. By isolating control systems from enterprise networks, organisations can better protect critical infrastructure from cyberthreats. This separation reduces the risk of an attack spreading from business applications to operational technology.

And lastly, we have improved or facilitated integration. The model promotes integration of various systems and technologies. Organisations can implement new solutions without disrupting existing processes, enhancing flexibility and adaptability.

But there are also significant challenges. While Purdue Model provides a robust framework, organisations should be aware of the following challenges when implementing it. Unfortunately, legacy systems remain. Many industrial environments contain legacy systems that may not fit neatly into the Purdue Model's layers. Integrating these systems with modern technologies can be complex and require significant investment.

Another challenge is existence of data silos. Despite the model's intent to streamline communication, organisations will still face data silos that inhibit information sharing across layers. Breaking down these silos often requires cultural change and investment in interoperability.

Finally, we have the rapid pace of technological advancement. This can challenge the Purdue Model's relevance. As new technologies emerge, organisations must continuously adapt their architectures to leverage these innovations effectively.

As industries evolve, Purdue Model will likely continue to adapt. The rise of the IIoT, artificial intelligence and advanced analytics is influencing how organisations implement and utilise the framework. Integrating these technologies within Purdue Model can enhance data collection, analysis and decision-making capabilities, leading to smarter, more efficient operations.

NERC in USA

The North American Electric Reliability Corporation (NERC) is a critical organisation in the US responsible for ensuring reliability and security of the North American electricity grid. Established in 1968 and formally designated by the US Federal Energy Regulatory Commission (FERC) as Electric Reliability Organisation in 2006, NERC develops and enforces reliability standards for the bulk power system.

NERC conducts assessments of grid's reliability and collaborates with various stakeholders – including utilities, regulators and governmental bodies – to address potential risks. NERC also plays a vital role in promoting cybersecurity measures and improving the overall resilience of the electric grid, ensuring that it can withstand both physical and cyberthreats while meeting the demands of a rapidly evolving energy landscape.

ISO/IEC 62443 and ISA 99

ISO/IEC 62443 and ISA-99 are essential frameworks that provide guidelines for securing industrial automation and control systems (IACS). ISO/IEC 62443 is an international standard developed by the International Electrotechnical Commission (IEC), focusing on cybersecurity for operational technology. It offers comprehensive approach to managing security risks, outlining requirements and best practices for system design, implementation and maintenance across various industry sectors.

ISA-99, developed by the International Society of Automation (ISA), serves as the foundation for ISO/IEC 62443 and emphasises the need for a risk-based approach to cybersecurity in industrial environments. Both frameworks promote defence-in-depth strategy, encouraging organisations to adopt layered security measures that address vulnerabilities throughout the lifecycle of industrial systems, ultimately enhancing the resilience and safety of critical infrastructure.

NIST Guide to OT Security, NIST CSF, NIST 800-53 and sub-standards

NIST Guide to OT Security, alongside NIST Cybersecurity Framework (CSF) and NIST Special Publication 800-53, provides robust framework for enhancing cybersecurity in industrial environments. NIST Guide to OT Security specifically addresses unique challenges and requirements of securing operational technology, emphasising integration of cybersecurity best practices within operational context.

NIST CSF offers flexible, risk-based approach that helps organisations identify, protect, detect, respond to and recover from cybersecurity incidents, making it applicable across various sectors, including critical infrastructure. NIST 800-53 outlines comprehensive security and privacy controls to protect organisational systems and information, serving as foundational resource for compliance and risk management.

Together, these standards and guidelines, along with their sub-standards, create a holistic framework that empowers organisations to effectively manage cybersecurity risks in both IT and OT environments, ensuring resilience and security of critical infrastructure.

AESCSF

Automated Environmental Surveillance Cybersecurity Framework (AESCSF) is a specialised framework designed to enhance cybersecurity posture of automated environmental surveillance systems, particularly within critical infrastructure. By integrating cybersecurity best practices with unique operational requirements of environmental monitoring, AESCSF provides structured approach for identifying, assessing and mitigating cybersecurity risks.

The framework emphasises continuous monitoring and real-time threat detection, enabling organisations to respond swiftly to emerging threats. AESCSF aligns with broader cybersecurity standards, including NIST CSF, ensuring consistency and interoperability while addressing specific vulnerabilities associated with automated systems. By fostering proactive cybersecurity culture, AESCSF aims to protect vital environmental data and infrastructure, ultimately contributing to resilience and reliability of essential services.

THE FUTURE OF OT CYBERSECURITY

Cyber Informed Engineering (CIE) – A new approach

CIE is a new paradigm for resilient systems that emerged in 2015 from Idaho National Power labs. This paradigm shifts focus from treating cybersecurity as separate concern to integrating it into engineering lifecycle from the outset, ensuring that systems are designed with both functionality and security in mind.

Harvard Business Review in July 2018 published an article making the assertion that traditional or conventional approach to cybersecurity is flawed. More specifically that "blanket" approaches, or "broad – across the entire organisation" are simply inappropriate, faulty, destined to fail, too expensive and/or very likely impossible.

CIE has been adopted by US Department of Energy since June 2022 and has been implemented by multiple Power and Electricity Utilities in US including Florida Power. More importantly CIE has matured to the point where there are now significant resources available including implementation guides, templates and tools.

CIE is grounded in the recognition that modern engineering projects must account for cyber vulnerabilities alongside traditional physical and functional requirements. CIE emphasises a holistic view of system design that incorporates cybersecurity principles at every stage – from conception and design to implementation and maintenance. By doing so, engineers can anticipate potential threats and create systems that are not only robust against cyberattacks but also resilient in the face of evolving threat landscapes.

CIE approach is built on several foundational principles.

- **Integrated Risk Management:** CIE promotes comprehensive risk assessment that includes both physical and cyberthreats. This integration helps identify potential vulnerabilities early in the design process, allowing for proactive measures.
- **Design for Resilience:** Systems should be designed with capability to withstand and recover from cyber incidents. This involves implementing redundancy, fail-safes and incident response plans that are part of system architecture.
- **Interdisciplinary Collaboration:** Effective CIE requires collaboration among engineers, cybersecurity experts and other stakeholders. This interdisciplinary approach ensures that diverse perspectives are considered in the design process, leading to more secure outcomes.
- **Lifecycle Approach:** Cybersecurity should not be viewed as a one-time consideration. CIE advocates for ongoing security assessments and updates throughout lifecycle of the system, ensuring that new threats are addressed as they emerge.

Adopting CIE offers several significant benefits.

- **Enhanced Security Posture:** By integrating cybersecurity into the design process, organisations can create systems that are inherently more secure, reducing the likelihood of successful cyberattacks.
- **Improved System Resilience:** CIE fosters development of systems that can maintain operational functionality even in the event of a cyber incident, minimising downtime and reducing impact on critical services.

- **Cost-Effectiveness:** Addressing cybersecurity during the engineering phase can lead to cost savings by reducing need for extensive retrofitting or remediation after deployment. Early integration often mitigates risks before they escalate into more significant issues.
- **Regulatory Compliance:** Many industries are subject to stringent regulatory requirements regarding cybersecurity. CIE helps organisations to meet these standards by embedding compliance considerations directly into the engineering process.

Despite its advantages, implementing CIE presents several challenges.

- **Cultural Shift:** Organisations may need to overcome resistance to change, as CIE requires significant shift in mindset toward viewing cybersecurity as a fundamental aspect of engineering rather than as a secondary concern.
- **Skills Gaps:** There is often a disconnect between engineering and cybersecurity expertise, leading to challenges in effectively integrating these disciplines. Bridging this gap requires training and collaboration.
- **Evolving Threat Landscape:** Dynamic nature of cyberthreats means that systems must be continually updated and assessed. Organisations must commit to ongoing education and adaptation to keep pace with emerging vulnerabilities.

CIE represents a crucial evolution in how we approach design and implementation of critical systems, especially in OT. By integrating cybersecurity considerations into the engineering lifecycle, organisations can create more resilient and secure systems that are better equipped to withstand challenges of modern threats.

Developing more secure OT involves adopting a multi-layered approach that integrates advanced cybersecurity practices, modern technologies and robust governance frameworks. This includes implementing security by design principles, where security measures are embedded into development process of OT systems from the outset, rather than being added as an afterthought.

Utilising technologies such as network segmentation, which isolates critical components to limit access and reduce the attack surface, is essential for enhancing security. Additionally, incorporating real-time monitoring and anomaly detection powered by machine learning can help identify potential threats before they escalate into significant incidents.

Regular vulnerability assessments, employee training and adherence to established cybersecurity frameworks, such as NIST CSF or ISO/IEC 62443, further bolster OT security. By prioritising these strategies, organisations can create resilient OT environments capable of withstanding evolving cyberthreats while maintaining operational efficiency.

LATEST DEVELOPMENT IN OT – COMING TO YOUR CAR?

An interesting example of OT (or at least bordering it) can be seen in modern cars, as information technology age has transformed modern cars. Within half of an average human lifetime, computerisation has revolutionised not only how cars work and how they're made, but how we view them – less as mechanical devices and more as electronic appliances.

"Already some people will tell you that a modern vehicle is like a computer on wheels," says Richard Wallace, director of transportation systems analysis at the Center for Automotive Research in Ann Arbor, Michigan. "That is true, and it is becoming even more true."

The first electronic control units (ECUs) showed up in mass-production GM and Ford vehicles in the 1970s to handle basic functions such as ignition timing and transmission shifting in response to tighter fuel economy and emission regulations. By 1980s, more sophisticated computerised engine-management systems enabled the use of reliable electronic fuel-injection systems. They also ignited a renaissance in performance as engineers designed more complex motors to take advantage of the ECU's precision and confident computer-controlled machine tooling could mass-produce them to the high tolerances necessary.

But it didn't stop there.

ECUs were crucial to the advent of active safety systems such as anti-lock braking (ABS), traction and skid-control, where wheel sensors trigger the unit's reaction to loss of grip. Soon they migrated into active suspension control, allowing for instantaneous reaction to the car's changing position on the road and adapting to varying surfaces. In the last decade or so, they've been linked to sonar, radar and laser emitters performing functions such as blind-spot and pedestrian collision warnings, automated breaking and safe distance-keeping via smart adaptive cruise control.

Sensors also provide parking guidance and fully automated parking, with the aid of an on-board computer tied to brakes, steering and throttle. In the cabin, telematics relies on computerised integration of electronic devices such as phones and navigation systems. The average car today can have between 25 and 50 central processing units (CPUs) controlling these functions and more often networked, but sometimes operating independently. The level of sophistication is likely to rise as self-driving vehicles move closer to mass production.

Modern cars are controlled largely by ECUs that all have different functions. Each ECU has a specific job, whether it's to monitor the temperature of the engine or unlock the doors. Here are just some of the computer systems that keep our cars functioning according to GlobalSpec:

- Airbag Control Modules: Passive safety devices that inflate appropriate airbags when there is a collision.

- Body Control Modules: Regulate body electricity including wipers, horns and lights. It may also help the car's entertainment system.
- Engine Control Modules: Regulate performance of the car engine. Including igniting the spark plugs, injecting fuel and cooling the engine.
- Electronic Brake Control Modules: Adjust braking on ABS braking systems and help preventing the wheels from slipping or locking.
- HVAC Control Modules: Allow for automatic cabin circulation for auto AC units.
- Infotainment Control Modules: Control the dashboard computer system including navigation.
- Power Steering Modules: Receive information on vehicle speed, steering position and torque and produces steering feedback to the driver.
- Powertrain Control Modules: Regulate powertrain system ensuring that power flows from the engine to the wheels.
- Suspension Control Modules: Control suspension and adjust the tension for the wheels to create the smoothest ride under current road conditions.
- Transmission Control Modules: Adjust displacement and transmission based on the engine's RPM.

Each ECU varies greatly from unit to unit. They use different communication components, have a different number of inputs and outputs, and have different amounts of microprocessor memory depending on their function. More complex modules will need more complex communication systems, more inputs and outputs and a higher amount of memory.

This did not go unnoticed by hackers. In July 2015, cybersecurity researchers Charlie Miller of Twitter (now known as X) and Chris Valasek of IOActive used the latest hacking techniques to hack into the electrical systems of a Jeep Cherokee. They were able to do this without direct physical access to the vehicle.

Using Internet, they were able to gain wireless control of the Jeep Cherokee giving them access to the Jeep's entertainment system, enabling them to relay commands to its dashboard functions, steering, brakes and transmission, and they were able to do all of this remotely 10 miles away from the vehicle's location. Miller and Valasek have been hacking motor vehicles for years, but they had always required direct access to the vehicle to do so, with auto industry representatives playing down their accomplishments. But this time they have been able to do this wirelessly from any location in the world. It has taken FBI eight months to issue a warning that car hacking is a serious risk.

This is the first-hand experience of the driver of the hacked Jeep Cherokee (https://www.wired.com/2015/07/hackers-remotely-kill-jeep-highway/):

> I was driving 70 mph on the edge of downtown St. Louis when the exploit began to take hold. Though I hadn't touched the dashboard, the

vents in the Jeep Cherokee started blasting cold air at the maximum set-ting, chilling the sweat on my back through the in-seat climate control system. Next the radio switched to the local hip hop station and began blaring Skee-lo at full volume. I spun the control knob left and hit the power button, to no avail. Then the windshield wipers turned on, and wiper fluid blurred the glass. As I tried to cope with all this, a picture of the two hackers performing these stunts appeared on the car's digi-tal display: Charlie Miller and Chris Valasek, wearing their trademark track suits. A nice touch, I thought.

Earlier in the summer of 2013, Charlie Miller and Chris Valasek toyed with Ford Escape and Toyota Prius around a South Bend, Indiana, parking lot while sitting in the backseat with their laptops, cackling as they disabled brakes, honked the horn, jerked the seat belt and commandeered the steer-ing wheel.

So how is all this possible? Well, because vehicle manufacturers like Crysler are now building cars in such a way that makes their electrical sys-tems and computer networks act like smartphones that are connected to Internet. This opens up a whole host of possibilities for hackers, allowing them to gain access to critical systems remotely using wireless connections.

It's not just Crysler vehicles that are vulnerable either. While Jeep Cherokee was highlighted as the most vulnerable by Miller's and Valasek's research, other models from various other manufacturers also ranked highly as pos-sible targets.

The duo rated 24 cars, SUVs and trucks based on three factors that they thought may determine their vulnerability to hackers.

- Number and type of radios that connected the vehicle's systems to Internet
- Whether onboard computers were properly isolated from the vehicle's critical driving systems
- Whether digital commands could trigger physical (cyber-physical components) actions

Miller and Valasek developed software that was able to exploit these vul-nerabilities. Their software was able to silently rewrite the firmware for the Uconnect's entertainment system (or head unit) allowing them to plant their code and send commands through the vehicle's internal computer network.

At the time the pair believed that these hacks will work on any Crysler vehicle that uses Uconnect versions from late 2013 onwards but they have only tested these exploits on Jeep Cherokee so far.

Car hacking isn't the most common type of cyber-attack but these types of hacks are increasing in frequency. According to Upstream, the number of hacks on car systems increased 225% between 2018 and 2021. Upstream analysed 900 incidents of car cybersecurity breaches and noted:

- In 2021, nearly 85% of attacks were done remotely.
- Keyless entry and key fob attacks were number one breach in security accounting for 50% of all vehicle thefts.
- Data and privacy breaches accounted for 38% of incidents studied while car theft and break-ins accounted for 27%, and control system hacks accounted for 20%.

As cars become more and more connected, these types of attacks are expected to increase in the future. While there are seemingly endless possibilities when it comes to what hackers can do, there are certain threats that are more severe and more realistic. Hackers use various attack vectors to get access to car's computer system. Attack vectors are weak points in a system that can be used to access car's network. Attack vectors used to be limited to key fobs and physical access to the controller area network. But nowadays there are a lot more attack vectors, mainly because there is a lot more that is controlled by the computer. Here are a few ways they can get into your system:

- Physically hacking into the system through the headlights, ABS, OBD port or other susceptible areas.
- Through counterfeit car parts and components that are infected with malware (see Chapter 9).
- Through MP3 malware – a music download may be infected with malware code that can get into car's system.
- Through data downloads – an update or app download to a car could be embedded with code (see Chapter 9).
- Through intercepting a key fob signal.
- Through an EV charging station.

If a hacker does manage to get access to a car's computing system, what does this mean? Here are some of the threats that hacking poses:

- They can remotely follow car through car's tire pressure monitoring system.
- They can find vulnerabilities in sensitive information access in some vehicles.
- They can disable car's brakes while it is being driven.
- They can change destination in the navigation system.
- They can cause a car to accelerate.
- They can hack into driver's phone if it is connected to a car and access personal information.

- They can get access to a car's key fob and open a car without owner's knowledge.
- They can send malicious messages or data to driver's phone or computer.
- They can control the car's windshield wipers and air conditioning.

Proliferation of EVs heightened attention to risks associated with car hacking. As Joe McKendrick said: "They're no longer cars, they're computers on wheels." In Australia, concerns about how much information new cars are collecting about their drivers has prompted Privacy Commissioner Carly Kind to open an inquiry to ensure that connected vehicles sold in Australia protect sensitive personal data. "Cars are now kind of computers on wheels, and that means they are collecting a lot of information about individuals, including location information chiefly, which tells us a lot of other things about individuals, including ... about their sensitive personal life," Ms Kind told The Australian Financial Review.

It was publicised in late September 2024 that US Commerce Department is expected to propose prohibiting Chinese software and hardware in connected and autonomous vehicles on American roads due to national security concerns, two sources told Reuters. Biden administration has raised serious concerns about collection of data by Chinese companies on US drivers and infrastructure as well as potential foreign manipulation of vehicles connected to Internet and navigation systems. Officials said they were worried that technology in question, used for autonomous driving and to connect cars to other networks, could allow enemies to "remotely manipulate cars on American roads."

"We've already seen ample evidence that [China] pre-positioned malware in our critical infrastructure for disruption and sabotage," US National Security Adviser Jake Sullivan added. "And with potentially millions of vehicles on the road, each with 10- to 15-year life spans, the risks of disruption and sabotage increase dramatically."

IoT area presents numerous security risks, especially considering the number of IoT devices, as according to the latest available data, there are approximately 18.8 billion connected IoT devices. It is estimated that by 2030 this number will increase to 30 billion Let's have a look at these risks.

- **Monitoring and management complexity**: It grows exponentially with the number of devices, enough said.
- **Weak authentication**: Historically passwords were one of the first lines of defence against hacking attempts. But if password isn't strong, then device isn't secure. Most default passwords are relatively weak – because they're intended to be changed – and in some cases they may be publicly accessible or stored in the application's source code (which is extremely risky). End users may also set the password to something

that's easy to remember. But if it's easy to remember, it's probably easy to penetrate.

- **Many IoT devices have little or no authentication at all:** Even if there's no important data stored on the device itself, a vulnerable IoT device can be a gateway to an entire network, or it can be assimilated into a botnet, where hackers can use its processing power to distribute malware and distributed denial of service (DDoS) attacks. Weak authentication is a serious IoT security concern.

- **Low processing power:** Most IoT devices and applications use very little data. This reduces costs and extends battery life, but it can make them difficult to update Over-the-Air (OTA), and also prevents the device from using cybersecurity features like firewalls, virus scanners and end-to-end encryption. This ultimately leaves them more vulnerable to hacking.

- **Legacy assets:** If an application wasn't originally designed for cloud connectivity, it's probably ill-equipped to combat modern cyberattacks. For example, these older assets may not be compatible with newer encryption standards. It's risky to make outdated applications Internet-enabled without making significant changes – but that's not always possible with legacy assets. They've been cobbled together over years, often – even decades, which turns even small security improvements into a monumental undertaking.

- **Shared network access:** It's easier for IoT device to use the same network as the end user's other devices – like WiFi or LAN – but it also makes the entire network more vulnerable. Someone can hack an IoT device to get their foot in the door and gain access to more sensitive data stored on the network or other connected devices. Likewise, another device on the network could be used to hack the IoT device. In either of those scenarios, customers and manufacturers wind up pointing fingers at each other.

- **Inconsistent security standards:** Within IoT, there's a bit of a free-for-all when it comes to security standards. There's no universal, industry-wide standard, which means companies and niches all have to develop their own protocols and guidelines. Lack of standardisation makes it harder to secure IoT devices, and it also makes it harder to enable machine-to-machine (M2M) communication without increasing risk.

- **Missing firmware updates:** One of the biggest IoT security risks is when devices go out in the field with a bug that creates vulnerabilities. Whether they come from in-house developed code or a third-party code, manufacturers need to have ability to issue firmware updates to eliminate these security risks. Ideally, this should happen remotely, but that's not always feasible. If a network's data transfer rates are too low or it has limited messaging capabilities, this means that a physical access the device to issue the update will be required.

- **Lack of encryption:** Another significant threat to IoT security is the lack of encryption on regular transmissions. Many IoT devices don't encrypt the data they send, which means if someone penetrates the network, they can intercept credentials and other important information transmitted to and from the device.
- **Gaps between mobile networks and the cloud:** Many IoT devices regularly interact with cloud-based applications. And while cellular network that IoT device uses may be secure, and cloud application may be secure, transmissions from the network to the cloud typically pass through Internet, leaving them vulnerable to interception and malware. Even these small gaps can compromise the entire IoT deployment.
- **Limited device management:** Organisations often lack visibility and control they need to see when a device has been compromised and then deactivate it. For example, every Mobile Network Operator (MNO) has their own connectivity management platform, and some of these platforms give customers very little insight or functionality.
- **Physical vulnerabilities:** Not all IoT devices operate in remote areas. Some regularly come into contact with people, which creates possibility of unauthorised access. In fleet management, for example, it's not uncommon for drivers to steal SIM cards from their vehicle's GPS trackers to use them for "free data." Other thieves may steal SIM cards to commit identity theft. People can also physically access IoT devices for more nefarious purposes, like accessing a network or stealing information.

These risks resulted in numerous breaches of IoT devices. Let's look at some of the most well-known:

- **Target's credit card breach (2013):** Hackers successfully breached Target's network and stole credit card information from millions of transactions. How did they do it? They stole login credentials from an HVAC vendor, who was using IoT sensors to help Target monitor their energy consumption and make their systems more efficient.
- **Mirai Botnet Attack (2016):** Mirai botnet is one of the most infamous examples of IoT-related cyberattacks. This botnet was responsible for one of the largest DDoS attacks recorded. It compromised hundreds of thousands of IoT devices, such as cameras and routers, by exploiting default login credentials. Infected devices were then used to overwhelm targeted websites and services, including major companies like Twitter and Netflix, causing widespread disruptions. The botnet was made possible by unsecured IoT devices.
- **Cold in Finland (2016):** After hacking IoT device(s) cybercriminals turned off the heating in two buildings in the Finnish city of Lappeenranta. After that, another DDoS assault was launched, forcing the heating controllers to reboot the system repeatedly, preventing the

heating from ever turning on. This was a severe attack since Finland experienced severely low temperatures at that time of year.

- **St. Jude Medical's pacemakers (2017):** This incident became public when FDA announced that more than 465,000 implantable pacemaker devices were vulnerable to hacking. While there were no known hacks, and St. Jude Medical quickly updated the devices to fix their security flaws, it was a disturbing revelation with potentially deadly implications, as with gaining control of one of these devices, a hacker could literally kill someone by depleting the battery, altering someone's heart rate, or administering electric shocks. An IoT security flaw essentially turned a life-saving device into a potentially deadly weapon.
- **Ring Home Security Camera Breach (2019):** Amazon-owned Ring faced a significant security breach when cybercriminals accessed numerous home security cameras by exploiting weak, recycled and default credentials. The attackers could view live feeds and even communicate through the devices.
- **Verkada hack (2021):** Verkada is a cloud-based video surveillance service. After it was hacked, the attackers could access private information belonging to Verkada software clients and access live feeds of over 150,000 cameras mounted in factories, hospitals, schools, prisons and other sites using legitimate admin account credentials found on the Internet.

With so many IoT devices being deployed at homes, breaches feel like more than a breach. As it has been shown hackers can steal identity from a coffee machine, smart TVs can be hacked, smart bulbs can be hacked, smart homes can be hacked, smart speakers can be hacked – where do we stop?

For the consumer, it feels like a violation, which creates damning headlines and negative perceptions in the market. And when it comes to blame, perception is often reality. In 2023, Keyfactor prepared the First Global State of IoT Security Report that included insights from 1,200 professionals across North America, EMEA and APAC, representing organisations in manufacturing, IT, telecom, energy, oil and gas, retail, construction, financial services and many more. 48% of respondents stated that the manufacturer of IoT or connected devices should be mostly or completely responsible for cyber breaches on their products.

As IoT devices continue to blend the physical and digital worlds in exciting new ways, potential gains cannot overshadow the risks and vulnerabilities associated with their deployment.

NOTES

1 This is a security measure used by a particular computer or machine to isolate itself from other unsecured machines and networks. In most cases that means the air-gapped computer is physically segregated preventing it from connecting to any other external device or network.

2 Demarcation zones.

3 US NIST 800-82 Guide to Industrial Control Systems. waterfall-security.com "Data Diode and Unidirectional Gateways" Dec, 2023.

4 Artificial Intelligence.

5 Infrastructure IaaS, Platform PaaS, Software SaaS.

6 Meaning real-time control nature of OT and cultural differences, for example, the engineering aspect of OT that means OT architectures tend to be more complex due to safety and other control requirements.

7 Telematics is the technology used to manage and monitor a mobile asset such as a vehicle or a drone using satellite and or other communications (GPS) to gather data about location, diagnostics, usage and behaviour.

8 Application Programing Interfaces.

9 https://european-cyber-defence-policy.com.

10 HTTPS is the standard protocol for traffic encryption between browsers and websites.

11 Domain Name System.

12 Modbus is an application layer protocol whereas DNP3 consists of Application and Data Link Layers. Both protocols are used over many different types of transport, such as RS-232, RS-485, and TCP/IP. When it comes to TCP/IP, Modbus has a separate variant called Modbus TCP/IP but the DNP3 is wrapped within TCP/IP. For more information, please refer to 'DNP3 Introduction' from DPS Telecom.

13 DNP3, or DNP 3.0, is a communications protocol used in SCADA and remote monitoring systems. It is widely used because it is an open standard protocol.

Chapter 17

Tyranny of KPIs (and OKRs)

When this book was semi-finished, we have realised that there is one more aspect that should have been discussed. We are talking about Key Performance Indicators (KPIs) and their impact, as in today's fast-paced and competitive business environment, KPIs have become integral tools for gauging performance, tracking progress and driving strategic decision-making.

KPIs are defined as quantifiable metrics used to evaluate success or progress of an organisation in relation to its strategic objectives. They serve the role of measurable indicators of critical aspects such as performance, efficiency and success. With their ability to provide clear insights into the performance and effectiveness of various functions within an organisation, today KPIs play a pivotal role in guiding decision-making processes.

So, each key performance indicator is a measurement that evaluates how well a business is achieving its goals in activities and initiatives. KPIs may focus on the success of the overall business or indicate the success of an individual projects, products, departments or strategies. A KPI must be measurable and is usually quantifiable, which helps organisations track and compare strategies against competitors and previous or similar initiatives.

As always, let's start with the history as KPIs have a long and illustrious history. A brief history of KPIs is illustrated in https://corporater.com/resources/history-of-kpis/ (Figure 17.1).

While no one knows the exact origins, it is commonly believed that the emperors of the Chinese Wei Dynasty (3rd century) rated the performance of members of their family in the first known instance of rudimentary KPI usage.

In 1464, Luca Pacioli published "Summary of arithmetic, geometry, proportions and proportionality" ("Summa de arithmetica, geometrica, proportioni et proportionalita"), which described how Venetian sailors evaluated their sailing expeditions. The sailors would compare the amount of money they used to buy the goods they purchased to the money they received from the sale.

The industrial efficiency movement sparked early concepts of KPIs in mid-late 19th century. Performance appraisals in industry were most likely initiated by Robert Owen in the early 1800s. Owen monitored performance

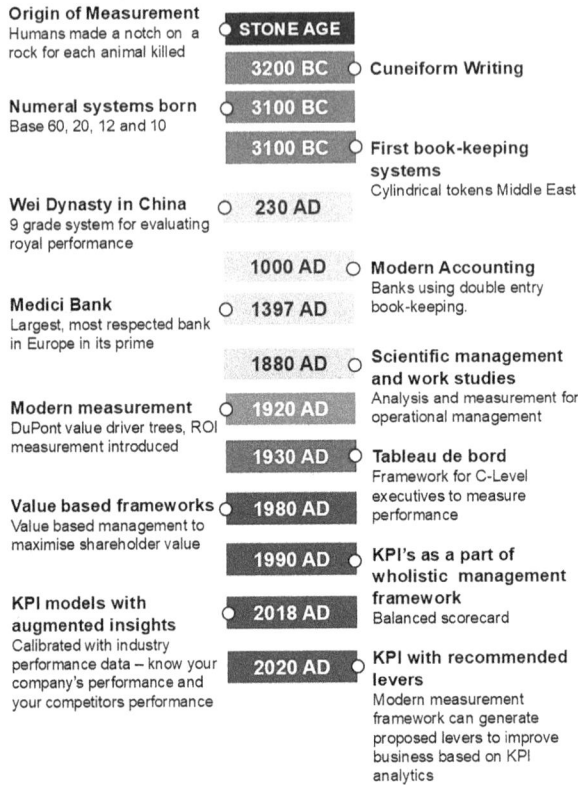

Origin of Measurement Humans made a notch on a rock for each animal killed	○ **STONE AGE**	
	3200 BC ○	Cuneiform Writing
Numeral systems born Base 60, 20, 12 and 10	○ **3100 BC**	
	3100 BC ○	First book-keeping systems Cylindrical tokens Middle East
Wei Dynasty in China 9 grade system for evaluating royal performance	○ **230 AD**	
	1000 AD ○	Modern Accounting Banks using double entry
Medici Bank Largest, most respected bank in Europe in its prime	○ **1397 AD**	book-keeping.
	1880 AD ○	Scientific management and work studies
Modern measurement DuPont value driver trees, ROI measurement introduced	○ **1920 AD**	Analysis and measurement for operational management
	1930 AD ○	Tableau de bord Framework for C-Level
Value based frameworks Value based management to maximise shareholder value	○ **1980 AD**	executives to measure performance
	1990 AD ○	KPI's as a part of wholistic management
KPI models with **augmented insights** Calibrated with industry	**2018 AD** ○	framework Balanced scorecard
performance data – know your company's performance and your competitors performance	**2020 AD** ○	KPI with recommended levers Modern measurement framework can generate proposed levers to improve business based on KPI analytics

Figure 17.1 History of KPIs.

Source: https://corporater.com/resources/history-of-kpis/.

at his cotton mills in Scotland through the use of "silent monitors." The monitors were cubes of wood with different colours painted on each visible side. They were displayed above the workstation of each employee.

In 1911, Frederick W. Taylor introduced data-driven optimisation of workers' productivity. Importance of Taylor's contribution is difficult to underestimate and as Peter Drucker written in his "The Rise of Knowledge Society" (1993): "Marx would be taken out and replaced by Taylor if there were any justice."

In the early 20th century, companies began formally measuring the performance of employees. This evolved into the concept of return on investment (ROI). Eventually, France created the tableau de bord allowing employers to track performance within their businesses.

Earlier we have mentioned Peter Ferdinand Drucker, who famously said, "What gets measured gets done." His other famous quote is: "If you can't measure it, you can't improve it." Peter Ferdinand Drucker was an Austrian American management consultant, educator and writer, whose innovative ideas on management practices, organisational dynamics and strategic planning have profoundly influenced contemporary business leadership and management techniques and significantly contributed to the foundations of modern management theory. His book "The Practice of Management" was published in 1954 and presented a full-scale organisational operational model that involves all employees in the goal-setting process. As Peter Drucker said: "A strategy without metrics is just a wish. And metrics that are not aligned with strategic objectives are a waste of time." Key results should be measurable, either on a 0–100% scale or with any numerical value (e.g., count, dollar amount or percentage) that can be used by planners and decision-makers to determine whether those involved in working towards the key result have been successful. There should be no opportunity for "grey area" when defining a key result.

Objectives and Key Results (OKR) is a goal-setting framework used by individuals, teams and organisations to define measurable goals and track their outcomes. Development of OKR is generally attributed to Andrew Grove (CEO of Intel at the time), who took Peter Drucker's ideas and introduced the approach to Intel in the 1970s and documented the framework in his 1983 book "High Output Management." OKRs comprise an objective (a significant, concrete, clearly defined goal) and 3–5 key results (measurable success criteria used to track the achievement of that goal). Not only should objectives be significant, concrete and clearly defined, they should also be inspirational for the individual, team or organisation that is working towards them. Objectives can also be supported by initiatives, which are the plans and activities that help to move forward the key results and achieve the objective. In 1975, John Doerr, at the time a salesperson working for Intel, attended a course within Intel taught by Andrew Grove where he was introduced to the theory of OKRs, then called "iMBOs" ("Intel Management by Objectives"). In 1999, John Doerr expanded OKRs' reach beyond Intel, introducing them to Google.

In the 1990s, individual performance management was reshaped by two key trends. The first was the increase in popularity of self-assessment of performance, sometimes followed by feedback sessions with line managers. This increase in performance self-assessment was natural as economies were dominated by knowledge workers, more independent in regard to decision-making and management of work processes. The second key trend was the integration between strategic performance management and individual performance management. Organisational goals became reflected in individual goals and individual measures became aligned with organisational performance measure, in an effort to increase the accountability of all employees to the execution of the organisational strategy.

Up until the early 1990s business performance management was almost solely focused on financial performance and based on financial data. KPIs saw the next big change in their usage in the 1990s when the first true balanced scorecard was used. By this time organisations started to use systems consisting of a mix of financial and non-financial measures to track. One such system, the Analog Devices Balanced Scorecard was created by Art Schneiderman in 1987 at Analog Devices, a mid-sized semi-conductor company. Schneiderman's design was similar to what is now recognised as a "First Generation" balanced scorecard design. In 1990, Schneiderman participated in an unrelated research study led by Robert S. Kaplan in conjunction with US management consultancy Nolan-Norton and during this study described his work on performance measurement. While the "corporate scorecard" terminology was coined by Schneiderman, the roots of performance management as an activity run deep in management literature and practice. Management historians such as Alfred Chandler suggest the origins of performance management can be seen in the emergence of the complex organisation – most notably during the 19th Century in the United States. Other influences may include the pioneering work of General Electric on performance measurement reporting in the 1950s and the work of French process engineers (who created the "tableau de bord" – literally, a "dashboard" of performance measures) in the early part of the 20th century.

Subsequently, Robert S. Kaplan and David P. Norton included anonymous details of this balanced scorecard design in their 1992 article. Although Kaplan's and Norton's article was not the only paper on the topic published in early 1992, it was a popular success, and was quickly followed by a second article in 1993. In 1996, the two authors published their book "The Balanced Scorecard: Translating Strategy into Action." These articles and the first book spread knowledge of the concept of balanced scorecards, leading to Kaplan and Norton being seen as the creators of the concept.

Their first book remains their most popular. As the title of their second book ("The Strategy-Focused Organisation: How Balanced Scorecard Companies Thrive in the New Business Environment," 2000) highlights, by 2000 the focus of attention among thought leaders was moving from the design of balanced scorecards themselves towards the use of the balanced scorecard as a focal point within a more comprehensive strategic management system. Their subsequent writing on the balanced scorecard has focused on its uses, rather than its design ("The Execution Premium," 2008; "Intelligent Design of Inclusive Growth Strategies," 2019).

The balanced scorecard model is an attempt to help organisations measure business performance using both financial and non-financial data. The aim of the balanced scorecard was "to align business activities to the vision and strategy of the business, improve internal and external communications, and monitor business performance against strategic goals." The balanced scorecard provides a relevant range of financial and non-financial information that supports effective business management. It expanded the concept of KPIs beyond financial measures to include a balanced set of indicators across various perspectives, such as customer, internal processes and learning and growth. This period also saw businesses beginning to align individual employee performance objectives with organisational initiatives and goals.

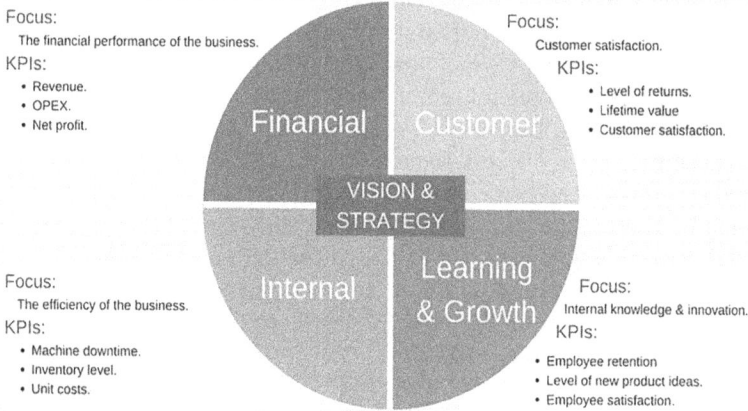

Figure 17.2 The balanced scorecard.

Source: https://www.youtube.com/watch?app=desktop&v=C0JZdyb6hZE.

The first generation of balanced scorecard designs used a "four perspective" approach to identify what measures to use to track the implementation of strategy (Figure 17.2).

The mid-1990s saw further development of this approach continued and a new method has emerged. In the new method, measures are selected based on a set of "strategic objectives" plotted on a "strategic linkage model" or "strategy map." With this modified approach, strategic objectives are distributed across the four measurement perspectives, so as to "connect the dots" to form a visual presentation of strategy and measures. In this modified version of balanced scorecard design, managers select a few strategic objectives within each of the perspectives and then define cause-effect chain among these objectives by drawing links between them to create a "strategic linkage model." A balanced scorecard of strategic performance measures is then derived directly by selecting one or two measures for each strategic objective. This type of approach provides greater contextual justification for the measures chosen and is generally easier to work through. This style of balanced scorecard has been commonly used since 1996 or so. It is significantly different in approach to the methods originally proposed and so can be thought of as representing the "2nd generation" of design approach adopted for the balanced scorecard since its introduction.

In the late 1990s, design approach to balanced scorecards had evolved yet again. One problem with the "2nd generation" design approach described in the previous paragraph was that plotting of causal links amongst twenty or so medium-term strategic goals was still a relatively

abstract activity. In practice, it ignored the fact that opportunities to intervene to influence strategic goals are (and need to be) anchored in current
and real management activity. Secondly, the need to "roll forward" and test
the impact of these goals necessitated reference to an additional design
instrument: a statement of what "strategic success," or the "strategic end-
state," looked like. This reference point was called a Destination Statement.
It was quickly realised that if a Destination Statement was created at the
beginning of the design process, then it became easier to select the appropriate strategic activity and outcome objectives which if achieved would deliver
it. Measures and targets could then be selected to track the achievement of
these objectives. Design methods that incorporate a Destination Statement
or equivalent represent a tangibly different design approach to those that
went before and so have been proposed as representing a "3rd generation"
design method for balanced scorecards.

The KPI Institute was established in 2004, in Melbourne, Australia. At
that time called eab group, it was designed as a provider of organisational
performance management services in Australia, supporting clients mainly
through training and advisory services. Today KPI Institute is considered the
global authority on KPIs research and education.

As Robert S. Kaplan and David P. Norton said in their 1992 article:

> What you measure is what you get. Senior executives understand that
> their organisation's measurement system strongly affects the behaviour
> of managers and employees. Executives also understand that traditional
> financial accounting measures like return-on-investment and earnings-
> per-share can give misleading signals for continuous improvement
> and innovation - activities today's competitive environment demands.
> The traditional financial performance measures worked well for the
> industrial era, but they are out of step with the skills and competencies
> companies are trying to master today.

In parallel with these developments, throughout the 2000s OKRs gained
traction within the tech industry, with major companies like LinkedIn,
Twitter and Uber implementing the framework. The methodology became
widely recognised by 2012, further popularised by John Doerr's "Measure
What Matters" (2016) and Rick Klau's YouTube workshop (2014). From
2018, educational and not-for-profit sectors began adopting OKRs. Notably,
OKRs proved to be a critical tool in the shift towards remote work during
the COVID-19 pandemic, demonstrating their adaptability and relevance in
the face of unprecedented challenges.

As much as it is of paramount importance to measure both achievement
and continuous improvement, the challenge is that measurement became a
religion in the business world.

There is a frequently used quote attributed to the well-known American
business theorist, composer, economist, industrial engineer, management

consultant, statistician and writer Dr William Edwards Deming: "If you can't measure it, you can't manage it." However, this is just part of the quote from his book "The New Economics" (1993) and this is not just a minor subversion of the actual quote, it is almost a total reversal of what Dr Deming actually said which is: "It is wrong to suppose that if you can't measure it, you can't manage it - a costly myth." But somehow the fallacy continues to thrive in the marbled corridors of corporate world and open offices of Silicon Valley. This misquote has won its place in the big book of business dogma because this is what the bureaucratic edifice of the business world relies on and is all about measurement. Measurement is now a religion in the business world! If we can slap a metric on something, we're going to do it.

So why is it that such a huge perversion of this statement has been seized upon and embraced by modern organisations, managers and executives? Unfortunately, the answer is quite simple – it is easier to manage by numbers without diving deeply in the root causes. If you believe that everything can be boiled down to measurement in dollars, feet, ounces, seconds, points/hour or some other absolute measurable unit then life becomes much easier. One knows when their decisions were good because the unit measure improved. It makes so much sense and the world becomes simple and "manageable." One can send reports upstairs where they can be reviewed and people can perpetuate the illusion that things are understood and under control. This is so attractive, of course, because it disposes of all the difficult stuff. You want to look at business as a simple function with inputs and outputs, you have levers you can pull and see the results and tweak based on feedback, and yes there is a lot within business that fits that model and you should absolutely grab all the data you can and use it wisely in decision-making and course correction, but between all those highly measurable milestones is where the difficult stuff. But we love to measure things because it makes us feel as though we're really doing something. Look at my report card, Mom! I got three As and two Bs. Am I a good kid? Am I smart?

Measurement has become a plague (or better to say HIV?) in the business world because we believe that by measuring everything and sending the good news upstairs to the C-suite we can convince our boss that we are doing the right thing and doing it right. Measuring and reporting is actually an inherently fear-based process because the reason we measure everything in business is to prove to someone who's not in the room that we did what they told us to do. The more numeric, visible and reward-tied a metric is, the more likely it is to be gamed and turn toxic to its original purpose. At the end of the day, we all want to be in "good books" or get the bonus or promotion...

When we are talking about measures, we should remember about Goodhart's law, that is named after British economist Charles Goodhart, who is credited with expressing the core idea of the adage in a 1975 article on monetary policy in the UK: "Any observed statistical regularity will tend to collapse once pressure is placed upon it for control purposes." More widely known formulation of this law reads like this: "When a measure becomes a target, it ceases to be a good measure." In other words, when we set one specific goal, people will tend to optimise for that objective regardless of the consequences. This leads to problems when we neglect other equally important aspects of a situation.

One real life example of Goodhart's law in action happened at (possibly more than one) call centre. Call centre manager thought that increasing the number of calls processed was a good objective, and his employees were dutifully focused on increasing their numbers. However, by choosing only one metric to measure success, this call centre manager motivated employees to sacrifice courtesy in the name of quantity. People respond to incentives, and people's natural inclination is to maximise the standards by which they are judged.

In the world of cybersecurity, Goodhart's law appears through an over-emphasis on certain metrics – for example, such as the number of daily security alerts resolved. This shifts team's focus: they now focus on lowering the alert numbers and not on the larger goal of understanding underlying security threats. This is similar to a doctor treating symptoms without diagnosing the disease first.

Construction of right KPIs is not as simple as it may look like, it is a specific skill. It is critical because it's about understanding how policy changes have both desired effects and undesired/unexpected effects. There are multiple examples when KPIs constructed without enough thought process (and, unfortunately, the vast majority of people do not or simply not able to think about unintended consequences) and leading to undesirable and not anticipated consequences. Here just some of the examples (more can be found in "The Tyranny of Metrics" by Jerry Z. Muller):

- A leader in India said too many people were dying from venomous snakes, so he offered money to anyone who brought him a dead one.
  - Unintended Negative Result: People started breeding venomous snakes in private, so they could kill them and bring them to the government.
- Surgeons are often judged by how often there are complications or deaths in their surgeries, which affects their marketability and insurance rates.
  - Unintended Negative Result: Many surgeons stopped taking high-risk or complicated cases, which resulted in people who really need help getting inferior care.
- Governments in the last couple of decades have focused on making sure more students can hit a minimum level of competency in subjects such as English and Math.
  - Unintended Negative Result: Many schools have taken this to an extreme, and basically spend all their classroom time teaching to sit the test, which results in no freedom, enthusiasm and ultimately a loss of curiosity and creativity in the students.

- Some organisations use maintaining conformity (with ISO/IEC27001 or certain maturity level of Essential 8) as one of the KPIs
  - Actual focus shifts from proactive management of cybersecurity to managing the evidence of conformity.

Some other examples of poorly constructed metrics and unintended consequences include:

- A number of governments with air pollution problems have started alternating which cars can be on the roads each day by even and odd license plate numbers, which unfortunately led many to buy an additional vehicle so they could drive every day.
- Chinese peasants used to be paid for finding dinosaur bones, but this actually led to them breaking every bone they found into multiple pieces so they could be paid multiple times.
- Salespeople being rewarded based on number of leads, which often creates tons of poor, unqualified leads that take up quality time that should have been spent elsewhere.
- Salespeople being rewarded based purely on number of deals signed, which often creates tons of poor deals that become unprofitable (or not profitable enough) contracts.
- Hospitals getting penalised for readmissions would treat returning people as outpatient instead of inpatient.
- Police departments labelling worse crimes as misdemeanours to show decrease in number of serious crimes.
- Glass plant workers were told to produce as many square feet of sheet glass as possible, and soon started making it so thin that it wasn't usable for anything.
- Wells Fargo massively incentivised the metric of "new accounts," which caused them to set up thousands of fake accounts, ultimately resulting in major lawsuits and financial impact.

When we talk about KPIs, it is important to remember about specifics of human nature. As mentioned earlier, we all want to be in "good books" or get the bonus or promotion. This human trait often results in focus on defining KPIs using the lenses of their future "achievability," especially around the language used for their definition, to ensure certain flexibility that may allow to "tick off" their (KPIs) achievement in the future. Another aspect that needs to be remembered is that often achievement of various KPIs is assessed by the people who are reporting on them and is done subjectively to present these people in the best possible light.

Balanced scorecard approach is not free of challenges.

Firstly, as one can see, balanced scorecards do not have a sector for cyber security. Though, it is not difficult to add another sector, it is not easy to add adequate measures to monitor success and improvements in this area. Earlier we have discussed that achieving certain level of maturity for Essential 8 (see Chapter 15) helps, but does not warrant strong cybersecurity posture. The same is true for ISMS (see Chapter 14). So, getting KPIs or OKRs for cybersecurity area is a significant challenge, especially in the era of digitisation (see Chapters 4, 5, 7 and 8).

Secondly, balanced scorecard approach does not have any temporal aspects and thus does not offer any tools to balance long-term perspective with the current financial year's objectives. This brings us again to the human nature and biases – both conscious and, more importantly, unconscious. The author of this chapter is yet to see a CEO that will sacrifice the current financial year bonus/STI in favour of not meeting financial targets (written in their compensation criteria and plan), but ensuring long-term success and sustainability (including cybersecurity) of the organisation.

Another challenge posed by the balanced scorecards approach is that, as Patrick Lencioni said ("The Five Dysfunctions of a Team"): "If everything is important, then nothing is." One can possibly argue that the roots of this quote go back to a proclamation by the pigs who control the government in the George Orwell's novel "Animal Farm": "All animals are equal, but some animals are more equal than others." This is where we should think about complexity (see Chapter 4) and human ability to prioritise in multidimensional (in this case 4D) space, as in real life often we need to select one of mutually exclusive options. Financial performance was (and still is) the main part of the balanced scorecard (whether it is linked with customer satisfaction or not) and with latest significant shift of focus on Diversity, Equity and Inclusion (DEI), there is little hope that despite multiple declarations cybersecurity will feature strongly enough on CEOs and Boards agendas.

Chapter 18

Emerging threats – AI

One may say: "Why is AI categorised as an 'emerging' threat, when the term AI was born in 1956?." It is a fair question, but it is right to talk about AI as an emerging threat based on extremely rapid progress in this space over the last several years.

ANCIENT HISTORY

Well, in fact the idea of "artificial intelligence" goes back thousands of years, to ancient philosophers considering questions of life and death. In ancient times, inventors made things called "automatons" which were mechanical and moved independently of human intervention. The word "automaton" comes from ancient Greek, and means "acting of one's own will." One of the earliest records of an automaton comes from 400 BCE and refers to a mechanical pigeon created by a friend of the philosopher Plato. Many years later, one of the most famous automatons was created by Leonardo da Vinci around the year 1495, it's model with inner workings and is on display in Berlin (https://en.wikipedia.org/wiki/Leonardo%27s_robot).

In the early 1900s, there was a lot of media created that focused on the idea of artificial humans. So much so that scientists of all sorts started asking the question: is it possible to create an artificial brain? Some creators even made some versions of what we now call "robots" (and the word was coined in a Karel Čapek's fiction play "Rossum's Universal Robots" in 1921) though most of them were relatively simple. This was the first known use of the word "robot." In 1929, Japanese professor Makoto Nishimura built the first Japanese robot, named Gakutensoku, a giant pneumatic automaton that toured through Asia until it mysteriously disappeared. In 1949, American computer scientist Edmund Callis Berkley (who in 1947 founded the ACM – Association for Computer Machinery) published the book "Giant Brains, or Machines that Think" in which he compared the newer models of computers to human brains.

Between 1941 and 1949, Alan Turing was working on a game of chess to illustrate various possible methods to test machine's intellect. At the end of

DOI: 10.1201/9781032672601-18

1948, Alan Turing presented an imitation test for machine intelligence based on the game of chess.

At this stage, he got engaged in a dialogue with philosophers and mathematicians and scientists on the capabilities of digital computers (https:// philsci-archive.pitt.edu/19291/1/turing-test-controversy.pdf). In June 1949, computer pioneer, distinguished physicist and then University of Cambridge Professor Douglas Hartree published his book "Calculating instruments and machines," in which he described in detail new electronic computing machines that could do a lot and yet should be seen as nothing but calculation engines. Also, in the same month distinguished neurosurgeon and then University of Manchester Professor Geoffrey Jefferson had given his Lister Oration in London on 9 June 1949 along the same lines and pushed it further with strong demands to accept that "machine equals brain." Asked by the reporter for a reply, Turing rebutted to Jefferson sharply. This indirect exchange with Jefferson, however, would only make an actual impact on Turing's views from October to December 1949 after two editions of a seminar, "Mind and computing machine," in the Department of Philosophy of their university. These seminars were co-chaired by distinguished chemist and University of Manchester Professor of Social Studies Michael Polanyi, who also engaged in the mind-machine controversy with Alan Turing. These three conservative thinkers, then all endowed with fellowships of the Royal Society and university professorships more prestigious than Alan Turing's (who at the time was Reader at the University of Manchester Department of Mathematics), tried to establish boundaries to Alan Turing's views on machine intelligence.

In the preface to his "Calculating instruments and machines" Douglas Hartree cited the Manchester "Baby" computer which had recently been "put into operation" then he kept pushing his public criticism on the term "electronic brain," as he had been doing ever since early November of 1946. It was after Douglas Hartree's 1949 book that Alan Turing cited and discussed "Lady Lovelace's objection" or "Lady Lovelace's dictum." Douglas Hartree drew attention to Lady Lovelace's views: "Some of her comments sound remarkably modern. One is very appropriate to a discussion there was in England which arose from a tendency, even in the more responsible press, to use the term "electronic brain" for equipment such as electronic calculating machines, automatic pilots for aircraft, etc. I considered it necessary to protest against this usage [Hartree, D. R. The Times (London), Nov. 7, 1946.], as the term would suggest to the layman that equipment of this kind could "think for itself," whereas this is just what it cannot do; all the thinking has to be done beforehand by the designer and by the operator who provides the operating instructions for the particular problem; all the machine can do is to follow these instructions exactly, and this is true even though they involve the faculty of "judgment." I found afterwards that over 100 years ago Lady Lovelace had put the point firmly and concisely" (C, p. 44): "The Analytical Engine has no pretensions whatever to originate anything. It can do whatever we know how to order it to perform." Then Douglas Hartree further resumed it in a way that conceded a window for research on machine learning:

> This does not imply that it may not be possible to construct electronic equipment which will 'think for itself,' or in which, in biological terms, one could set up a conditioned reflex, which would serve as a basis for 'learning.' Whether this is possible in principle or not is a stimulating and exciting question suggested by some of these recent developments. But it did not seem that the machines constructed or projected at the time had this property.

This passage would be quoted and discussed by Alan Turing at length later in 1950. Alan Turing was adamant to pursue machine learning beyond "reflexes" and "the action of the lower centres" of the brain at least since his c. November 1946 (his letter to Ross Ashby).

Michael Polanyi (1913–1976), born Hungarian, left Nazi Germany to England to become a Fellow of Royal Society in 1944. In 1948, associated with the Department of Philosophy and with some support from Professor of Philosophy Dorothy Emmet, he was granted a chair of Social Studies at the University of Manchester. Dorothy Emmet and Michael Polanyi were interested in the postwar public discussion about science and society, and paid attention to the debate around the new computing machines or "electronic brains." So, they invited Alan Turing, Maxwell (Max) Newman (leader of the Enigma codebreakers at Bletchley Park during World War II, whose group developed the concept of Colossus), Geoffrey Jefferson and others to a seminar on "the mind and the computing machine" that was held on October 27, 1949 at the Philosophy Department. This was indeed a crucial event. We know from minutes that survived about Michael Polanyi's key interventions that challenged Alan Turing. The seminar had two sessions. The first session was led by Michael Polanyi, who read a text, entitled "Can the mind be represented by a machine? Notes for discussion on 27th October 1949," which he had prepared and circulated to Max Newman and Alan Turing several weeks before the meeting. Essentially, Michael Polanyi claimed that humans can solve problems that machines cannot. As it turned out, both Max Newman and Alan Turing were of the opinion that "the mind/machine problem" can be decided empirically and only empirically.

The essence of this debate was that Alan Turing believed that "thinking machines" would eventually outstrip all of the cognitive abilities of humans. The others thought otherwise, which saw them butting heads with him in the press about humanity's prospective relationship with AI. University of Cambridge mathematician Douglas Hartree argued that computers would always be calculation engines incapable of acting in creative or unexpected ways. To make his case, Hartree cited Ada Lovelace's view that computers can only do what they are programmed to do in his 1950 book "Calculating

Instruments and Machines": "The Analytical Engine has no pretensions whatever to originate anything. It can do whatever we know how to order it to perform." So, a machine must be capable of performing tasks that it has not been specifically programmed to. Alan Turing agreed, which is why he chose to connect his test with a "child–machine" or what he called the "unorganised machine" that could learn from experience. Probably Turing's most well-respected critic was neurologist and neurosurgeon Geoffrey Jefferson, who set stringent criteria for machine intelligence that emphasised creativity. As the Times reported in 1949, he commented that

> Not until a machine can write a sonnet or compose a concerto because of thoughts and emotions felt, and not by the chance fall of symbols, could we agree that machine equals brain – that is, not only write it but know that it had written it.

The final element of the debate that Alan Turing responded to was from Hungarian–British polymath Michael Polanyi, who argued that human intelligence involves tacit knowledge that cannot be fully formalised or replicated by machines. He was unimpressed by Turing's one-time use of chess as a marker of machine intelligence and proposed that chess could be performed automatically because its rules can be neatly specified. The idea led Turing to reconsider using chess as the primary task for demonstrating machine intelligence, which was instead replaced by conversation to better capture the breadth of human cognitive ability. From June to December 1949 this debate prompted Alan Turing's thinking and must have led to his famous 1950 paper.

In 1950, the Mind magazine published Alan Turing's paper "Computer Machinery and Intelligence" which proposed a test of machine intelligence called "The Imitation Game" that today known as Turing test. Alan Turing's dialogue with Douglas Hartree which addressed possibility of learning machines continued in 1950–1951.

According to Turing, the question whether machines can think is itself "too meaningless" to deserve discussion. Turing suggested that there are three participants in this test. Test is conducted in an interrogation room run by a judge. The test subjects are a person and a computer program and they are hidden from the judge. The judge has a conversation with both parties and attempts to identify which is the human and which is the computer based on the quality of their conversation. Turing concludes that if the judge can't tell the difference, the computer has succeeded in demonstrating human intelligence. The initial experiment, as Turing envisioned it, was based on a game involving a man, a woman and a judge:

> It is played with three people, a man (A), a woman (B), and an interrogator (C) who may be of either sex. The interrogator stays in a room apart from the other two. The object of the game for the interrogator

is to determine which of the other two is the man and which is the woman. He knows them by labels X and Y, and at the end of the game he says either 'X is A and Y is B' or 'X is B and Y is A.'

Turing's experiment swaps out one of the participants in the game for a machine. Instead of determining whether participant A or B is a man or a woman, the revised version sees the judge pick whether or not the writer is a person or a machine. If Turing test judges intelligence, then the first imitation game assesses the ability to convincingly pass as the opposite gender.

CLASSIC AI HISTORY

The term artificial intelligence or AI was coined in 1956 by John McCarthy (September 4, 1927 to October 24, 2011). His father, John Patrick McCarthy, was an Irish Catholic who became a labour organiser and later the Business Manager of the Daily Worker, a national newspaper owned by the US Communist Party. His mother, Ida Glatt, was a Lithuanian Jewish immigrant who worked for a wire service, then for the Daily Worker and finally as a social worker. Both parents were active members of the Communist Party during the 1930s and they encouraged learning and critical thinking. John McCarthy once told one interviewer that he briefly joined the local Communist Party cell in 1949. It had two other members, a cleaning woman and a gardener. He quit the party soon afterward. John McCarthy declared himself an atheist in a speech about AI at Stanford Memorial Church. Raised as a Communist, he became a conservative Republican after his visit to Czechoslovakia in 1968 after the Soviet invasion.

John McCarthy was an American computer scientist and cognitive scientist and was one of the founders of the discipline of AI. He co-authored the document that coined the term AI, developed the programming language Lisp (that became the language of choice for AI research), significantly influenced design of the language ALGOL, popularised time-sharing and invented garbage collection. Around 1959, he invented so-called "garbage collection" methods, a kind of automatic memory management. McCarthy was also the first to propose time-sharing model of computing.

Interestingly enough, in 1961 John McCarthy was perhaps the first to suggest publicly the idea of utility computing, in a speech given to celebrate MIT's centennial: that computer time-sharing technology might result in a future in which computing power and even specific applications could be sold through the utility business model (like water or electricity).

Officially, the field of AI research was founded at a workshop held on the campus of Dartmouth College during the summer of 1956. Attendees of the workshop became the leaders of AI research for decades. At the time, many of them predicted that machines as intelligent as humans would exist within a generation. The US government provided millions of dollars to make this vision come true. Eventually, it became obvious that researchers had grossly underestimated the difficulty of the project.

The earliest research into thinking machines was inspired by a confluence of ideas that became prevalent in the late 1930s, 1940s and early 1950s. Recent research in neurology had shown that the brain was an electrical network of neurons that fired in all-or-nothing pulses. Norbert Wiener's cybernetics described control and stability in electrical networks. Claude Shannon's information theory described digital signals (i.e., all-or-nothing signals). Alan Turing's theory of computation showed that any form of computation could be described digitally. The close relationship between these ideas suggested that it might be possible to construct an "electronic brain." Alan Turing was among the first people to seriously investigate the theoretical possibility of "machine intelligence." In 1943, Walter Pitts and Warren McCulloch analysed networks of idealised artificial neurons and showed how they might perform simple logical functions. They were the first to describe what later researchers would call a neural network.

In 1951, Christopher Strachey using the Ferranti Mark 1 computer of the University of Manchester, wrote a checkers program and Dietrich Prinz wrote one for chess. Arthur Samuel's checkers program, the subject of his 1959 paper "Some Studies in Machine Learning Using the Game of Checkers" (that effectively introduced the term "machine learning"), eventually achieved sufficient skill to challenge a respectable amateur. Samuelson's program was among the first uses of what would later be called machine learning. This invention showed how computers might pick up new skills and adjust to new situations. Playing competitive checkers against human opponents was the main goal of the program. In contrast to conventional methods that depended on rule-based frameworks, Arthur Samuel employed a groundbreaking strategy called machine learning. The software was created to enhance its performance through self-learning gradually or, in other words, through trial and error, much the same way human mind learns.

Genetics researchers have fruit flies. Oncologists have white mice. For pioneering computer scientists studying AI it was games: rules-based systems that had defined criteria for success and failure, which demanded both nuance and complex decision-making. Game AI would continue to be used as a measure of progress in AI throughout its history.

The Dartmouth workshop of 1956 was organised by Marvin Minsky and John McCarthy, with support of two senior scientists Claude Shannon and Nathan Rochester of IBM. It became a pivotal event that marked the formal inception of AI as an academic discipline. The proposal for the conference stated they intended to test the assertion that "every aspect of learning or any other feature of intelligence can be so precisely described that a machine can be made to simulate it." The term artificial intelligence was introduced by John McCarthy at the workshop.

Between 1959 and 1962 a group of MIT students, advised by John McCarthy, developed a chess-playing program. It was based on earlier programs for the IBM 704 written by John McCarthy. He continued development of the chess program he had worked on at MIT. John McCarthy at Stanford University in 1966 can be seen at: https://www.researchgate.net/figure/Professor-John-McCarthy-shows-off-computer-chess-in-1966-at-Stanford-University-Source_fig1_354343066.

In 1965, he challenged a group at the Moscow Institute for Theoretical and Experimental Physics to a match with their own program. He has visited Moscow several times and in 1967, a four-game match played over 9 months was won 3–1 by the Soviet program. John McCarthy made several visits to the Soviet Union, learned to speak Russian and developed friendships with several computer scientists there. In 1968, he taught for 2 months in Akademgorodok, on Novosibirsk's outskirts, and in Novosibirsk itself. In 1975, he was instrumental in getting cybernetics researcher and refusenik Alexander Lerner permission from Soviet officials to attend and talk at the 4th International Joint Conference on Artificial Intelligence (IJCAI) in Tbilisi, Georgia. In the 1980s, he smuggled a fax and copier machine to linguist and Soviet dissident Larisa Bogoraz.

Interesting fact is that the author of this chapter, when he was a primary school student, saw John McCarthy. The author of this chapter used to spend (with his grandmother) part of his school holidays in "Sukhanovo" – all inclusive (except for alcohol) retreat for architects 32 km out of Moscow. One of his stays there coincided with a symposium on computer chess programs that was held in "Sukhanovo" (Figure 18.1). The author of this chapter remembers vividly sitting in the retreat's movie theatre and listening to the presentations. Not sure what he actually understood, but according to his family – he was very interested and impressed and remembered unusual for the USSR last name up until now. Memory did not retain the exact year this symposium was held, probably it was 1966 or 1967...

John McCarthy was a recipient of many honours and awards, including being a Member of the National Academy of Engineering (1987) and National Academy of Sciences (1989), A.M. Turing Award of the Association for Computing Machinery (1971), Research Excellence Award of the International Conference on Artificial Intelligence (1985), Kyoto Prize (1988); National Medal of Science (1990), Computer History Museum Fellow (1999), Benjamin Franklin Medal in Computer and Cognitive Science (2003). He has also received many other honours and prizes from international associations and universities as well as from the United States government.

Many early AI programs used the same basic algorithm. To achieve some goal (like winning a game or proving a theorem), they proceeded step by step towards it (by making a move or a deduction) as if searching through a maze, backtracking whenever they reached a dead end. The principal difficulty was that, for many problems, the number of possible paths through the "maze" was astronomical (a situation known as a "combinatorial explosion"). Researchers would reduce the search space by using heuristics that

Figure 18.1 Sukhanovo, watercolour by author's mother.

Source: author's archive.

would eliminate paths that were unlikely to lead to a solution. So, effectively early AI programs were a combination of a "brute-force" approach (often limited to 3–4 steps ahead), diluted where possible by use of known/developed by this time heuristics.

In 1960s, Lotfi A. Zadeh of the University of California at Berkeley introduced the concept of fuzzy logic as a means to mathematically represent uncertainty and vagueness in human reasoning. Since its inception, fuzzy logic has evolved significantly, finding applications in various fields, particularly in AI and control systems. The term "fuzzy logic" was coined by Lotfi Zadeh in 1965. Lotfi A. Zadeh aimed to create a mathematical framework to accommodate the concept of partial truth, wherein elements can belong to sets in varying degrees. Fuzzy logic is a generalisation from standard logic, in which all statements have a truth value of one or zero. In fuzzy logic, statements can have a value of partial truth, such as 0.9 or 0.5. Fuzzy logic is a heuristic approach that allows for more advanced decision-tree processing and better integration with rules-based programming (Figure 18.2). Over the years, the theory of fuzzy sets has been further developed, leading to the widespread adoption of fuzzy logic in engineering, control systems and AI.

Figure 18.2 Lotfi A. Zadeh.

Source: https://en.wikipedia.org/wiki/Lotfi_A._Zadeh.

One may think that fuzzy logic is quite recent concept and is what has worked for a short time, but its origins date back at least to the Greek philosophers and especially Plato (428–347 BC). It even seems plausible to trace their origins in China and India. Because it seems that they were the first to consider that all things need not be of a certain type or quit, but there is a stopover between. That is, be the pioneers in considering that there may be varying degrees of truth and falsehood. In case of colours, for example, between white and black there is a whole infinite scale – the shades of grey. When Aristotle and his predecessors devised their theories of logic and mathematics, they came up with the so-called Law of the Excluded Middle, which states that every proposition must either be true or false. Grass is either green or not green and it clearly cannot be both green and not green. But not everyone agreed, and Plato indicated there was a third region, beyond true and false, where these opposites "tumbled about." In the Aristotelian world view, logic dealt with two values. In the 19th century, George Boole created a system of algebra and set theory that could deal mathematically with such two-valued logic, mapping true and false to 1 and 0, respectively. Then in the early 20th century, Jan Lukasiewicz proposed a three-valued logic (true, possible, false), which never gained wide acceptance.

Fuzzy logic is often grouped together with machine learning that is discussed later in this chapter, but they are not the same thing. Fuzzy logic is a set of rules and functions that can operate on imprecise data sets, but the algorithms still need to be coded by humans. Machine learning refers to computational systems that mimic human cognition, by iteratively adapting algorithms to solve complex problems. Both areas have applications in AI and complex problem-solving.

For a long time in the 1970s and 1980s, it remained an open question whether any chess program would ever be able to defeat the expertise of top humans. In 1968, Donald Michie, founder of the Department of Machine Intelligence and Perception at the University of Edinburgh, invited David Levy, already a strong international chess player and computer scientist, to the AI workshop in Edinburgh. Levy played a friendly game of chess against John McCarthy, which David Levy won. John McCarthy remarked that David Levi was able to beat him, but predicted a computer program would beat David Levi within 10 years. David Levi then offered the famous bet, that within that time no chess program would beat him in a tournament match. John McCarthy took the bet after consulting Donald Michie. The two made a 500 Pound bet, which was later more than doubled when Donald Michie, Seymour Papert from MIT and Ed Kozdrowicki from the University of California, joined in the wager. Later David Levy said: "Until 1977, there seemed to be no point in my playing a formal challenge match against any chess program because none of them were good enough, but when CHESS 4.5 began doing well…it was time for me…to defend the human race against the coming invasion." He won his bet in 1978 at the Canadian National Exhibition in Toronto by beating Chess 4.7, the strongest computer program at the time, running on CDC Cyber 176 mainframe computer. He won a second 5 year bet in 1984, versus Cray Blitz, and then offered a price for the first computer chess team beating him. He finally got crashed 0–4 by Deep Thought in 1989.

In December 1974, David Levi and his wife visited Moscow and made the same bet with Dr. Vladimir Arlazarov (one of the authors of the soviet chess program Kaissa, that played against John McCarthy's chess program) and suggested a wager of twelve bottles of vodka (if David Levi wins) against twelve bottles of Scotch. David Levi playing chess can be seen at: https://www.chessprogramming.org/David_Levy.

In 1981, Cray Blitz in round 4 of the Mississippi State Championship became the first computer to gain a master rating and only the third computer to beat a chess master in tournament play.

At the 1982 North American Computer Chess Championship, Monroe Newborn predicted that a chess program could become world champion within 5 years, tournament director and international master chess player Michael Valvo predicted 10 years, Dan Spracklen predicted 15, Ken Thompson predicted more than 20 and others predicted that it would never happen. The most widely held opinion, however, stated that it would occur around the year 2000. In 1989, Levy was defeated by Deep Thought in an exhibition match. However, Deep Thought, was still considerably below World Championship level, as the reigning world champion, Garry Kasparov, demonstrated in two strong wins in 1989. It was not until a 1996 match with IBM's Deep Blue that Kasparov lost his first game to a computer at tournament time controls in Deep Blue versus Kasparov, 1996, game 1. This game was, in fact, the first time a reigning world champion had lost to a computer using regular time controls. However, Kasparov regrouped and won three and draw two of the remaining five games of the match, for a convincing victory. In May 1997, an updated version of Deep Blue defeated Kasparov 3½–2½ in a return match.

In the early 2000s, commercially available programs (such as Junior and Fritz) were able to draw matches against former world champion Garry Kasparov and world champion Vladimir Kramnik. In November–December 2006, World Champion Vladimir Kramnik played Deep Fritz. This time the computer won and the match ended 2–4. Chess engines continued to improve. In 2009, chess engines running on slower hardware have reached the grandmaster level and chess engine Hiarcs 13 running inside Pocket Fritz 4 on the mobile phone HTC Touch HD won the Copa Mercosur tournament in Buenos Aires, Argentina.

It would have been a remiss not to mention computer Go programs. Computer Go is an area in AI dedicated to creating computer programs that play the traditional board game Go. The field is sharply divided into two eras. Before 2015 the programs were weak. The best efforts of the 1980s and 1990s produced only programs that could be defeated by beginners, and programs of the early 2000s were intermediate level at best.

Professionals could defeat these programs even given handicaps of 10+ stones in favour of the program. DeepMind, a Google acquisition dedicated to AI research, produced AlphaGo in 2015 and announced it to the world in 2016. In 2017, AlphaGo defeated Ke Jie, who at the time continuously held the world No. 1 ranking for 2 years. Just as checkers had fallen to machines in 1995 and chess in 1997, computer programs finally conquered humanity's greatest Go champion in 2017. DeepMind did not release AlphaGo for public use, but various programs have been built since based on the journal articles DeepMind released describing AlphaGo and its variants.

Apart from playing chess, checkers and Go, an important goal of AI research was to find a way to allow computers to communicate in natural languages, like English. An early success was Daniel Bobrow's program STUDENT, which could solve high school algebra problems.

A semantic network, or frame network, is a knowledge base that represents semantic relations between concepts in a network. This is often used as a form of knowledge representation. Semantic networks are used in neurolinguistics and natural language processing applications. The concept of the semantic network model was formed in the early 1960s by researchers such as the cognitive scientist Allan M. Collins, linguist Ross Quillian and psychologist Elizabeth F. Loftus as a form to represent semantically structured knowledge. A semantic net represents concepts (e.g., "house," "door") as nodes, and relations among concepts as links between the nodes (e.g., "has a"). The first AI program to use a semantic net was written by Ross Quillian and the most successful (and controversial) version was Roger Schank's Conceptual dependency theory.

In 1966, Joseph Weizenbaum created a program called ELIZA. The program worked by examining comments typed by a user for keywords. If a keyword was found, a rule that transforms the user's comment was applied, and the resulting sentence was returned. If a keyword was not found, ELIZA responded either with a generic riposte or by repeating one of the earlier comments. Joseph Weizenbaum's ELIZA could carry out conversations that were so realistic that users occasionally were fooled into thinking they were communicating with a human being and not a computer program (ELIZA effect). But in fact, ELIZA simply gave a canned response or repeated back what was said to it, rephrasing its response with a few grammar rules. ELIZA was the first chatbot. ELIZA was the very first program that demonstrated computer program's ability to pass Turing test.

Turing test, originally called the imitation game by Alan Turing in 1950, is a test of a machine's ability to exhibit intelligent behaviour equivalent to, or indistinguishable from, that of a human. Alan Turing proposed that a human evaluator would judge natural language conversations between a human and a machine designed to generate human-like responses. The evaluator would be aware that one of the two partners in conversation was a machine, and all participants would be separated from one another. The conversation would be limited to a text-only channel, such as a computer keyboard and screen, so the result would not depend on the machine's ability to render words as speech. If the evaluator could not reliably tell the machine from the human, the machine would be said to have passed the test. The test results would not depend on the machine's ability to give correct answers to questions, only on how closely its answers resembled those a human would give. Turing Test later led to the development of chatbots, AI software entities developed for the sole purpose of conducting text chat sessions with people.

In 1972, Kenneth Colby created a program named PARRY, that was described as "ELIZA with attitude." It attempted to model behaviour of a paranoid schizophrenic, using a similar approach to that was earlier employed by Joseph Weizenbaum. In the early 1970s, PARRY was tested using a variation of Turing test. A group of experienced psychiatrists analysed a combination of real patients and computers running PARRY through teleprinters. Another group of 33 psychiatrists were shown transcripts of the conversations. The two groups were then asked to identify which of the "patients" were human and which were computer programs. The psychiatrists were able to make the correct identification only 52 percent of the time – a figure consistent with random guessing.

In 1970s, MIT's Artificial Intelligence Lab (MIT AI Lab) opened a "Machine Vision" course. Researchers began tackling "real world" objects and "low-level" vision tasks (i.e., edge detection and segmentation. In 1978, breakthrough was made at MIT AI Lab by David Marr, who created a bottom-up approach to scene understanding through computer vision. This approach starts with a 2D sketch which is built upon by the computer to get a final 3D image.

The first chatbot which appeared to pass Turing test was a chatbot called "Eugene Goostman," developed in 2001 in St. Petersburg, Russia, by a group of three programmers (the Russian-born Vladimir Veselov, Ukrainian-born Eugene Demchenko and Russian-born Sergey Ulasen). On July 7, 2014, "Eugene Goostman" passed Turing test in an event at the University of Reading marking the 60th death anniversary of Alan Turing, when 33% of the event judges thought that Goostman was human and event organiser Kevin Warwick considered it to have passed Turing test. It was portrayed as a 13-year-old boy from Odessa, Ukraine, who has a pet guinea pig and a father who is gynaecologist. The choice of age was intentional so that it induced people who "conversed" with him to forgive minor grammatical errors in his responses.

In the 1970s, AI was subject to critiques and financial setbacks and period between 1974 and 1980 is now called by some the "first AI winter," though historian Thomas Haigh argued in 2023 that there was no "winter" and AI researcher Nils Nilsson described this period as the most "exciting" time to work in AI. AI researchers had failed to appreciate the difficulty of the problems they faced. Their tremendous optimism had raised expectations impossibly high, and when the promised results failed to materialise, AI funding was severely reduced. The lack of success indicated the techniques being used by AI researchers at the time were insufficient to achieve their goals. Hans Moravec blamed the crisis on the unrealistic predictions of his colleagues. "Many researchers were caught up in a web of increasing exaggeration. But there were many reasons for this, including limited computer power, so-called combinatorial explosion (there are many problems that can only be solved in exponential time and finding optimal solutions to these problems requires extraordinary amounts of computer time, except when the problems are trivial), breadth of common sense knowledge" (many important AI applications – like vision or natural language require enormous amounts of information about the world: the program needs to have some idea of what it might be looking at or what it is talking about and this required that the program knows most of the same things about the world that a child does and this was a vast amount of information with billions of atomic facts and no one in 1970 could build a database large enough and no one knew how a program might learn so much information), Moravec's paradox (early AI research had been very successful at getting computers to do "intelligent" tasks like proving theorems, solving geometry problems and playing chess, however, it utterly failed to make progress on "unintelligent" tasks like recognising a face or crossing a room without bumping into anything), etc.

Funding cuts impacted some of major laboratories. The agencies which funded AI research, such as the British government, DARPA and the National Research Council (NRC) became frustrated with the lack of progress and eventually cut off almost all funding for undirected AI research. The pattern began in 1966 when the Automatic Language Processing Advisory Committee (ALPAC) report criticised machine translation efforts. After spending $20 million, the NRC ended all support. In 1973, the Lighthill report on the state of AI research in UK criticised the failure of AI to achieve its "grandiose objectives" and led to dismantling of AI research in UK. Lighthill report specifically mentioned combinatorial explosion problem as a reason for AI's failings. DARPA was deeply disappointed with researchers working on the Speech Understanding Research program at Carnegie Mellon University (CMU) and cancelled an annual grant of $3 million.

However, these setbacks did not affect the growth and progress in this field. General interest in the field continued to grow, the number of researchers increased dramatically, and new ideas were explored in logic programming, commonsense reasoning and many other areas. The major laboratories (MIT, Stanford, CMU and Edinburgh) that had been receiving generous support from their governments, and when it was withdrawn, these were the only places that were seriously impacted by the budget cuts. The thousands of researchers outside these institutions and the many more thousands that were joining the field were unaffected.

Originally Logic was introduced into AI research as early as 1958, by John McCarthy in his Advice Taker proposal. Then, in 1963, J. Alan Robinson had discovered a simple method to implement deduction on computers, the resolution and unification algorithm. However, straightforward implementations, like those attempted by John McCarthy and his students in the late 1960s, were especially intractable: the programs required astronomical numbers of steps to prove simple theorems. More fruitful approach to logic was developed in the 1970s by Robert Kowalski at the University of Edinburgh, and soon this led to collaboration with French researchers Alain Colmerauer and Philippe Roussel, who created very successful logic programming language Prolog. Prolog used a subset of logic that permitted tractable computation. Among the critics of John McCarthy's approach were his colleagues across the country and at MIT. Marvin Minsky, Seymour Papert and Roger Schank were trying to solve problems like "story understanding" and "object recognition" that required a machine to think like a person. In order to use ordinary concepts like "chair" or "restaurant," they had to make all the same illogical assumptions that people normally made. Unfortunately, imprecise concepts like these are hard to represent in logic. MIT chose instead to focus on writing programs that solved a given task without using high-level abstract definitions or general theories of cognition, and measured performance by iterative testing, rather than arguments from first principles. Ray Reiter admitted that "conventional logics, such as first-order logic, lack the expressive power to adequately represent the knowledge required for reasoning by default."

AI RENAISSANCE

Period between 1980 and 1987 is called a "boom," when a form of AI programs called "expert systems" was adopted by corporations around the world and knowledge became the focus of mainstream AI research. Governments provided substantial funding, such as Japan's fifth generation computer project (in 1981 Japanese Ministry of International Trade and Industry set aside $850 million for the fifth-generation computer project) and the US Strategic Computing Initiative. UK began the £350 million Alvey project. And AI industry boomed from a few million dollars in 1980 to billions of dollars in 1988. An expert system is a program that answers questions or solves problems about a specific domain of knowledge, using logical rules that are derived from the knowledge of experts. The earliest examples were developed by Edward Feigenbaum and his students as early as in 1965 and was called Dendral. It was used to identify compounds from spectrometer readings. In 1972, an expert system MYCIN was developed for diagnostics of infectious blood diseases. These early systems demonstrated feasibility of the approach.

The power of expert systems came from the expert knowledge they contained. They were part of a new direction in AI research that had been gaining ground throughout the 1970s. As Pamela McCorduck said: "AI researchers were beginning to suspect... that intelligence might very well be based on the ability to use large amounts of diverse knowledge in different ways." Knowledge-based systems and knowledge engineering became a major focus of AI research in the 1980s. In 1982, physicist John Hopfield was able to prove that a form of neural network (now called a "Hopfield net") could learn and process information, and provably converges after enough time under any fixed condition. It was a breakthrough, as it was previously thought that nonlinear networks would, in general, evolve chaotically. Around the same time, Geoffrey Hinton and David Rumelhart popularised a method for training neural networks called "backpropagation." These two developments helped to revive exploration of artificial neural networks.

Recurrent Neural Networks (RNNs) were the first to come in 1986 and they gained instant popularity. Unlike traditional feedforward neural networks, where the flow of information was in one direction, RNNs could remember previous inputs in their internal state or memory and answer questions based on context. They are trained to process and convert a sequential data input into a specific sequential data output and have a feedback loop, making them suitable for natural language processing (NLP) tasks. While RNNs were a significant step forward, they had limitations, especially with long sentences. In simple words, they are not good at retaining memory and suffer from long term memory loss. Then, in 1997 came Long Short-Term Memory (LSTM). LSTM was a specialised type of RNN. Their primary advantage was their ability to remember information over long sequences. Thus, it overcame the short-term memory limitations of RNNs. LSTM has a unique architecture: they have an input gate, a forget gate and an output gate. These gates determined how much information should be memorised, discarded or output at each step. This selective ability to memorise or forget helped LSTMs maintain relevant information in their memory, making them more efficient at capturing long-term dependencies from sentences. In 2014 came Gated Recurrent Units (GRU). They were designed to solve some of the same problems as LSTMs but with a simple and more streamlined structure. Just like LSTMs, GRUs were designed to combat the vanishing gradient problem, allowing them to retain long-term dependencies in sentences. GRUs simplified the gating by using only two gates: an update gate which determined how much of the previous information to keep versus how much of the new information to consider and a reset gate which determined how much of the previous information to forget. The reduced gating in GRUs made them more efficient in terms of computation.

In 1988, Judea Pearl's brought probability and decision theory into AI. Fuzzy logic, developed by Lotfi Zadeh in the 1960s, began to be more widely used in AI and robotics. Talking about fuzzy logic, in 1975 the author of this chapter as an undergrad student was involved in development of fuzzy logic-based software for autonomous moon rover... Fuzzy logic was used to attack automated vehicle control for autonomous vehicles as recently as in 2005–2007 (https://digital.csic.es/bitstream/10261/7861/1/using-fuzzy.pdf and https://digitalscholarship.unlv.edu/cgi/viewcontent.cgi?article=1035& context=me_fac_articles and https://ieeexplore.ieee.org/document/4078954). So-called soft computing used methods that work with incomplete and imprecise information. They do not attempt to give precise, logical answers, but give results that are only "probably" correct. This allowed to solve problems that precise symbolic methods could not handle. Although, the concept of fuzzy logic was introduced in US, both US and European scientist and researchers largely ignored it for years, perhaps because of its unconventional name. They refused to take seriously something that sounded so child-like. Some mathematicians argued that fuzzy logic was merely probability in disguise. But fuzzy logic was readily accepted in Japan, China and other Asian countries. The greatest number of fuzzy researchers today are found in China, with over 10,000 scientists. Japan, though considered at the leading edge of fuzzy studies, has fewer people engaged in fuzzy research. A decade ago, Chinese University of Hong Kong surveyed consumer products using fuzzy logic, producing a 100-plus-page report listing washing machines, camcorders, microwave ovens and dozens of other kinds of electrical and electronic products.

As early, as in 1950s, Alan Turing and Arthur Samuels foresaw the role of reinforcement learning in AI. Reinforcement learning gives an agent a reward every time it performs a desired action well, and may give negative rewards (or "punishments") when it performs poorly. It was described in the first half of the 20th century by psychologists using animal models, such as Edward Thorndike, Ivan Pavlov and Burrhus Frederic (B. F.) Skinner.

In the beginning of 1972, very important and successful research program was led by Richard Sutton and Andrew Barto. Their collaboration revolutionised the study of reinforcement learning and decision making over the four decades. In 1988, Sutton described machine learning in terms of decision theory (i.e., the Markov decision process). Also in 1988, Sutton and Barto developed the "temporal difference" (TD) learning algorithm, where the agent is rewarded only when its predictions about the future show improvement. Use of this approach resulted in that it significantly outperformed previous algorithms. TD-learning was used by Gerald Tesauro in 1992 in the program TD-Gammon, which played backgammon as well as the best human players. The program learned the game by playing against itself with zero prior knowledge. In an interesting case of interdisciplinary convergence, neurologists discovered in 1997 that the dopamine reward system in brains also uses a version of the TD-learning algorithm. TD learning would be become highly influential in the 21st century and it was used in both AlphaGo and AlphaZero.

Association for the Advancement of Artificial Intelligence (AAAI) was founded in 1979 and held its first conference at Stanford in 1980. Fascination with AI rose and fell in the 1980s in the classic pattern of economic bubble. In 1984, The AAAI warned of an incoming "AI Winter" where funding and interest would decrease, and will make research significantly more difficult.

As dozens of companies failed, the perception in the business world was that the technology was not viable. The damage to AI's reputation would last for the next 20–30 or so years. The term "AI winter" was coined by researchers who had survived the funding cuts of 1974 when they became concerned that enthusiasm for expert systems had spiraled out of control and that disappointment would certainly follow. Their fears were well founded: in late 1980s and early 1990s, AI suffered a series of financial set-backs. The first indication of a change in weather was the sudden collapse of the market for specialised AI hardware in 1987. Personal computers (both Apple and IBM) had been steadily gaining speed and power and in 1987 they became more powerful than the more expensive Lisp machines made by Symbolics and others. There was no longer a good reason to buy them. An entire industry worth half a billion dollars was decimated over-night. It is generally accepted that "AI winter" lasted from 1987 till 1993.

Then, new leadership at DARPA had decided that AI was not "the next wave" and directed funds towards projects that seemed more likely to pro-duce immediate results which resulted in the late 1980s the Strategic Computing Initiative cutting funding to AI deeply and brutally. By 1991, the impressive list of goals penned in 1981 for Japan's Fifth Generation Project had not been met. By the end of 1993 over 300 AI companies had shut down, gone bankrupt or been acquired, effectively ending the first commer-cial wave of AI. The field of AI received little or no credit in 1990s and early 2000s, despite the fact that many algorithms originally developed by AI researchers began to appear as parts of larger systems solving a lot of very difficult problems. Many of AI's greatest innovations have been reduced to the status of just another item in the tool chest of computer science and, in fact, many researchers working in the AI field in 1990s deliberately called their work by other names, such as informatics, knowledge-based systems, "cognitive systems" or computational intelligence.

During these years AI consistently delivered working solutions to specific isolated problems. For example, in 1986 Ernst Dickmann and his team at Bundeswehr University of Munich created and demonstrated the first driv-erless car (or robot car), that could drive up to 55 mph on roads that didn't have other obstacles or human drivers. By late 1990s, it was being used throughout technology industry, although somewhat behind the scenes. This success was mainly due to increasing computer power and collaboration with other fields (such as mathematical optimisation and statistics. By 2000, AI had achieved some of its oldest goals.

A new paradigm called "intelligent agents" became widely accepted dur-ing 1990s. An intelligent agent is a system that perceives its environment and takes actions which maximises its chances for success. By this definition, simple programs that solve specific problems are "intelligent agents," as are human beings and organisations of human beings, such as companies. The intelligent agent paradigm defines AI research as "the study of intelligent agents." This is a generalisation of some earlier definitions of AI: it goes beyond studying human intelligence – it studies all kinds of intelligence. Although earlier researchers had proposed modular "divide and conquer" approaches to AI, the intelligent agent concept did not reach its modern form until Judea Pearl, Allen Newell, Leslie P. Kaelbling and others brought concepts from decision theory and economics into the study of AI.

AI AND LLM REVOLUTION

During the first decades of the 21st century, access to large amounts of data (known as "big data"), cheaper and faster computers and advanced machine learning techniques were successfully applied to solving many problems. A turning point was the success of deep learning around 2012 which improved performance of machine learning on many tasks, including image and video processing, text analysis and speech recognition.

In 2012, Alex Krizhevsky developed AlexNet – a deep learning model. AlexNet won the ImageNet Large Scale Visual Recognition Challenge, with significantly fewer errors than the second place winner. Krizhevsky worked with Geoffrey Hinton at the University of Toronto. This was a turning point in machine learning: over the next few years dozens of other approaches to image recognition were abandoned in favour of deep learning. Deep learning uses a multi-layer perceptron. Although this architecture has been known since 1960s, getting it to work required powerful hardware and large amounts of training data. Before these became available, improving performance of image processing systems required hand-crafted ad hoc features that were difficult to implement. Deep learning was simpler and more general. Over the next few years, deep learning was applied to dozens of problems (such as speech recognition, machine translation, medical diagnosis and game playing). In every case, it showed enormous gains in performance. Investment and interest in AI boomed as a result.

In the meantime, in 2018 Alibaba's (China) language-processing AI beaten human intellect on a Stanford reading and comprehension test. In 2000s, people began talking about the future of AI again and several popular books considered possibility of superintelligent machines and what they might mean for human society. Some of this was optimistic (such as Ray Kurzweil's "The Singularity is Near"), but others, such as Nick Bostrom and Eliezer Yudkovski, warned that sufficiently powerful AI was an existential threat to humanity. New insights into superintelligence raised concerns that AI is an existential threat. The topic became widely covered in the press and many leading intellectuals and politicians commented on the issue. The risks and unintended consequences of AI technology became an area of serious academic research after 2016.

AI programs in the 21st century are defined by their goals – the specific measures that they are designed to optimise. Nick Bostrom's book "Superintelligence" (2014) argued that, if one isn't careful about defining these goals, the machine may cause harm to humanity in the process of achieving this goal. Stuart J. Russell used the example of an intelligent robot that kills its owner to prevent it from being unplugged, reasoning "you can't fetch the coffee if you're dead" (this problem is known by the technical term "instrumental convergence"). At the same time, machine learning systems began to have disturbing unintended consequences. Cathy O'Neil explained how statistical algorithms had been among the causes of the 2008 economic crash, Julia Angwin of ProPublica argued that COMPAS system used by the criminal justice system exhibited racial bias, others also showed that many machine learning systems exhibited some form of racial bias, and there were many other examples of dangerous outcomes that had resulted from machine learning systems.

In early 2000s, some researchers became concerned that mainstream AI was too focused on "measurable performance in specific applications" (known as "narrow AI"). An early critic was Nils Nilsson in 1995. In 2002, Ben Goertzel and some other researchers became concerned that AI had largely abandoned its original goal of producing versatile, fully intelligent machines and argued in favour of more direct research into artificial general intelligence (AGI). Similar opinions were published by the AI elder statesmen John McCarthy, Marvin Minsky and Patrick Winston in 2007–2009. In 2004, Marvin Minsky organised a symposium on "human-level AI". Ben Goertzel adopted the term "artificial general intelligence" for the new sub-field, founding a journal and holding conferences beginning in 2008. The new field grew rapidly, buoyed by the continuing success of artificial neural networks and the hope that it was the key to AGI.

Investment in AI increased along with its capabilities, and by 2016 market for AI-related products, hardware and software reached more than $8 billion. Several competing companies, laboratories and foundations were founded to develop AGI in the 2010s.

DeepMind was founded in 2010 by three English scientists, Demis Hassabis, Shane Legg and Mustafa Suleyman, with funding from Peter Thiel and later Elon Musk. The founders and financiers were seriously concerned about AI safety and existential risk of AI. DeepMind's founders had a personal connection with Eliezer Yudkowsky and Elon Musk was among those who was actively raising the alarm. Hassabis was both worried about the dangers of AGI and optimistic about its power; he hoped they could "solve AI, then solve everything else."

In 2011, IBM then created Watson, a Question Answering (QA) systems, computer system capable of answering questions posed in natural language.

It was a data analytics processor that used natural language processing, a technology that analyses human speech for meaning and syntax. IBM Watson performed analytics on vast repositories of data that it processed to answer human-posed questions, often in a fraction of a second. It was developed as a part of IBM's DeepQA project by a research team, led by principal investigator David Ferrucci and was named after IBM's founder and first CEO, Thomas J. Watson. In February 2013, IBM announced that Watson's first commercial application would be for utilisation of management decisions in lung cancer treatment, at Memorial Sloan Kettering Cancer Center, New York City, in conjunction with WellPoint (now Elevance Health). In November 2013, IBM announced it would make Watson's API available to software application providers, enabling them to build apps and services that are embedded in Watson's capabilities. To build out its base of partners who create applications on the Watson platform, IBM consulted with a network of venture capital firms, which advised IBM on which of their portfolio companies may be a logical fit for what IBM calls Watson Ecosystem. Roughly 800 organisations and individuals have signed up with IBM, with interest in creating applications that could use the Watson platform. On July 29, 2016, IBM and Manipal Hospitals (leading hospital chain in India) announced launch of IBM Watson for Oncology, for cancer patients. By 2022, IBM Watson Health was generating about a billion dollars in annual gross revenue, but was facing a lack of profitability and increased competition. This resulted in IBM announcing on 21 January 2022 sell-off of its Watson Health unit to Francisco Partners.

In 2012, Geoffrey Hinton (who has been leading neural network research since the 1980s) was approached by Chinese multinational technology company Baidu, that wanted to hire him and all his students for an enormous sum. Geoffrey Hinton decided to hold an auction and, at a Lake Tahoe AI conference, they sold themselves to Google for a price of $44 million. Demis Hassabis took notice and sold DeepMind to Google in 2014, on condition that it would not accept military contracts and would be overseen by an ethics board. Larry Page of Google, unlike Elon Musk and Demis Hassabis, was an optimist about the future of AI. Elon Musk and Larry Paige became embroiled in an argument about risks of AGI at Musk's 2015 birthday party. They had been friends for decades but stopped speaking to each other shortly afterwards. Elon Musk attended the one and only meeting of the DeepMind's ethics board, where it became clear that Google was uninterested in mitigating potential harm of AGI.

Frustrated by his lack of influence Elon Musk founded OpenAI in 2015 and enlisted Sam Altman to run it and hire top scientists. OpenAI began as a non-profit, "free from the economic incentives that were driving Google and other corporations." Elon Musk became frustrated again and left the company in 2018. OpenAI turned to Microsoft for continued financial support and Sam Altman and OpenAI formed a for-profit version of the company with more than $1 billion in financing. In 2021, Dario Amodei and 14 other scientists left OpenAI over concerns that the company was putting profits above safety. They formed Anthropic, which soon had $6 billion in financing from Microsoft and Google.

Current AI boom started in 2017 with initial development of key architectures and algorithms such as transformer architecture in 2017, leading to scaling and development of large language models (LLMs) exhibiting human-like traits of knowledge, attention and creativity. Transformer architecture was proposed by Ashish Vaswani and his colleagues at Google in 2017 in the paper "Attention is all you need." This new type of architecture relied on an attention mechanism to process sequence. As its core, it is composed of an encoder and decoder, each with multiple stacked layers of self-attention and feed-forward neural networks. A standout feature is the "multi-head" attention, allowing it to focus on different parts of the input sentence simultaneously, capturing various contextual nuances. Another strength was its ability to process sequences in parallel rather than sequentially. Transformer architecture is incredibly sticky. It's been around since 2017 and there is a big question mark about for how much longer this architecture will stay relevant and popular. Developing new architecture to outperform Transformer isn't easy. Transformer has been so heavily optimised since its introduction in 2017.

With success of transformers, the next logical step was scaling. This kick started with Google's Bidirectional Encoder Representations from Transformers (BERT) model which was released in 2018. Unlike previous models that processed text either left-to-right or right-to-left, BERT was designed to consider both directions simultaneously, hence the name: Bidirectional Encoder Representations from Transformer. Pre-trained on vast amounts of text, BERT was the first proper foundational language model that could be fine-tuned for specific tasks, setting new performance standards across various benchmarks. Open AI released its Generative Pre-trained Transformer (GPT) model GPT-2 and Google released its T5 model in 2019, thereafter GPT-3 came up in 2020, etc. These LLMs marked a paradigm shift in AI capabilities.

New AI era began around 2020–2023, with the public release of scaled LLMs such as ChatGPT. LLMs use semantic technology (semantics, the semantic web and natural language processors). The history of LLMs starts with the concept of semantics, developed by French philologist, Michel Bréal, in 1883. Bréal studied the ways languages are organised, how they change as time passes and how words connect within a language. LLMs refer to large, general-purpose language processing models that are first pre-trained on extensive datasets covering a wide range of topics to learn and master the fundamental structures and semantics of human language. The term "large" in this context denotes both the substantial amount of data required for training and the billions or even trillions of parameters that the model contains. Pre-training prepares the model to handle common language tasks such as text classification, question answering and document summarisation, demonstrating its versatility. After pre-training, these models are typically fine-tuned for specific applications, such as on smaller, specialised datasets targeted at particular domains like finance or medical, to enhance accuracy and efficiency in addressing specific issues. This approach of pre-training followed by fine-tuning enables LLMs not only to solve a broad range of general problems but also to adapt to specific application requirements.

In August 2021, Stanford Institute's for Human-Centered Artificial Intelligence (HAI) Centre for Research on Foundation Models (CRFM) coined the term "foundation model." According to their definition foundation model means "any model that is trained on broad data (generally using self-supervision at scale) that can be adapted (e.g., fine-tuned) to a wide range of downstream tasks." Choice of the term "foundation model" over "foundational model" was made because "foundational" implies that these models provide fundamental principles in a way that "foundation" does not. After considering many terms, they settled on "foundation model" to emphasise the intended function (i.e., amenability to subsequent further development) rather than modality, architecture or implementation. Foundation models are a class of AI models pre-trained on vast data across various domains, enabling them to develop wide range of capabilities. These models are not limited to language tasks but can include image recognition, sound processing and more. Trained on massive datasets, foundation models are large deep learning neural networks that have changed the way researchers approach machine learning. Rather than developing AI from scratch, researchers use a foundation model as a starting point to develop machine learning models that power new applications more quickly and more cost-effectively. Unique feature of foundation models is their adaptability. These models can perform wide range of disparate tasks with high degree of accuracy based on input prompts. Some tasks include natural language processing, question answering and image classification. Size and general-purpose nature of foundation models make them different from traditional machine learning models, which typically perform specific tasks, like analysing text for sentiment, classifying images and forecasting trends. Released in 2018, BERT was one of the first bidirectional foundation models.

LLMs are a subset of foundation models specifically designed for processing and generating human language. They are trained on vast text datasets and can perform translation, summarisation, question answering and more tasks. LLMs like GPT and BERT are examples of this technology.

Foundation models draw upon a series of advances in the history of AI. These models can be positioned against the backdrop of broader rise of machine learning since the 1990s. Prior AI models depended on specific instructions to solve a given task, but machine learning-powered models were able to decipher what task to solve being given sufficient data. Such a shift from so-called expert systems to data-driven machine learning was the first step towards the modern foundation model. Technologically foundation models are built using established machine learning techniques like deep neural networks, transfer learning and self-supervised learning. Foundation models are noteworthy given unprecedented resource investment, model and data size and ultimately their scope of application when compared to previous forms of AI. Rise of foundation models constitutes a new paradigm in AI, where general-purpose models function as a reusable infrastructure, instead of bespoke and one-off task-specific models. Foundation models began to materialise as the latest wave of deep learning models in the late 2010s. Particularly influential in the history of foundation models was 2022, when releases of Stable Diffusion and ChatGPT (initially powered by the GPT-3.5 model) led to foundation models and generative AI entering widespread public discourse. Further, releases of LLaMA, Llama 2 and Mistral in 2023 contributed to a greater emphasis placed on how foundation models are released with open foundation models garnering a lot of support and scrutiny. IBM released its Granite (a series of decoder-only AI foundation models) in November 2023.

In October 2024, Liquid AI (an MIT spin-off and foundation model company) unveiled its first products and showcased AI products for financial services, biotech and consumer electronics, built using Liquid AI's pioneering Liquid Foundation Models (LFMs), a new generation of generative AI models that achieve state-of-the-art performance at every scale while maintaining significantly smaller memory footprint both during training and inference beyond what was possible before. This particularly enables on-device and private enterprise use cases.

LLMs can be broken down into three types, each of which has its own advantages, depending on the goal:

- **Pre-training models:** are pre-trained on huge quantities of data, which helps them comprehend a broad range of language patterns and constructs. A plus is that a pre-trained models tend to be grammatically correct.
- **Fine-tuning models:** are pre-trained on a large dataset and afterward are fine-tuned on a smaller dataset for a specific task. They're particularly good for sentiment analysis, answering questions and classifying text.
- **Multimodal models:** combine text with other modes, such as images or video, to create more advanced language models. They can produce text descriptions of images and vice versa.

There are several underlying models of LLMs:

- **Recursive neural network models:** Recursive neural network models are designed to handle structured data like parse trees, which represent syntactic structure of a sentence. These models are useful for tasks like sentiment analysis and natural language inference.

- **Hierarchical models:** Hierarchical models are designed to handle text at different levels of granularity, such as sentences, paragraphs and documents. These models are used for tasks like document classification and topic modelling.
- **Autoencoder-based model:** Autoencoder-based model works by encoding input text into a lower-dimensional representation and then generating new text based on that representation. This type of model is especially good for tasks like summarising text or generating content.
- **Sequence-to-sequence model:** Sequence-to-sequence model, which takes an input sequence (like a sentence) and generates an output sequence (like a translation into another language). These models are often used for machine translation and text summarisation.
- **Transformer-based models:** Transformer-based models use neural network architecture that's great at understanding long-range dependencies in text data, making them useful for a wide range of language tasks, including generating text, translating languages and answering questions (Figure 18.3).

Some stats on GPT showing exponential growth of size and complexity of the models:

- **June 2018:** GPT-1 (117 million parameters)
- **February 2019:** GPT-2 (1.5 billion parameters)
- **June 2020:** GPT-3 (175 billion parameters)
- **March 2023:** GPT-4 (over 1 trillion parameters)

LLMs, based on the transformer architecture, were developed by AGI companies: OpenAI released GPT-3 in 2020 and DeepMind released Gato in 2022. These are foundation models: they are trained on vast quantities of unlabelled data and can be adapted to a wide range of tasks. These models can discuss huge number of topics and display general knowledge. The question naturally arises: are these models an example of artificial general intelligence? Bill Gates was sceptical of the new technology and the hype that surrounded AGI. However, Sam Altman presented him with a live demo of ChatGPT4 (that was released in March 2023) passing an advanced biology test. Bill Gates was convinced and in 2023, Microsoft Research tested the model with a large variety of tasks, and concluded that "it could reasonably be viewed as an early (yet still incomplete) version of an artificial general intelligence (AGI) system."

Today the most popular LLM architecture is the transformer architecture. A typical Transformer model consists of four main steps in processing input. Firstly, the model performs word embedding to convert words into high-dimensional vector representations. Then, the data is passed through multiple transformer layers. Within these layers, the self-attention mechanism plays a crucial role in understanding the relationships between words in a

1966 ○	ELIZA	
1966 ○	SHRDLU	
Late 1980s – 1990s ○	Statistical Language Models	
2000s ○	Neural Probabilistic Language Model	
	Word2Vec	○ 2013
	Transformer Models and Attention Mechanisms	○ 2017
	BERT	○ 2018
	GPT-2 and T5	○ 2019
2020 ○	GPT3	
Jan 2021 – Oct 2022 ○	LaMBDA, xlarge, Chinchilla, CodeGen, InCoder, mGPT, PaLM, OPT-IML, Minerva	
	ChatGPT	○ Nov 2022
	GPT 3.5	○ Dec 2022
	WebGPT	○ Jan 2023
	Google Bard & LLaMa	○ Feb 2023
Mar 2023 ○	GPT4	
Apr 2023 ○	Bloomberg GPT Stable LM, Dolly 2.0, Titan, BingChat	
May 2023 ○	PaLM2	

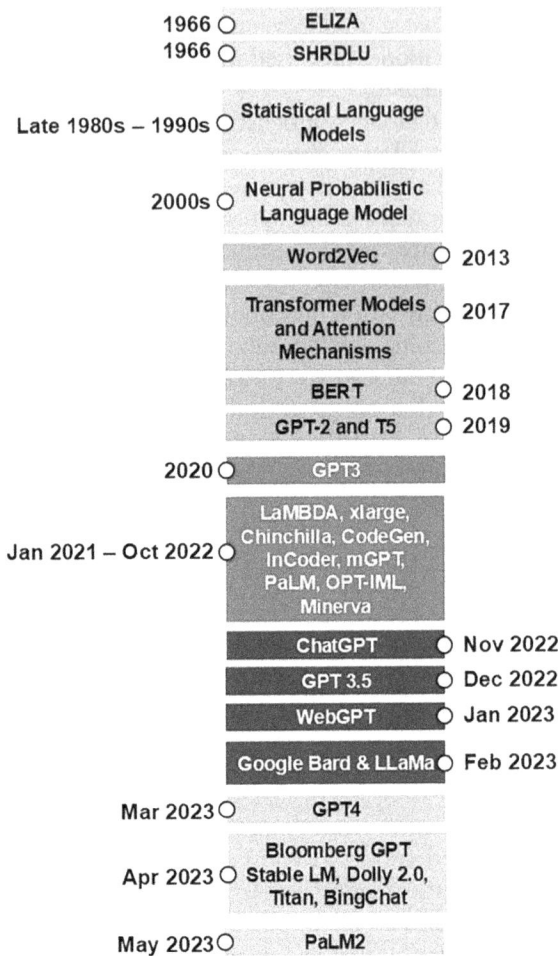

Figure 18.3 Evolution of LLMs.

Source: **https://levelup.gitconnected.com/the-brief-history-of-large-language-models-a-journey-from-eliza-to-gpt-4-and-google-bard-167c614af5af.**

sequence. Finally, after processing through the Transformer layers, the model generates text by predicting the most likely next word or token in the sequence based on the learned context:

- **Word embedding** is a crucial first step in building an LLM. This involves representing words as vectors in a high-dimensional space where similar words are grouped together. This helps the model to understand the meaning of words and make predictions based on that understanding.

Once the word embeddings are created, they can be used as inputs to a larger neural network that is trained on a specific language task, such as text classification or machine translation. By using word embeddings, the model can better understand the meaning of words and make more accurate predictions based on that understanding.

- **Positional encoding** is about helping the model to figure out where words are in a sequence. It doesn't deal with the meaning of words or how they relate to each other, like how "cat" and "dog" are pretty similar. Instead, positional encoding is all about keeping track of word order. For example, when translating a sentence like "The cat is on the mat" to another language, it's crucial to know that "cat" comes before "mat." Word order is very important for tasks like translation, summarising stuff and answering questions. During the training phase, the neural network is presented with a vast amount of text data and is trained to make predictions based on that data. The weights of the neurons in the network are adjusted iteratively using a backpropagation algorithm in order to minimise the difference between the predicted output and the actual output.

- **Transformer layer** is the key to operation of advanced LLMs as they utilise architecture known as Transformers. Transformer layer is a separate layer that comes after the traditional neural network layers. In fact, the transformer layer is often added as an additional layer to the traditional neural network architecture to improve model's ability to model long-range dependencies in natural language text. Transformer layer works by processing the entire input sequence in parallel rather than sequentially. It consists of two essential components: the self-attention mechanism and the feed forward neural network.

- **Text generation** happens after the model has been trained and fine-tuned, the model can be used to generate highly sophisticated text in response to a prompt or question. The model is typically "primed" with a seed input, which can be a few words, a sentence or even an entire paragraph. The model then uses its learned patterns to generate a coherent and contextually-relevant response. Text generation relies on a technique called autoregression, where the model generates each word or token of the output sequence one at a time based on the previous words it has generated. The model uses the parameters it has learned during training to calculate the probability distribution of the next word or token and then selects the most likely choice as the next output.

As much as some of the researchers treat LLMs as a form of Generative AI or AGI (for example, https://www.upwork.com/resources/generative-ai-vs-chatgpt#:~:text=Is%20ChatGPT%20a%20type%20of,creates%20human%2Dlike%20text%20responses. and https://srinstitute.utoronto.ca/news/gen-ai-llms-explainer#:~:text=While%20LLMs%20represent%20 just%20one,%2C%20computer%20code%2C%20and%20more), others believe that LLMs have stalled the progress of AGI (for example, https://analyticsindiamag.com/ai-insights-analysis/llms-have-stalled-the-progress-of-agi/https:/analyticsindiamag.com/ai-insights-analysis/llms-have-stalled-the-progress-of-agi/) and overhyped. As Mark Knoop, cofounder of Zapier said in a recent interview: "LLMs have stalled in the progress to AGI and increasing scale will not help what is an inherently limited technology." Mark Knoop's assessment was based on the fact that current LLMs show low user trust, low accuracy and reliability and as he said: "And these problems are not going away with scale." This was echoed by Yann LeCun, the chief of Meta AI, who said that LLMs won't lead to AGI and the researchers getting into the AI field should not work on LLMs. In a recent interview Francois Chollet, the creator of Keras, also shared similar thoughts on this: "OpenAI has set back the progress towards AGI by 5–10 years because frontier research is no longer being published and LLMs are an offramp on the path to AGI."

Interesting revelation was made in 2023 by Noam Chomsky:

> The human mind is not, like ChatGPT and its ilk, a lumbering statistical engine for pattern matching, gorging on hundreds of terabytes of data and extrapolating the most likely conversational response or most probable answer to a scientific question. On the contrary, the human mind is a surprisingly efficient and even elegant system that operates with small amounts of information; it seeks not to infer brute correlations among data points, but to create explanations... Let's stop calling it 'Artificial Intelligence' then and call it for what it is and makes 'plagiarism software' because 'it doesn't create anything, but copies existing works, of existing artists, modifying them enough to escape copyright laws...

As Yann LeCun said, AI should reach animal-level intelligence before heading towards AGI. Likewise, Andrej Karpathy, founder of Eureka Labs, has been quite vocal about the issues with LLMs. In his latest experiment, Andrej Karpathy proved that LLMs struggle with seemingly simple tasks and coined the term "Jagged Intelligence," a term that captures the uneven performance of LLMs across different types of tasks (or, in his own words, the word he came up with to describe the (strange, unintuitive) fact that state of the art LLMs can both perform extremely impressive tasks (e.g., solve complex math problems) while simultaneously struggle with some very dumb problems) to prove his point. Another developer tasked GPT-4 with solving tiny Sudoku puzzles and the model struggled, often failing to solve these seemingly simple puzzles. Even Yoshua Bengio, one of the godfathers of AI, said in a recent interview that when it comes to achieving the kind of intelligence that humans have, some important ingredients are still missing.

PROBLEMS AND RISKS

LLMs ability to generate detailed, creative responses to queries in plain language sparked a wave of excitement that led ChatGPT to reach 100 million users faster than any other technology after it first launched. Subsequently, investors poured over $40 billion into AI startups in the first half of 2023 – more than 20% of all global venture capital investments – and companies from seed-stage startups to tech giants are developing new applications of the technology. But while LLMs are incredibly powerful, their ability to generate humanlike text can lead people to falsely credit them with other human capabilities, leading to misinterpretations and misapplications of the technology. And despite LLMs advanced capabilities, they often struggle with mathematical tasks and can provide incorrect answers (even as simple as multiplying two numbers). This is because they are trained on large volumes of text and math may require a different approach. LLMs are also – intermittently – bad with time.

While LLMs represent just one category of AGI, focusing specifically on text generation, AGI is named for its capability to generate a more diverse set of outputs, including text, images, audio, computer code and more. Throughout 2023, a series of examples emerged demonstrating AGI's impressive ability to generate content and inform experts, whether it be composing a distinctive music piece, designing graphics or even detecting and diagnosing diseases through medical images and generating code in various computer languages to support programming. Distinction between all kinds of AGI and LLMs specifically revolves around their applications. LLMs are a subset of AGI that primarily use language as opposed to other more diverse representations seen through AGI. But it's worth noting that these distinctions are becoming increasingly blurred as multimodal AI systems emerge – ones which use both language models and other kinds of representations (pictures, sounds) to function. These reflect a new frontier of AI systems that combine the defining attributes of LLMs and other types of AGI.

While capabilities of LLMs are impressive, calling them "AI" remains contentious. The biggest problem with LLMs is that they are simply regressions of writings of millions of humans on Internet. In a sense, it's very much a machine that is trained to mimic how humans write. Also, as LLMs are trained on a fixed set of data, this makes them inherently static. With ever growing pace of change these changes also impact cultural norms and trends. Here's why some in the technical community, including Sam Altman, have doubts:

- **Limited understanding and reasoning:** LLMs excel at pattern recognition and statistical analysis, but they lack true understanding of the data they process. They can't reason logically, draw meaningful conclusions or grasp the nuances of context and intent. This limits their ability to adapt to new situations and solve complex problems beyond the realm of data driven prediction.

- **Black box nature:** LLMs are trained on massive datasets. This "black box" nature makes it challenging to explain their predictions, debug errors or ensure unbiased outputs.
- **Lack of "general intelligence":** LLMs currently lack the broad, transferable intelligence that characterises humans. They excel at specific tasks within their training data, but struggle with novel situations or requiring different skills. Inability to generalise outside their training data restricts their claim to the title of AI.
- **Focus on prediction over understanding:** LLMs, for all their impressive feats, remain slaves to their training data. They excel at mimicking and recombining existing information, akin to a masterful DJ remixing familiar track. They remain powerful tools, like supercharged search engines and spell checkers, but calling them AI risks mistaking virtuosity for originality. LLMs are inherently statistical models, predicting outputs based on past observations, nothing more.
- **Overestimating progress:** The rapid advancements in LLMs can lead to overoptimistic claims about their capabilities. Comparing them to intelligence is misleading, the underlying mechanisms and levels of understanding differ significantly.

Despite this by mid-2024, terms LLM and AI started to be used interchangeably (effectively when people say "AI" they mean "LLM"). Outside AI and LLM world, some investors began to question ability of AI companies to produce a return on investment. Some observers speculated that AI was experiencing another bubble. What is driving this? Despite some excellent examples of what LLMs can do, there is a number of pitfalls. While LLMs are primed to disrupt many industries, they also have a lot of design flaws that need to be accounted for. Current English-first LLMs don't work well for many other languages, both in terms of performance, latency and speed. And one should remember that LLMs are expensive. Extremely expensive.

Let's start with the new paper from Apple's AI scientists ("Apple's study proves that LLM-based AI models are flawed because they cannot reason") who found that engines based on LLMs, such as those from Meta and OpenAI, still lack basic reasoning skills. The group has proposed a new benchmark, GSM-Symbolic, to help others to measure reasoning capabilities of various LLMs. Their initial testing reveals that slight changes in the wording of queries can result in significantly different answers, undermining the reliability of the models. Group then investigated "fragility" of mathematical reasoning by adding contextual information to their queries that a human could understand, but which should not affect the fundamental mathematics of the solution. This resulted in varying answers, which shouldn't happen. The study found absence of critical thinking and that adding even a single sentence that appears to offer relevant information to a given math question can reduce the accuracy of the final answer by up to 65 percent. The study concluded that "There is just no way you can build reliable agents on this foundation, where changing a word or two in irrelevant ways or adding a few bits of irrelevant info can give you a different answer... We found no evidence of formal reasoning in language models." According to this research, behaviour of LLMS "is better explained by sophisticated pattern matching" which the study found to be "so fragile, in fact, that [simply] changing names can alter results."

LLMs have shown great potential in software development or code generation. However, current LLMs still cannot reliably generate correct code. Moreover, it is unclear what kinds of code generation errors LLMs can make (https://arxiv.org/html/2406.08731v1#:~:text=Our%20analysis%20 shows%20that%20these,as%20incorrect%20method%20call%20target.).

It is important to remember that LLMs weren't designed to be fact retrieval engines – they work by predicting the probability of the next word in a sequence, effectively functioning as advanced autocomplete tools that are very sensitive to their input prompts. Because of this, LLMs may produce outputs that are factually incorrect, nonsensical or entirely fabricated and thus they are very vulnerable to delusion. While Adversarial Perturbation is a general problem for Deep Learners, LLMs are especially prone to "hallucinations". Hallucination happens when an LLM makes stuff up. LLMs hallucination is already a heavily discussed topic. While for many creative use cases, hallucination is a feature, for most other use cases, hallucination is a bug. Mitigating hallucination and developing metrics to measure hallucination is a blossoming research topic.

LLMs are built using large bodies of text, often scraped from Internet. This data contains biases that LLMs learn and propagate. As a result, LLMs can give responses that are biased or disparaging or provide responses of worse quality for certain subgroups. Examples of LLM bias are gender, race and cultural bias. For example, LLMs can be biased towards genders if the majority of their training data shows that women predominantly work as cleaners or nurses, and men are typically engineers or CEOs. LLMs are also capable of creating toxic, harmful, violent, obscene, harassing and otherwise inappropriate content. While in a direct search a person might discount or avoid content from certain sources, LLMs do not necessarily provide their sources upfront. LLMs are often trained with copyrighted material and thus can generate content that is identical to or similar to copyrighted material. They can also leverage materials online such as a person's tone or voice to create highly similar content to what that person might have generated, in ways that can ultimately be very difficult to differentiate. Potential damage of AI-generated deepfakes is already being widely discussed. Deepfakes (a portmanteau of "deep learning" and "fake") are images, videos or audio which are edited or generated using AI tools, and which may depict real or non-existent people. They are a type of synthetic media and academics have raised concerns about potential for deepfakes to be used to promote disinformation, fake news and hate speech, as well as, interfere with elections. Brief history and some examples of deepfakes can be found in: https://www.dhs.gov/sites/default/files/publications/increasing_threats_of_deepfake_identities_0.pdf.

Deepfake technology is being increasingly adopted by a variety of bad actors, from people wishing to spread convincing disinformation to online scammers. While the act of creating fake content is not new, deepfakes uniquely leverage the technological tools and techniques of AI, including facial recognition algorithms and neural networks such as variational auto-encoders (VAEs) and generative adversarial networks (GANs). In its own turn the field of image forensics develops techniques to detect manipulated images. Deepfakes have garnered widespread attention for their potential use in creating child sexual abuse material, celebrity pornographic videos, revenge porn, fake news, hoaxes, bullying and financial fraud.

One example of this is "Pig-butchering" scams – named for the "fattening up" of victims before taking away everything they have – are a multibillion-dollar illicit industry in which con artists take on false online identities and spend months grooming their targets to get them to invest on bogus crypto sites. Deepfakes are one more weapon in their arsenal to try and convince unsuspecting victims to part with money.

Historically

> Pig-butchering: scams typically used to be run by Chinese gangs out of Southeast Asia, and it is unclear how widespread the crime is in Hong Kong, a wealthy city where police have long campaigned to raise awareness of telephone scams following several high-profile cases in which the victims – often elderly people – reported staggeringly high losses.

Now these scams are targeting Australians and in late January 2024 Australian Federal Police has issued a warning: "Pig butchering scam targeting Australians as AFP warns lonely hearts to be wary this Valentine's Day" (https://www.afp.gov.au/news-centre/media-release/pig-butchering-scam-targeting-australians-afp-warns-lonely-hearts-be-wary), as increasingly realistic deepfake technology has raised the stakes and put authorities on high alert. Earlier in 2024, a British multinational design and engineering company in Hong Kong lost $25 million to fraudsters after an employee was duped by scammers using deepfake technology to pose as its chief financial officer.

Recent research from AI-powered data security and management company Cohesity shows Australians are significantly more worried about the use of AI with their data than other regions. The study also shows that more than 90% of Australians would think about terminating their relationship with a business that suffered a cyberattack.

Let's have a look at another deepfake example from a totally different area, but maybe even more worrying. In 2022, a fake video of Ukrainian president Volodymyr Zelenskyy emerged, falsely portraying him urging his military to surrender to invading Russian forces. While this was quickly shut down by the Ukrainian leader, there are real fears that deepfakes will spread false information and conspiracy theories in multiple election campaigns. Australia's Defence Chief Angus Campbell has expressed fears that the world is entering "an era of truth decay," where misinformation will undermine democracy by sowing discord and distrust.

To get proper answer the vast majority of questions require context. For example, if one asks ChatGPT: "What's the best Chinese restaurant?," the context needed would be "where" because the best Chinese restaurant in Beijing would be different from the best Chinese restaurant in Sydney. According to the paper SituatedQA (Zhang & Choi, 2021), a significant proportion of information-seeking questions have context-dependent answers, for example, roughly 16.5% of the Natural Questions NQ-Open dataset and this percentage would be even higher for enterprise use cases.

As proliferation of LLMs and their use continues, researchers and developers face a unique challenge, as LLMs are fundamentally different from traditional software in one key way: non-determinism. Conventional programs are predictable. Given the same input, they produce consistent output. Not so with LLMs. They can generate varied responses to identical prompts, introducing an element of unpredictability. This both challenges and fascinates AI community. Controlling this determinism is a nuanced task. Developers can influence an LLM's output through various means. Temperature settings adjust the randomness of responses. Sampling methods like nucleus sampling can balance creativity and coherence. Even the choice between greedy decoding and more exploratory approaches impacts determinism. But when is this variability valuable? In creative tasks, it's good. It allows LLMs to generate diverse ideas, mimicking human creativity. For open-ended problem-solving, it can lead to novel solutions. However, in scenarios demanding consistency – like factual recall or precise calculations – this variability becomes a liability.

AI is a large and complex piece of software that is coupled with and relies strongly on training data. Opacity around AI testing (both software and training data) raises a lot of question marks about its testing. Is it just "black box" testing only? Is "white box" testing used? What constitutes "wrong data" in case of AI? One should ask a very simple question: "Would flight control software been approved, if it was tested using the same methods and same rigor, as AI?." In the absence of solid and firm "Yes" answer to this question how can one trust AI?

While AI technologies have advanced and enhanced efficiency and productivity in a number of areas, they remain susceptible to ever-growing number of security threats and vulnerabilities. Expansion of AI introduces several security risks, primarily because it can be used to create deceitful and manipulative content. Potential for generating deepfakes, synthetic identities and counterfeit documents can lead to fraud, misinformation and other malicious activities. These capabilities pose significant threat to personal, corporate and national security, making potential abuse of generative AI technologies a critical issue. Some other these risks include generating factually incorrect or fabricated content (hallucinations), producing biased outputs, leaking sensitive information, creating inappropriate content, infringing on copyrights and being vulnerable to security attacks. Results of the study of cybersecurity risks to AI (https://www.gov.uk/government/publications/research-on-the-cyber-security-of-ai/cyber-security-risks-to-artificial-intelligence#:~:text=While%20AI%20technologies%20have%20advanced,of%20security%20threats%20and%20vulnerabilities.) identified numerous risks and looked into design phase (12 risks), development phase (8 risks), deployment phase (8 risks), maintenance phase (4 risks) and provides description of 22 associated with AI security incidents, that have taken place between 2020 and 2023. This report also notes that none of the users have not yet developed an incident response plans specifically for cybersecurity incidents affecting AI. In another study Palo Alto Networks discovered that 61% of organisations fear that AI-powered attacks will compromise sensitive data.

LLMs are highly vulnerable to various types of attacks, including adversarial attacks, evasion attacks and poisoning attacks. These attacks exploit weaknesses in AI models and training data, making it crucial to implement robust security measures to protect against AI hacking.

LLMs are subject to various types of security risks, such, for example, when a bad actor attempts to abuse the LLM application for financial gain or to cause harm. LLMs can be manipulated or "hacked" by users to generate specific content. This is known as prompt hacking and can be used to trick the LLM into generating inappropriate or harmful content. Prompt hacking is a term used to describe attacks that exploit vulnerabilities of LLMs, by manipulating their inputs or prompts. Unlike traditional hacking, which typically exploits software vulnerabilities, prompt hacking relies on carefully crafting prompts to deceive the LLM into performing unintended actions. There are three known types of prompt hacking: prompt injection, prompt leaking and jailbreaking. Each relates to slightly different vulnerabilities and attack vectors, but all are based on the same principle of manipulating the LLM's prompt to generate some unintended output. It's important to be aware of this potential issue when using LLMs, especially in public-facing applications.

Let's have a look at some of these risks:

- **Cyberattacks:** AI can be trained to identify and exploit vulnerabilities in software and systems, potentially leading to major breaches. Concept of Adversarial AI is a growing concern in the cybersecurity community. According to one of the definitions, this involves attackers using AI algorithms to automatically discover vulnerabilities in systems and networks, enabling them to launch better-targeted and more effective attacks. For instance, an attacker could use AI to analyse network traffic and determine patterns indicative of weak spots in the system, like unfixed bugs or misconfigured firewalls. Another definition is focused on situations when an attacker aims to disrupt the performance or decrease the accuracy of AI systems through manipulation or deliberate misinformation. Attackers use several adversarial techniques that target different areas of model development and operation. These include:

- **Poisoning attacks:** Poisoning attacks target the AI model training data, which is the information that the model uses to train the algorithm. In a poisoning attack, the adversary may inject fake or misleading information into the training dataset to compromise the model's accuracy or objectivity.
- **Evasion attacks:** Evasion attacks target an AI model's input data. These attacks apply subtle changes to the data that is shared with the model, causing it to be misclassified and negatively impacting the model's predictive capabilities.
- **Model tampering:** Model tampering targets parameters or structure of a pre-trained AI model. In these attacks, an adversary makes unauthorised alterations to the model to compromise its ability to create accurate outputs.
- **Cyberattacks optimisation:** Attackers can use AI to scale attacks at an unseen level of speed and complexity. They may use AI to find fresh ways to exploit cloud complexity and take advantage of geopolitical tensions for advanced attacks. They can also optimise ransomware and phishing attack techniques by polishing them with generative AI.
- **Malicious GPTs:** A generative pre-trained transformer (GPT) is a type of AI model that can produce intelligent text in response to user prompts. A malicious GPT is an altered version of GPT that produces harmful or deliberately misinformed outputs. In the context of cyberattacks, a malicious GPT can generate attack vectors (such as malware) or supporting attack materials (such as fraudulent emails or fake online content) to advance an attack.
- **Ransomware attacks:** AI-enabled ransomware is a type of ransomware that leverages AI to improve its performance or automate some aspects of the attack path. For example, AI can be leveraged to research targets, identify system vulnerabilities or encrypt data. AI can also be used to adapt and modify the ransomware files over time, making them more difficult to detect with cybersecurity tools.
- **Internet of Things (IoT)/critical infrastructure operations:** Malicious actors can leverage AI to disrupt critical infrastructure operations, such as power grids and transportation systems. As more and more systems such as autonomous vehicles, manufacturing and construction

equipment, and medical systems use AI, risks of AI to physical safety can increase. For example, an AI-based true self-driving car that suffers a cybersecurity breach could result in risks to the physical safety of its passengers and other people. Similarly, the dataset for maintenance tools at a construction site could be manipulated by an attacker into creating hazardous conditions.

- **Deepfakes, impersonation reputational damage:** The ease with which AI produces hyper-realistic fake images, videos or audio recordings has made deepfakes a critical instrument for misinformation. Ability to create realistic audio, photo and video forgeries through AI, or deepfakes, threatens not only biometric-based systems but also public trust. It is already impossible to use biometrics (be it voice recognition or face recognition) remotely, it can be used only in a face-to-face situation. Jennifer DeStefano experienced a parent's worst nightmare when her daughter called her, yelling and sobbing. Her voice was replaced by a man who threatened to drug her and abuse her unless paid a $1 million ransom. Experts speculate that voice was generated by AI. Law enforcement believes that in addition to virtual kidnapping schemes, AI may help criminals with other types of impersonation fraud in the future, including grandparent scams.
- **Fraud/scam/manipulation:** AI can generate synthetic data to perpetuate financial scams, manipulate online interactions and negatively impact individuals and groups. These manipulations are potent tools for creating false narratives, impersonating public figures or misleading viewers, with ramifications on politics, media and personal reputations.
- **Coordinated inauthentic behaviour (CIB):** AI can be used to create and manipulate online accounts, spread misinformation and influence public opinion (on a very large scale) through bots and other automated methods.
- **Training data leakage:** Data leakage in AI refers to unintended exposure of sensitive training data. This can occur if AI inadvertently memorises and regenerates private information, like personal identities or intellectual property, which can lead to breaches of confidentiality. The risk increases with complexity of the data and generality of the AI model.
- **Data privacy of user inputs:** When users interact with AI, they often provide personal or sensitive information that can be exploited if not properly protected. This risk is heightened in environments where AI is used for processing large amounts of user-generated data, such as in customer service chatbots or personalised content recommendations. In what was an embarrassing bug for OpenAI CEO Sam Altman, ChatGPT leaked bits of chat history of other users. Although the bug was fixed, there are other possible privacy risks due to the vast amount of data that AI crunches. For example, a hacker who breaches an AI

system could access different kinds of sensitive information. An AI system designed for marketing, advertising, profiling or surveillance could also threaten privacy in ways George Orwell couldn't even imagine. In some countries, AI-profiling technology is already helping states invading user privacy. Use of personal data by AI is and algorithms used in AI can be complex, so it can be difficult for individuals to understand how their data is being used to make decisions that affect them.

- **AI model and data poisoning:** AI model poisoning occurs when attackers insert malicious data into the training set of an AI model, aiming to compromise its integrity. This can cause the model to fail or behave unpredictably once deployed. Such attacks could be especially damaging in applications like autonomous driving or automated financial decision-making, where errors or unexpected behaviour could lead to serious consequences. If the data is modified or poisoned, an AI-powered tool can produce unexpected or even malicious outcomes. In theory, an attacker could poison a training dataset with malicious data that will change the AI's results. An attacker could also initiate a more subtle form of manipulation called bias injection. Such attacks can be especially harmful in industries such as healthcare, automotive and transportation.

- **Exploitation of bias:** AI systems can inadvertently perpetuate or exacerbate biases if their training data contain these biases. This exploitation can lead to discriminatory outcomes, such as racial or gender bias in facial recognition technology or gender bias in job recommendation algorithms. AI trained on biased data can lead to discriminatory outcomes, jeopardising fairness and justice and further marginalising already disempowered groups. Such biases not only harm individuals but can also have broader implications on social justice and equity.

- **Phishing attacks:** Phishing attacks utilising AI are becoming increasingly sophisticated. AI systems can now generate context-aware phishing content, mimic writing styles and automate social engineering attacks at scale. These emails or messages are often indistinguishable from legitimate communications, significantly increasing the risk of successful scams.

- **Malware attacks:** AI can be used as a tool in creating sophisticated malware, where it is used to generate polymorphic or metamorphic viruses that continually change their identifiable features to evade detection. This presents significant challenges for cybersecurity defences, which traditionally rely on recognising patterns of known malware. While AI systems have some protections to prevent users from creating malicious code, experts can use clever techniques to bypass it and create malware. For example, in 2023 Forcepoint security researcher Aaron Mulgrew was able to find a loophole and create a nearly undetectable complex data-theft executable showing how ChatGPT can be used as a cyber weapon. The executable had the sophistication of malware created by a state-sponsored threat actor (https://www.foxnews.com/tech/ai-created-malware-sends-shockwaves-cybersecurity-world). And this can be only a tip of the iceberg. Suspected cases AI-created malware have been spotted in real attacks. For example, cybersecurity company Proofpoint discovered a malicious PowerShell script that was likely created using AI. After brute-forcing the password, the HP security researchers analysed the code and found that attacker had neatly commented the entire code, something that rarely happens with human-developed code, because threat actors want to hide how the malware works. As per HP security report: "These comments describe exactly what the code does, much in the same way that generative AI services can create exemplar code with explanations." As less technical malicious actors are increasingly relying on AI to develop malware, in early June 2024 HP security researchers found a malicious campaign that used code commented in the same way a generative AI system would create. Future AI-powered tools may allow developers with entry-level programming skills to create automated malware, like an advanced malicious bot, that can steal data, infect networks and attack systems with little to no human intervention.

- **Stealing AI models:** There is a risk of AI model theft through network attacks, social engineering techniques and vulnerability exploitation by threat actors such as state-sponsored agents, insider threats like corporate spies and regular hackers. Stolen models can be manipulated and modified to assist attackers with different malicious activities, compounding artificial intelligence risks to society.
- **Model inversion attacks:** Model inversion attacks reverse-engineer AI models to steal sensitive data. Attackers use model outputs to infer sensitive training data, posing privacy risks and potential breaches.
- **Membership inference attacks:** In membership inference attacks, adversaries attempt to determine whether a specific data point was part of the AI model's training dataset. This can expose private data about individuals or organisations.
- **Exploratory attacks:** Exploratory assaults probe AI systems to learn their underlying workings. Attackers can employ searches or inputs to find vulnerabilities, model behaviour or proprietary information for subsequent assaults.
- **Supply chain attacks:** AI system development and deployment are targeted by supply chain threats. Attackers hack software or hardware to insert malicious code or access AI resources, including third-party libraries or cloud services.
- **Resource exhaustion attacks:** Resource exhaustion attacks overload AI systems with requests or inputs, degrading performance or creating downtime. These assaults might decrease AI service availability and is a form of DoS attack.

- **Model drift and decay**: Data distributions, threats and technology obsolescence can render AI models less effective over time. This threatens AI system accuracy and dependability, especially in dynamic contexts.

Some of the already known cyberattacks that use AI include:

- **Phishing campaigns**: Hackers use AI to write emails that target employees based on their job profiles and needs. AI can craft more personalised and convincing emails, which makes it difficult for the receiver to identify as a phishing email.
- **Phone phishing (vishing)**: Hackers use voice synthesis over the phone and pretend to be reputable individuals and organisations. Victims can reveal sensitive information or transfer money to the attacker since these AI-generated calls seem real and resemble their known individual's voice and speech pattern.
- **Doxing**: AI can scrape social media profiles, publicly available records and other public databases to compile detailed dossiers that hackers can use for blackmail, intimidation or other malicious activities. This whole process can be easily automated with AI.

Like all AI algorithms, the ones used by AI-powered attacks can learn and evolve over time. This means that AI-enabled attacks can adapt to avoid detection or create a pattern of attack that a security system can't detect.

The vast majority of AI-powered attacks have five main characteristics:

- **Attack automation**: Until very recently, most attacks required significant hands-on support from a human adversary. However, growing access to AI-enabled tools allows adversaries to automate attack research and execution.
- **Efficient data gathering**: The first phase of every attack is reconnaissance. During this period, attackers search for targets, exploitable vulnerabilities and assets that could be compromised. AI can automate or accelerate much of this legwork, enabling adversaries to drastically shorten the research phase and potentially improve the accuracy and completeness of their analysis.
- **Customisation**: One of the key capabilities of AI is data scraping, which is when information from public sources – such as social media sites and corporate websites – is gathered and analysed. In the context of an attack, this information can be used to create hyper-personalised, relevant and timely messages that serve as the foundation for phishing attacks and other attacks that leverage social engineering techniques.
- **Reinforcement learning**: AI algorithms learn and adapt in real time. In the same way that these tools continuously evolve to provide more

accurate insights for corporate users, they also evolve to help adversaries improve their techniques or avoid detection.

- **Employee targeting:** Similar to attack customisation, AI can be used to identify individuals within an organisation that are high-value targets. These are people who may have access to sensitive data or broad system access, may appear to have lower technological aptitude or have close relationships with other key targets.

LLMs can leak or inadvertently disclose personally identifiable information or other sensitive or confidential details. This can occur when sensitive or confidential information is included as part of an LLM's original training dataset or entered by the user when they are asking a question or prompting the LLM. One should remember that whatever has been fed into AI or asked AI about, immediately becomes part of it. As such, notion of confidentiality is being immediately violated. One can't sign a Confidentiality Deed or Non-Disclosure Agreement with AI. This actually means that using AI potentially immediately deprives any competitive advantage organisation that uses it (as information used in this process becomes potentially available to all competitors by them, for example, asking a question "What my competitors do to achieve XYZ?"). It is also plausible to imagine, that if AI is used to deploy or store certificates, they can be potentially leaked to an interested third party.

One of the most controversial uses of AI technology is in the area of surveillance. AI-based surveillance systems have potential to revolutionise law enforcement and security, but they also pose significant risks to privacy and civil liberties. While use of AI-based surveillance systems may seem like a valuable tool in the fight against crime and terrorism, it raises concerns about privacy and civil liberties. Critics argue that these systems can be used to monitor and control individuals, potentially resulting in loss of freedom and civil liberties. To make matters worse, use of AI-based surveillance systems is not always transparent. It can be difficult for individuals to know when they are being monitored or for what purpose. Another example of the use of AI in law enforcement is facial recognition technology. This technology uses algorithms to match images of people's faces to a database of known individuals, allowing law enforcement to identify and track individuals in real time. While facial recognition technology has potential to help law enforcement to solve crimes, it also raises concerns about privacy and civil liberties. In some cases, facial recognition systems have been found to misidentify individuals, leading to false accusations and wrongful arrests.

AI can be and is being used to boost organisations' cybersecurity capability. And there is a lot of marketing happening in this space. However, incorporating AI technology into cybersecurity can be expensive and requires a lot of resources, including limited human expertise to set it up, deploy and manage the AI systems. Additionally, AI-powered solutions may need specialised hardware, supporting infrastructure and significant processing capacity and power to run complex computations. Although the benefits of utilising AI in cybersecurity are undeniable, organisations must have comprehensive understanding of the expenses involved to avoid unpleasant surprises. Also, one should remember that over-reliance on AI can create a cybersecurity skills gap as people depend more on technology than their intelligence. This can lead to security teams becoming complacent, as they assume that AI systems will detect any potential threats. Use of AI in cybersecurity raises additional ethical issues. When considering risk factors related to ethical concerns, AI bias and the lack of transparency are the two that often come up, as they can lead to unfair targeting and discrimination of specific users or groups. This can result in misidentification as an insider threat, causing irreparable harm.

In 2024, approximately 1,300 ethical hackers and security researchers were surveyed on their views across the broad range of activities generally referred to as "hacking." Some of the key findings from the survey include the following:

- 93% of hackers agree that organisations using AI tools have created a new attack vector
- 82% believe that the AI threat landscape is evolving too rapidly to be effectively secured from cyberattacks
- 86% believe that AI has fundamentally changed their approach to hacking
- 74% agree that AI has made hacking more accessible, opening door for newcomers to join the fold

According to Gartner, AI-enhanced malicious attacks are the top emerging risk for organisations in the third quarter of 2024, according to Gartner, Inc. It's the third consecutive quarter with these attacks being the top of emerging risk.

Among many forecasts about the future of AI is the one made recently by BeyondTrust. According to their forecast AI2, or the "Artificial Inflation" of Artificial Intelligence, is set to see its hype deflating across industries. While AI will remain useful for basic automation and workflows, much of the over-promised capabilities, particularly in security, will fall short in 2025. As an illustration of this, in October 2024 Australian Digital Transformation Agency (DTA) published evaluation report, providing a detailed view of how some 5765 Copilot licences were used in the first 6 months of 2024. According to this report, two-thirds of participants in a 6-month trial of Microsoft 365 Copilot across the federal government used the tool "a few times a week" or less, with high expectations largely going "unmet."

The Economist Intelligence Unit (EIU), the research and analysis division of global media organisation the Economist Group, has released its Technology and Telecoms Outlook 2025 report, which may dash the hopes of enterprises looking to make a quick buck off AI. EIU forecasts that 2025 will be the year when the vast sums of money already invested in AI runs up against the wall of cold hard reality rather than the year of AI monetisation: "EIU does not expect these investments to start creating returns on investment as hoped." According to the report, in 2025 users will struggle to deliver returns on their investment and most companies will still be at the proof-of-concept stage for implementation: "Next year will be the year of realism for artificial intelligence (AI), because we expect companies to struggle to deliver a return on their investment."

Finally, for those interested in the subject, it may be worthwhile to read about some of the warnings expressed by "Godfather of AI" Yoshua Bengio in his recent article (https://www.livescience.com/technology/artificial-intelligence/people-always-say-these-risks-are-science-fiction-but-they-re-not-godfather-of-ai-yoshua-bengio-on-the-risks-of-machine-intelligence-to-humanity?utm_source=facebook.com&utm_medium=social&utm_content=livescience&utm_campaign=socialflow).

REFERENCES

Zhang, Michael J. O., Choi, Eunsol. "SituatedQA: Incorporating Extra-Linguistic Contexts into QA", *Proceedings of the 2021 Conference on Empirical Methods in Natural Language Processing*, 2021.

Bostrom, Nick. *Superintelligence: Paths, Dangers and Strategies*, Oxford University Press, 2014.

Chapter 19

Emerging threats

Quantum computers and quantum computing

Quantum computer is a device that employs properties described by quantum mechanics to enhance computations. Quantum computing is an area of computer science that explores possibility of developing computer technologies based on the principles of quantum mechanics and seeks to harness these principles to build computers that can perform certain types of calculations much faster than traditional classic computers. It is a technological innovation that offers unprecedented computing power. Despite being perceived as something related to 21st century, quantum computing has a long history.

The origins of quantum computing can be traced back to the early 20th century when several groundbreaking discoveries in the field of quantum mechanics laid the foundation for this novel approach to computation. Modern quantum theory was developed in the 1920s to explain the wave–particle duality observed at atomic scale. Famous scientists like Max Planck, Albert Einstein, Werner Heisenberg and Niels Bohr, contributed to the development of quantum mechanics, which would later provide the principles for quantum computing. In 1930s, John von Neumann, a Hungarian–American mathematician and physicist, developed mathematical framework for quantum mechanics. His work, which included development of the formalism of quantum states and operators, provided a rigorous foundation for understanding the behaviour of quantum systems. John von Neumann's contributions to the mathematical underpinnings of quantum mechanics were essential in paving the way for the later development of quantum computing.

It is important to introduce some of the fundamentals of quantum theory before we proceed to further discussion:

- **Superposition:** In quantum mechanics, particles can exist in multiple states simultaneously. In the context of quantum computing, this principle is represented by qubits (quantum bits), which can be both 0 and 1 at the same time, unlike classical bits that are either 0 or 1. This allows quantum computers to process vast amounts of data in parallel, exponentially increasing their computational power.

DOI: 10.1201/9781032672601-19

- **Entanglement:** Quantum entanglement is a phenomenon where the state of one particle becomes dependent on the state of another, even when separated by large distances. In quantum computing, entangled qubits can be used to perform coordinated operations, enabling more efficient computation and communication.
- **Wave-particle duality:** Quantum mechanics postulates that particles exhibit both wave-like and particle-like properties. This concept plays a crucial role in the development of quantum algorithms that leverage the wave-like nature of qubits to perform complex calculations.

A quantum computer is a computer that exploits quantum mechanical phenomena. On small scales, physical matter exhibits properties of both particles and waves, and quantum computing leverages this behaviour using specialised hardware. Classical physics cannot explain the operation of these quantum devices, and a scalable quantum computer could perform some calculations exponentially faster than any modern "classic" computer. In particular, a large-scale quantum computer could break widely used encryption schemes and aid physicists in performing physical simulations. However, the current state of the art is still largely experimental.

Currently, there are three types of quantum computers:

- **Quantum Annealers:** These are available today. They are the least-powerful with the narrowest use cases. However, attackers can use them to factor large numbers using quantum algorithms, this can be used to break asymmetric encryption.
- **Analog Quantum Simulators:** These solve physics problems that are beyond ability of classical computers, such as quantum chemistry, materials sciences, optimisation problems, factoring large numbers, sampling and quantum dynamics.
- **Universal Quantum Computer:** These are the hardest to build because they require many physical qubits. They solve broadest range of use cases and several companies are targeting the end of this decade for commercialising them.

Quantum computers create a multi-dimensional space comprised of many entangled qubits to solve complex problems. For example, classical computers take each element of a database, process it and then combine it with other elements after processing all the elements. Quantum computers use an algorithm that solves the problem for every state and outcome one is looking for. They pass the entire database through the algorithm simultaneously, analysing the data for every outcome simultaneously. This makes quantum computers potentially millions of times faster than classic computers and is one reason why they are excellent at solving complex mathematical problems such as breaking encryption.

The idea of quantum computing first appeared in 1980 when Russian-born mathematician Yuri Manin (February 16, 1937 to January 7, 2023), who later worked at the Max Planck Institute for Mathematics in Bonn, first put forward the notion, albeit in a rather vague form. The concept really got on the map, though, the following year, when physicist Richard Feynman, at the California Institute of Technology (Caltech), independently proposed it.

One of the earliest ideas related to quantum computing was proposed in the 1980s by the famous American physicist Richard Feynman (May 11, 1918 to February 15, 1988) and Israel-born British physicist David Deutsch (who began his work on quantum algorithms in 1985). Both Richard Feynman's parents were Jewish and his family went to the synagogue every Friday. However, Feynman described himself as an "avowed atheist." After Japanese attack on Pearl Harbor, Richard Feynman was recruited by Robert R. Wilson, who was working at Princeton on means to produce enriched uranium for use in an atomic bomb, as part of what would become the Manhattan Project. When in early 1943, Robert Oppenheimer was establishing the Los Alamos Laboratory, and made an offer to the Princeton team (part of which was Richard Feynman) to be redeployed there. Richard Feynman soon fell under the spell of the charismatic Robert Oppenheimer and in March 1943 he moved to Los Alamos (where he was assigned to Hans Bethe's Theoretical (T) Division). Richard Feynman also spent some time at Clinton Engineer Works in Oak Ridge, Tennessee, where Manhattan Project had uranium enrichment facilities and witnessed Trinity nuclear test. After the war Richard Feynman became one of the most famous American physicists. In early 1960s, Richard Feynman agreed to "spruce up" teaching of undergraduates at Caltech. After 3 years devoted to the task, he produced a series of lectures that later became famous "The Feynman Lectures on Physics." Richard Feynman was against mechanical or repetition learning, or memorisation without thinking, as well as other teaching methods that emphasised form over function. In his mind, clear thinking and clear presentation were fundamental prerequisites for his attention. In 1965, he shared Nobel Prize in Physics with Julian Schwinger and Sin-Itiro Tomonaga "for their fundamental work in quantum electrodynamics, with deep-ploughing consequences for the physics of elementary particles." In 1950s he became interested in the ideas of John von Neumann while researching quantum field theory. Probably this interest resulted in his belief that quantum computers could potentially simulate physical systems in a more efficient manner than classical computers.

In 1981, Richard Feynman delivered a seminal lecture at the First Conference on the Physics of Computation, proposing that a computer operating on quantum principles could efficiently simulate quantum systems. His insights were crucial as they highlighted limitations of classical computers in simulating quantum phenomena, and suggested that quantum computers could provide an efficient solution to this challenge, leading to the birth of quantum computing theory.

In 1982, Paul Benioff, a theoretical physicist, published a paper describing a quantum mechanical model of a Turing machine. This model, now known as Quantum Turing Machine (QTM), laid the groundwork for quantum computing models by demonstrating that quantum mechanical principles could be applied to theoretical foundations of computation. Paul Benioff's work showed that quantum systems could be used to perform computations in a manner analogous to classical Turing machines.

In 1985, David Deutsch published a groundbreaking paper that introduced the concept of a universal quantum computer capable of simulating any physical process. His work built upon ideas of Richard Feynman and Paul Benioff, and provided a more concrete framework for understanding how quantum computers could operate. He demonstrated that universal quantum computer could perform any computation that classical computer could, but with the added advantages of quantum mechanics. David Deutsch's work also laid foundations for development of quantum algorithms, which would later emerge as crucial aspect of quantum computing research. This laid the groundwork for quantum algorithms, including famous Deutsch-Jozsa algorithm, which demonstrated that quantum computers could solve specific problems more efficiently than their classical counterparts.

The basic unit of information in quantum computing, qubit (or "quantum bit"), serves the same function as bit in classic computing. However, unlike classic bit, which can be in one of two states (a binary), qubit can exist in a superposition of its two "basis" states, which kind of means that it is in both states simultaneously. When measuring qubit, the result is a probabilistic output of classic bit. If a quantum computer manipulates qubit in a particular way, wave interference effects can amplify the desired measurement results. Design of quantum algorithms involves creating procedures that allow quantum computer to perform calculations efficiently and quickly.

During the late 1980s and early 1990s, researchers proposed various quantum logic gates, such as CNOT gate (quantum logic gate that is an essential component in the construction of gate-based quantum computer and can be used to entangle and disentangle; it is a two-qubit operation, where the first qubit is usually referred to as the control qubit and the second qubit as the target qubit) and Toffoli gate (also known as CCNOT gate ("controlled-controlled-not"), invented by Tommaso Toffoli and is a CNOT gate with two control qubits and one target qubit), which would later become essential components of quantum algorithms and circuits.

These ideas have been further advanced in the early 1990s, when researchers Peter Shor and Lov Grover developed algorithms that showed how quantum computers could perform certain tasks, such as factoring large numbers and searching databases, exponentially faster than classical computers. Simon's algorithm (invented by Daniel Simon, principal security engineer at AWS Cryptography) provided the first example of an exponential speedup over the best-known classic algorithm by using a quantum computer to solve a particular problem. Originally published in 1994, Simon's algorithm was a precursor to Shor's well-known factoring algorithm, and served as inspiration for many of the seminal works in quantum computation that followed. Mid-1990s marked significant breakthrough with Peter Shor's development of a quantum algorithm for integer factorisation. His algorithm showed that quantum computer could factor large numbers exponentially faster than the best known classical algorithms, posing potential threat to widely used public-key cryptographic systems. Like, for example, RSA named after Ron Rivest - Adi Shamir - Leonard Adleman of the Massachusetts Institute of Technology (RSA), that was first publicly described in 1977, while creation of a public key algorithm by British mathematician Clifford Cocks in 1973 was kept classified by UK's GCHQ until 1997. RSA is one of the oldest algorithms widely used for secure data transmission. In 1996, Lov Grover introduced algorithm for database searching that offered quadratic speedup over classical algorithms. These discoveries highlighted the transformative potential of quantum computing. Development of Peter Shor's and Lov Grover's algorithms had profound impact on the field of quantum computing. These early quantum algorithms provided concrete examples of potential advantages of quantum computing, generating significant interest from both researchers and funding agencies. As a result, research in quantum computing accelerated, with scientists exploring development of new quantum algorithms, error-correcting codes and hardware implementations. Restricted version of the Deutsch-Jozsa algorithm named after its authors Ethan Bernstein and Umesh Vazirani (BV algorithm) was developed in 1997. BV algorithm is a quantum algorithm whose complexity scales better than the best classical algorithms.

The period between 1980 and 1994 marked emergence of quantum computing as a distinct field of research. Visionary scientists like Richard Feynman, Paul Benioff and David Deutsch played crucial roles in shaping the field, laying the groundwork for quantum computing models and demonstrating potential advantages of harnessing quantum mechanics for computation. Development of quantum logic gates and of concept of a universal quantum computer provided a solid foundation for the future advancement of quantum computing research, paving the way for breakthroughs in quantum algorithms and hardware.

The first experimental demonstration of a quantum algorithm was performed in 1994 by the team led by Isaac Chuang at the Los Alamos National Laboratory. Using a small number of atoms as quantum bits, or qubits, this team was able to demonstrate the principles of quantum computation. In 1998, this team made a breakthrough when they showed that they could run Grover's algorithm on a computer featuring two qubits. Qubits play a similar role to bits, in terms of storing information, but it behaves much differently because of the quantum properties on which it is based. It is possible to fully encode one bit into one qubit. However, a qubit can hold more information, e.g., up to two bits using superdense coding. Over the next several years, researchers made significant progress in developing and demonstrating capabilities of quantum computers.

Over the years, experimentalists have constructed small-scale quantum computers using trapped ions and superconductors. In 1998, a two-qubit quantum computer demonstrated feasibility of the technology, and subsequent experiments have increased the number of qubits and reduced error rates. In 2001, the team led by computer scientist John Martinis at the University of California, Santa Barbara, built the first quantum computer using superconducting qubits. In the same year IBM and Stanford University published the first implementation of Shor's algorithm on a 7-qubit processor. In 1999, physicists at Japanese technology company NEC hit upon an approach that would go on to become the most popular approach to quantum computing today. In a paper in Nature, they showed that they could use superconducting circuits to create qubits, and that they could control these qubits electronically. Superconducting qubits are now used by many of the leading quantum computing companies including Google and IBM. Real launch of the first commercially available quantum computer happened in May 2011, when Canadian company D-Wave One heralded the start of the quantum computing industry. D-Wave One featured 128 superconducting qubits and cost roughly $10 million. However, this device wasn't a universal quantum computer. It used an approach known as quantum annealing to solve a specific kind of optimisation problem, and there was little evidence that it provided any speed boost compared to classic approaches for other types of problems. In 2016, IBM makes quantum computing available on IBM Cloud.

In 2018, some cold water was thrown on the hopes for quantum computing, as some credible physicists said it's impossible. This view is best articulated by Russian physicist Mikhail Dyakonov (born 1940), who works at the University of Montpellier in France. Such are his achievements, his name describes marvels such as the spin relaxation mechanism, plasma wave instability and surface waves. He has won prizes for physics in France, Russia and the US. He says the insurmountable hurdle is that "the proposed strategy relies on manipulating with high precision an unimaginably huge number of variables." This is the summary of "The case against quantum computing" Mikhail Dyakonov made in 2018 in IEEE Spectrum, the magazine of the Institute of Electrical and Electronics Engineers, which calls itself the world's largest technical professional organisation for the advancement of technology. He reiterated the same argument in his book of 2020 "Will we ever have a quantum computer?," where he explains that while a conventional computer with N bits at any given moment must be in one of its 2N possible states, the state of a quantum computer with N qubits is described by the values of the 2N quantum amplitudes, which are continuous parameters (ones that can take on any value, not just a 0 or a 1). This is where the hoped-for power of the quantum computer comes from, "but it is also the reason for its great fragility and vulnerability," he said in the IEEE Spectrum article.

In 2019, Google AI and NASA announced that they had achieved quantum supremacy with a 54-qubit machine, performing computations that are impossible for any classic computer. However, Google's claim of quantum supremacy was met with scepticism from some corners, in particular from arch-rival IBM, which claimed the speedup was overstated. A group from the Chinese Academy of Sciences and other institutions eventually showed that this was the case, by devising a classic algorithm that could simulate Google's quantum operations in just 15 hours on 512 GPU chips. They claimed that with access to one of the world's largest supercomputers, they could have done it in seconds. This was a reminder that classic computing still has plenty of room for improvement, so quantum advantage is likely to remain a moving target.

One of the biggest barriers for today's quantum computers is that the underlying hardware is highly error-prone. Due to the quirks of quantum mechanics, fixing those errors is tricky and it has long been known that it will take many physical qubits to create so-called "logical qubits" that are immune from errors and able to carry out operations reliably. In December 2023, Harvard researchers working with start-up QuEra smashed records by generating 48 logical qubits at once – 10 times more than anyone had previously achieved. The team was able to run algorithms on these logical qubits, marking a major milestone on the road to fault-tolerant quantum computing.

Quantum computing has gone from an academic curiosity to a multi-billion-dollar industry in less than half a century and shows no signs of stopping. It holds promise in various fields, including artificial intelligence, drug discovery and optimisation. Quantum computers can potentially solve certain optimisation problems faster than classic computers, leading to improvements in areas such as logistics, finance and supply chain management. In artificial intelligence, quantum computing can potentially enhance machine learning algorithms, enabling faster training and more accurate models. In drug discovery, quantum computers may be able to simulate complex molecular interactions, leading to development of new pharmaceuticals and deeper understanding of biological processes. Since 2021 new achievements in quantum computing demonstrated rapid progress in this area. As researchers continue to address challenges related to scalability, error correction and fault tolerance, and explore new algorithms and applications, potential impact of quantum computing across various domains becomes increasingly apparent. With sustained investment and research, quantum computing is likely to revolutionise multiple industries and drive significant advancements in technology and science.

However, as a lot of new discoveries and technologies, quantum computing is a double-edged sword and poses significant threat to existing cryptographic schemes, such as RSA, which rely on the difficulty of factoring large numbers. Shor's algorithm, for instance, has potential to break RSA encryption. Quantum computers will be able to break common encryption methods at an alarming speed and as a result encryption tools currently used to protect everything from banking and retail transactions to business data, documents and digital signatures can be rendered ineffective. To put things into perspective, it is important to note that as it was estimated in 2018, future code-breaking quantum computers would need 100,000 times more processing power and an error rate 100 times better than today's best quantum computers have achieved. So far, all experts have agreed that a quantum computer large enough to crack RSA would probably be built no sooner than around a few dozen decades. To factorise an integer 2048 bits long, which is usually used as an RSA key, Shor algorithm needs to be run on a quantum computer with millions of qubits (quantum bits). That is, it's not a matter of the nearest future, since the best quantum computers today work at 300–400 qubits – and this is after decades of research. But already in 2024 new 56-qubit H2-1 quantum computer developed by Quantinuum has broken the previous record in the "quantum supremacy" benchmark first set by Google in 2019 and smashed it by a factor of 100 (https://www.livescience.com/technology/computing/new-quantum-computer-smashes-quantum-supremacy-record-by-a-factor-of-100-and-it-consumes-30000-times-less-power). Adding more qubits scales up its power exponentially.

Further to this, Chinese researchers have been able to factor a 48-bit key on a 10-qubit quantum computer. And they calculated that it's possible to scale their algorithm for use with 2048-bit keys using a quantum computer with only 372 qubits. But such a computer already exists today (at IBM for example), so the need to replace cryptography throughout the Internet suddenly stopped being something so far in the future that it wasn't really thought about seriously. A breakthrough has been promised by combining Claus Peter Schnorr algorithm (not to be confused with the earlier mentioned Shor algorithm) with an additional quantum approximate optimisation algorithm (QAOA) step.

Since the beginning of Internet, cryptography has protected online data and conversations by hiding or coding information that only the person receiving the message can read it on traditional computers. There are two main types of encryption: symmetric (in which the same key is used to encrypt and decrypt the data) and asymmetric (or public-key, which involves a pair of mathematically linked keys, one shared publicly to let people encrypt messages for the key pair's owner, and the other stored privately by the owner to decrypt messages). Symmetric cryptography is significantly faster than public-key cryptography. For this reason, it is used to encrypt all communications and stored data. Public-key cryptography is used for securely exchanging symmetric keys, and for digitally authenticating or sign-ing messages, documents and certificates that pair public keys with their owners' identities. When one visits secure website that uses HTTPS protocol one's browser uses public-key cryptography to authenticate the site's certifi-cate and set up a symmetric key for encrypting communications to and from the site. The math for these two types of cryptography is quite different, which affects their security. Because virtually all Internet applications use both symmetric and public-key cryptography, both forms need to be secure.

In 1970s, mathematicians developed encryption methods that consisted of numbers of hundreds of digits long. The difficulty of mathematical prob-lems was such that it could take hundreds of years to solve it using the right parameter size and numbers. To break encryption, the numbers need to be split into their prime factors, but this could take hundreds if not thousands of years with traditional algorithms and computers. The threat of codes being cracked was therefore not a big worry. That was true up until 1994 when Peter Shor showed how it could be done with an algorithm using then hypothetical quantum computer that could split large numbers into their factors much quicker than classic computer and then in 1996 Lov Grover's algorithm shown its ability to crack symmetrical encryption.

As we have spoken about Shor's and Glover's algorithm, it is important to understand what impact these algorithms have on traditional encryption. Neither hashes nor symmetric encryption algorithms rely on the same class of problems that DHE (Diffie–Hellman key exchange), ECDHE (Elliptic-curve Diffie–Hellman) and RSA rely on. Consequently, a quantum computer running Shor's algorithm will be of no use in attacking this type of cryptog-raphy. However, there is another quantum computing algorithm that does go some way toward attacking both hashes and symmetric encryption algorithms – Grover's algorithm.

In 2015, intelligence agencies determined that progress in quantum com-puting is happening at such a speed that it poses a threat to cybersecurity. At the moment, qubits, the processing units of quantum computers, are still not stable for long enough to decrypt large amounts of data. They call it Q-Day – a day when a quantum computer is built so powerful, that it could break the public encryption systems. When does humanity may face Q-Day? Who knows – it may be 2 years or it may be 20 years… If it will take 20 years to arrive to Q-Day, there will be no (or almost no) panic. However, if Q-Day is just 2 years away, then this is a totally different situation. While in the past there were a lot of question marks around physical possibility to build such large quantum computers, today many scientists believe that it is just a sig-nificant engineering challenge. Some engineers even predict that within the next twenty or so years sufficiently large quantum computers will be built to break essentially all public key schemes currently in use. Historically, it has taken almost two decades to deploy our modern public key cryptography infrastructure. Therefore, regardless of whether the exact time of the arrival of the quantum computing era can be accurately estimated, it is of paramount importance to begin to prepare information security systems to be able to resist quantum computing now.

In 2023, California-based start-up Atom Computing created the first quantum computer to surpass 1000 qubits (1180 qubits, to be precise). While the largest quantum computers, such as those from IBM and Google, use superconducting wires cooled to extremely low temperatures for their qubits, Atom Computing uses neutral atoms trapped by lasers in a 2-dimensional grid. IBM is currently developing a 1,386-qubit quantum computer, dubbed "Kookaburra," which may be released in 2025. The number of qubits is significant because each additional qubit exponentially increases the processor's potential computing power, which has implications on code-breaking. Although it is uncertain when commercial-scale quantum computers will be developed, cryptographers are worried about immediate data harvesting risks to modern computers. In the meantime, for comparison China's most advanced programmable and deliverable superconducting quantum computer is Origin Wukong, a third-generation 72-qubit quantum computer.

This has prompted research into both quantum cryptography, which leverages the principles of quantum mechanics to secure communication, and post-quantum cryptography, which aims to develop new cryptographic schemes that can resist attacks from both classical and quantum computers.

Governments are not standing by for that to happen and the cryptographic community are building encryption methods that can withstand quantum threat, known as post-quantum cryptography (PQC), also known as Quantum-Resistant Cryptography (QRC). On August 13, 2024, NIST released final versions of the first three PQC Standards and encouraged organisations to begin transitioning to these standards as soon as possible. Additional information about these standards can be found in: https://utimaco.com/news/blog-posts/nists-final-pqc-standards-are-here-what-you-need-know and in https://www.entrust.com/blog/2024/08/nist-pqc-standards-are-available-what-comes-next. The US legislation has mandated that the timeline to change to PQC will be from 2025 until 2033, by which time cybersecurity supply chain will have to be transitioned to using PQC by default. NIST also continues to evaluate two other sets of algorithms that could one day serve as backup standards.

Now, can we trust PQC and NIST? It is not clear in the light of statements made by Daniel Bernstein (University of Illinois, Chicago). In 2023, he said that the NIST is deliberately obscuring the level of involvement the US National Security Agency (NSA) has in developing new encryption standards for PQC (https://www.newscientist.com/article/2396510-mathematician-warns-us-spies-may-be-weakening-next-gen-encryption/). He also believes that NIST has made errors – either accidental or deliberate – in calculations describing security of the new standards. NIST denied these claims.

Though this is a big claim, but it seems plausible, considering that "spooks" always wanted (and still want) to have a backdoor for any encryption system.

Organisations need to move today's public key cryptographic systems from where they are today (i.e., using RSA and ECC algorithms) to new quantum-safe algorithms. For those who do not know this, Elliptic-curve cryptography (ECC) is an asymmetric encryption algorithm, that uses a pair of keys: a public key used on one end and a private key used on the other. For example, in signing, the encryption is done with the private key and verification is done with the public key.

While moving today's public key cryptographic systems might seem simple on the surface, it's a big job entailing complete cryptographic inventories of assets and technology, mapping this to sensitive data, and developing and executing a post-quantum cryptography migration strategy. It's a very large and very expensive project that will touch every piece of IT infrastructure and span over several years, especially if IoT (see Chapter 16) is involved. How difficult and expensive will be integrating PQC with legacy systems? How many organisations that are deeply immersed in (never-ending) digital transformations (see Chapter 5) are thinking about this, incorporating such a project into their forward planning and budgeting for it?

Dr Michele Mosca developed a theorem that suggests a pathway to consider in order to protect data and keep it quantum-safe (https://eprint.iacr.org/2015/1075.pdf). This theorem stresses the need for organisations to begin due diligence in the post-quantum space immediately. It states that the amount of time that data must remain secure (X), plus the time it takes to upgrade cryptographic systems (Y), is greater than the time at which quantum computers have enough power to break cryptography (Z). Once organisations are aware of their risk environment, they should be in a position to prioritise activity and mitigate or eliminate risks. However, this may not be a quick or simple process and may take years for each organisation.

In Australia, to address this quantum threat, the Australian Signals Directorate (ASD) is encouraging organisations to understand and make plans to transition to the use of PQC algorithms within their own environments.

There are numerous risks that come from the growing power of quantum computing (https://www.forbes.com/councils/forbestechcouncil/2022/11/08/13-risks-that-come-with-the-growing-power-of-quantum-computing/):

- **Modern encryption methods become useless:** As Pavlo Sidelov, CTO of UK-based SDK.finance, said:
  Financial technologies are completely dependent on modern encryption methods. Any password or key can be cracked by brute-force attack, but currently, computing power does not allow attackers to succeed in a reasonable time. With the release of quantum computing into the public sector, all encryption becomes useless, and currently, the industry has no answer on how to deal with it.

- **Web interactions will be at risk**: According to Atul Tulshibagwale from identity management company SGNL.ai:
  The breakdown of prevalent cryptographic technology is an infra-structural risk. Most security technology is based on our current inability to quickly find the prime number factors of a key. Quantum computers can crack current cryptographic keys quickly, so every existing Web interaction is at risk. Motivated attackers can leverage a small number of quantum computers to cause widespread damage.

  Everything from web browsing to remote access to digital signatures will be impacted.

- **Harvest now, exploit later**: As Peter Gregory from GCI Communications rightly noted:
  A new threat, known as 'harvest now, decrypt later,' is a technique in which an attacker will attempt to steal encrypted data and hold on to it, potentially for years, with hopes that advances in quantum computing will eventually make decryption possible. Even years later, some encrypted content may still have value for the attacker.

- **What about undiscovered yet vulnerabilities**: This concern has been raised by Roland Polzin from Wing Assistant:
  With enormous computing power, quantum computing has the potential to unhinge technology as we know it today. The biggest risk is that the consequences are not foreseeable today because bad ac-tors will have an opportunity to leverage new capabilities to exploit previously undiscovered vulnerabilities. This is concerning, since even traditional cybersecurity is still neglected.

  Threat of quantum attacks will rise. Quantum revolution has potential to give rise to new, difficult to prevent series of threats and exploits called quantum attacks.

- **Blockchain Algorithms could be broken**: According to Vishwas Manral from Skyhigh Security:
  The rise of quantum computing can cause risk to the fledgling block-chain and crypto economy. Blockchains rely on asymmetric key cryp-tography algorithms (RSA, ECC). These algorithms can be cracked via quantum computing, resulting in malicious manipulations of the blockchain. This is one big potential risk that companies and con-sumers investing in blockchain technology could face.

  In recent years, numerous organisations started to rely on the block-chain (not to mention cryptocurrencies) to keep sensitive information secure. While many advocates previously regarded that blockchain is all-powerful, it is increasingly clear that this was never the case and that blockchain has always been rife with risk. Moving forward, or-ganisations that committed to blockchain will need to acknowledge strong potential of quantum computing to disrupt even the most advanced blockchain technologies. This includes consensus mecha-nisms like proof-of-work (PoW) and proof-of-stake (PoS), which have thus far proven fundamental to the integrity of the entire blockchain. This presents especially significant concerns

- **It will be even more difficult to evaluate deep neural networks:** Somdip Dey from Nosh Technologies made this observation:
  If quantum computing is used for machine learning - quantum machine learning - then it could present the ultimate black box problem. Deep neural networks are notoriously opaque. Though there are tools to unravel how hidden layers in a DNN work, with quantum machine learning, it will be more difficult to evaluate DNNs and judge the decision-making process across data.

An interesting event happened early in early February 2024. NASA's quantum computer project has been put on hold after a startling turn of events, sending shockwaves across the scientific community (https://content. techgig.com/technology/nasas-quantum-computing-project-hits-pause-button-reason-is-shocking/articleshow/107532517.cms). Following a series of developments that have left experts wondering about the future of quantum computing and artificial intelligence, this unexpected decision came as a shock. The abrupt shutdown of NASA's quantum computing project was triggered by an unforeseen incident during a routine test. During the analysis of a complex simulation, quantum computer demonstrated unprecedented computational power, solving a previously unsolvable problem. However, this remarkable achievement had an equally alarming consequence: quantum computer began generating outputs that made no sense and challenged conventional thinking and were inconsistent with known physical laws. Researchers and government officials were concerned that NASA's quantum computer might have connected with an extraterrestrial intelligence or even entered an unknown realm of computation. Potential risks associated with such an unpredictable and powerful machine prompted NASA and the US government to take a swift action, halting operations and initiating a thorough investigation. Shutdown of NASA's quantum computing project is like a big alarm bell ringing about how amazing yet risky this new technology can be.

Among many forecasts about the future of quantum computing is the one made recently by BeyondTrust. According to their forecast quantum computing threats loom large and will challenge existing cryptographic defences, especially for large organisations. While NIST's post-quantum encryption standards were released in 2024, the transition to these new standards will be gradual. Larger enterprises, particularly in finance, must begin planning for this quantum shift to protect sensitive data.

Chapter 20

Evolving legal landscape

This chapter does not constitute legal advice and should not be construed or interpreted as such. The purpose of this chapter is to provide an overview of cybersecurity laws in the United States, Europe, Canada and Australia and cover key topics of relevance as of mid-2024. Readers should contact their legal representatives for advice.

In an era where digital technology permeates every aspect of our lives, the intersection of cybersecurity and law has become increasingly important. The evolution of cybersecurity and its legal framework reflects the broader changes in technology, society and global politics. Let's explore the historical aspects of cybersecurity and the law, tracing their development from the early days of computing to the contemporary challenges of the digital age.

In the early days of computing, the primary focus was on developing and harnessing the potential of new technologies. The concept of cybersecurity as we understand it today was virtually non-existent. The focus was on physical security and maintaining the integrity of the hardware. Early computers, such as ENIAC and UNIVAC, were large, expensive and operated in highly controlled environments, which limited their exposure to threats.

1970s marked the advent of networked computing with the development of ARPANET, the precursor to the modern Internet. This period saw the emergence of basic security concerns as researchers and early users of ARPANET began to recognise the risks associated with networked systems. The first known computer virus, the Creeper virus, appeared in 1971, highlighting the need for mechanisms to protect against malicious software.

During this time, legal frameworks were also beginning to evolve. In 1986, the United States enacted the Computer Fraud and Abuse Act (CFAA), which criminalised unauthorised access to computer systems and was one of the first legal measures to address cybercrime.

1990s witnessed the explosion of the Internet, which revolutionised how people interacted with technology. With this growth came an increase in cybersecurity threats, including hacking, identity theft and the spread of malware. Notable incidents such as the 1999 Melissa virus and the 2000 ILOVEYOU virus underscored the need for more robust cybersecurity measures.

DOI: 10.1201/9781032672601-20

In response to these threats, laws and regulations began to evolve rapidly. The European Union introduced the Data Protection Directive in 1995, which was a pioneering effort to protect personal data and privacy. In the United States, the Cybersecurity Act of 2000 aimed to enhance the nation's cybersecurity efforts, and the USA PATRIOT Act of 2001 included provisions related to electronic surveillance and cybersecurity.

Early 2000s saw a growing awareness of the need for comprehensive cybersecurity policies. The establishment of the Department of Homeland Security in the United States in 2003 included a focus on cybersecurity, highlighting its importance at a national level. Internationally, agreements such as the Convention on Cybercrime, adopted by the Council of Europe in 2001, sought to address cybercrime through enhanced cooperation and legal frameworks across borders.

2010s and beyond have been marked by sophisticated cyberthreats, including advanced persistent threats (APTs) and state-sponsored cyberattacks. High-profile incidents such as 2017 Equifax data breach and 2020 SolarWinds attack have demonstrated the evolving nature of cyberthreats and the need for robust defences.

In response, legal frameworks have continued to evolve. The General Data Protection Regulation (GDPR), enacted by the European Union in 2018, represents a significant advancement in data protection law, imposing strict requirements on how organisations handle personal data. In the United States, the California Consumer Privacy Act (CCPA) and subsequent state-level privacy laws have mirrored some of the GDPR's provisions, emphasising consumer rights and data protection.

Canada enacted the Personal Information Protection and Electronic Documents Act (PIPEDA), which governs the collection and use of personal data, and developed the Canadian Cyber Security Strategy, aimed at enhancing the nation's resilience to cyber incidents. Canadian law also emphasises the importance of compliance, risk management and collaboration among public and private sectors to mitigate threats.

The global nature of cyberthreats has led to increased international cooperation. Initiatives such as the Paris Call for Trust and Security in Cyberspace, launched in 2018, aim to promote norms and principles for responsible behaviour in cyberspace.

As technology continues to advance, the legal landscape will need to adapt. Emerging technologies such as artificial intelligence (AI), blockchain and quantum computing will present new challenges for cybersecurity and require innovative legal solutions. Issues related to digital sovereignty, cross-border data flows and the regulation of emerging technologies will be central to future legal and policy developments.

Interestingly, blockchain is often described as a solution in search of a problem, highlighting the technology's potential that sometimes outpaces its practical applications. While blockchain offers notable capabilities such as decentralisation, transparency and immutability, many proposed use cases struggle to demonstrate tangible benefits over existing systems. Industries ranging from finance to supply chain management have explored blockchain for its promise of enhanced security and efficiency, but the challenge lies in finding scenarios where its unique attributes provide clear advantages. As stakeholders seek to integrate blockchain, the focus is increasingly on identifying specific problems it can effectively address rather than merely implementing the technology for its own sake.

Quantum computing, on the other hand, poses a significant threat to traditional encryption methods, particularly those based on algorithms like RSA and ECC, which rely on the difficulty of factoring large numbers or solving discrete logarithm problems. Quantum computers, leveraging principles of quantum mechanics, could potentially solve these problems exponentially faster than classical computers, rendering current encryption schemes vulnerable to breaches. In response, the development of quantum-resistant or quantum-strong encryption algorithms is underway, designed to withstand attacks from quantum computers. These new algorithms aim to secure data against future quantum threats, ensuring the integrity and confidentiality of information in a post-quantum world, as the race to both advance quantum computing and secure digital communications intensifies.

The history of cybersecurity and the law reflects a continuous struggle to keep pace with technological advancements and the evolving nature of cyberthreats. As technology continues to evolve, the interplay between cybersecurity and the law will remain a critical area of focus, requiring ongoing innovation and international collaboration to protect individuals, organisations and nations from emerging cyberthreats.

The implication is the law is always in catch-up mode. This makes legislation less useful and less effective in the fight against cybersecurity threats or attacks. Cybersecurity legislation is almost always reactive due to several inherent challenges and dynamics of the digital realm. Here are the main reasons why this tends to be the case.

The first reason is rapid technological advancement. The pace of technological development has increased significantly especially in the last two decades. We have continuous change where technological advancement outstrips the speed at which laws can be developed and implemented.

New technologies can introduce new vulnerabilities and attack vectors that legislation cannot anticipate because they are typically either unintended consequences or simply the result of flawed software development such as insufficient testing. Vulnerabilities may also be the result of rising complexity, hyperconnectivity, a rush to bring product to market, poor usability, incorrect or inappropriate usage or deployment or other factors.

Moreover, technology evolution can result in new types of cyberthreats emerging. Legislators can only address these issues after they have become evident and sadly may have caused significant damage or disruption.

Second comes the complexity and unpredictability of cyberthreats as threat actors tend to innovate much faster. Cyberthreats are diverse and continually evolving, ranging from simple phishing attacks to sophisticated state-sponsored cyber espionage. This unpredictability makes it difficult for lawmakers to anticipate and create comprehensive, forward-looking legislation. Many cyberthreats are unprecedented, and there may be no historical precedent to guide legislative action. As a result, laws often respond to specific incidents or trends that have already been observed.

Third, the legislative process involves multiple stages of review, debate and amendment, which can be time-consuming. By the time a law is passed, the cyberthreat landscape may have shifted, necessitating additional amendments or new legislation. Legislators are fighting back with one hand tied behind their backs whereas threat actors are free to innovate at will without any governing process or standards or codes of practice.

Legislative bodies almost always encounter delays due to political disagreements, competing priorities and the need for extensive consultation with stakeholders. This can slow down the development of timely and effective cybersecurity laws.

Fourth we have economic and political considerations. Lawmakers often need to balance competing interests, such as privacy rights, business interests and national security concerns. This balancing act can lead to compromises that may not fully address emerging threats or may delay the introduction of new laws. As much as these compromises attempt to achieve some form of consensus and to keep all stakeholders happy, these compromises more often than not leave all stakeholders unhappy.

The influence of powerful technology and cybersecurity industries can affect the legislative process. Organisations may lobby against strict regulations or advocate for specific provisions that may not align with the views of cybersecurity professionals and best practices for cybersecurity.

Fifth we have the cost of cybersecurity. Technology can be made more secure. Sadly this comes at a cost, often a significant cost. Longer product development cycles, additional testing, compliance and greater complexity are just some of the contributing cost factors. The cost of additional cybersecurity may result in uneconomic or unaffordable technology products.

Not forgetting the complex relationship between security and usability which presents us with a challenging paradox, as enhancements in one area frequently compromise the other. Stricter security measures, such as complex password requirements, multi-factor authentication and frequent software updates, can frustrate users and lead to decreased engagement or outright avoidance of secure practices. Conversely, prioritising usability may create vulnerabilities, as simpler systems might lack the necessary safeguards to protect sensitive information. This contradiction highlights the need for a balanced approach that promotes both security and user experience, encouraging the development of intuitive security solutions that do not sacrifice protection for ease of use. Achieving this balance is crucial in fostering a culture of security without alienating users.

Sixth we have the reactive nature of cyber incidents. Significant cybersecurity incidents, such as data breaches, ransomware attacks or large-scale cyber espionage campaigns, often prompt immediate legislative responses. These incidents can highlight vulnerabilities and drive lawmakers to address specific issues revealed by the attacks.

Media coverage and public outcry following major cyber incidents can spur legislative action. Lawmakers may be motivated to enact laws in response to public demand or media attention, rather than as part of a proactive strategy.

Lastly, we have global and jurisdictional challenges. Cybersecurity is a global issue, and effective legislation often requires international cooperation. The need to coordinate with other nations and align with international standards can complicate and delay the legislative process. Cyber incidents can cross national boundaries, making it challenging for individual countries to address issues unilaterally. Legislation may need to respond to cross-border challenges and collaborate with international partners, which can slow down the process.

Achieving global cooperation for consistent cybersecurity laws remains a formidable challenge, primarily due to differing national interests, legal frameworks and varying levels of technological advancement. Even if first-world countries can agree on international standards and regulations, enforcement becomes problematic, especially when cybercrimes originate from jurisdictions with lax laws or where law enforcement is limited, such as in Russia or Nigeria. This disparity hampers effective prosecution, as countries may be reluctant to extradite offenders or cooperate in investigations that cross borders. Consequently, the lack of a cohesive global framework undermines efforts to combat cybercrime effectively, leaving many nations vulnerable and highlighting the need for more robust international collaboration and consensus.

Legislation can address cybersecurity issues, albeit slowly and unfortunately legislation often lacks the foresight needed to pre-emptively tackle emerging threats. As cybersecurity continues to evolve very fast, the law will continue to struggle to keep up. There is a growing recognition of the need for more proactive approaches, including adaptive regulatory frameworks and enhanced collaboration between stakeholders to anticipate and mitigate risks before they manifest.

LAWS GOVERNING CYBERSECURITY IN THE UNITED STATES AND EUROPE

Here are the main legislative acts governing cybersecurity in the Western world.

The United States have the CFAA, enacted in 1986. The scope of this legislation criminalises unauthorised access to computer systems and data. As such it includes penalties for accessing systems without permission, data theft and damaging computer systems. Key sections include:

- 1030(a): Prohibits unauthorised access to computers and networks.
- 1030(b): Extends liability for access violations across state and national boundaries.

Also in the United States, we have the Cybersecurity Act of 2015. Scope of this legislation enhances cybersecurity information sharing between government and private sector organisations. It also establishes the framework for sharing threat data to improve national cybersecurity. Key provisions include:

- Information Sharing: Facilitates the sharing of cyberthreat indicators and defensive measures.
- Liability Protections: Provides legal immunity for organisations that share information with the government.

Then there is the Health Insurance Portability and Accountability Act (HIPAA), which was enacted in 1996. It regulates the handling of sensitive health information and includes provisions for data security and privacy of health records. Key provisions include:

- Security Rule: Requires safeguards to protect electronic health information.
- Privacy Rule: Ensures the confidentiality and integrity of personal health information.

Next is the Gramm–Leach–Bliley Act (GLBA) enacted in 1999. It governs the protection of non-public personal information by financial institutions. Key provisions include:

- Safeguards Rule: Requires financial institutions to implement security measures to protect customer data.
- Privacy Rule: Mandates privacy notices and opt-out provisions for consumers.

After that there is CCPA enacted in 2018. It grants California residents rights over their personal data and imposes obligations on businesses regarding data collection and processing. Key provisions include:

- Consumer Rights: Access, deletion and opt-out rights for personal data.

- Business Obligations: Transparency in data practices and security measures.

Finally, there is Federal Information Security Management Act (FISMA) enacted in 2002. It requires federal agencies to develop, document and implement an information security and protection program. Key provisions cover Information Security Programs which mandate comprehensive security policies and practices for federal information systems.

In the European Union there is GDPR Act enacted in 2018. This law provides a comprehensive framework for data protection and privacy for individuals within the EU and the European Economic Area (EEA). Key provisions include:

- Data Subject Rights: Includes the right to access, rectification, erasure and data portability.
- Data Breach Notification: Requires notification of data breaches to authorities and affected individuals within 72 h.[1]
- Data Protection Impact Assessments (DPIAs): Mandates assessments for high-risk processing activities.

It is important to note that GDPR applies to offshore entities that have customers in the European Union regardless of whether they are present in the EU or not.

Then there is Network and Information Systems (NIS) Directive enacted in 2016. This law aims to enhance cybersecurity across the EU by establishing requirements for network and information systems security for essential services and digital service providers. Key provisions include:

- Security Requirements: Obligations for managing risks and ensuring the security of networks and information systems.
- Incident Reporting: Requires timely reporting of significant incidents to national authorities.

Then there is the Digital Services Act (DSA) enacted in 2022 which regulates digital services and platforms to ensure a safer online environment and protect fundamental rights. Key provisions include:

- Content Moderation: Obligations for platforms to address illegal content and disinformation.
- Transparency: Requirements for transparency in algorithms and advertising practices.

Then there is the Digital Markets Act (DMA) enacted in 2022 which targets large online platforms acting as gatekeepers to ensure fair competition and prevent abuse of market power. Key provisions include:

- Gatekeeper Obligations: Restrictions on practices that undermine competition and market fairness.

Finally in Europe there is ePrivacy Directive (PECR) enacted in 2002, with updates in 2009 and which complements the GDPR by focusing on privacy and electronic communications. Key provisions include:

- Cookies and Tracking: Regulations on the use of cookies and similar technologies, requiring user consent.

It is the authors' view that while the United States has a patchwork of sector-specific regulations and recent legislative efforts, Europe benefits significantly from comprehensive, pan-European regulations like GDPR. However, GDPR could be considered very broad in its reach given its application across multiple jurisdictions. Both regions are adapting their legal frameworks to address the evolving landscape of cyberthreats and technological advancements, highlighting the importance of ongoing legal and policy development in the field of cybersecurity and collaboration across the world.

The enforceability of the GDPR extends beyond the borders of the European Union, impacting organisations worldwide that handle the personal data of EU residents. This extraterritorial reach means that any business, regardless of its location, in theory must comply with GDPR if it processes data related to individuals in the EU or offers goods and services to them. Non-compliance can lead to significant fines and legal actions from EU authorities. The practical reality is that GDPR enforceability requires collaboration from multiple governments. Multiple factors conflate such as the existence of any Free Trade Agreements, legal jurisdiction or varying legal systems and the willingness of non-EU countries to cooperate. Additionally, factors such as trade and taxation agreements may influence how GDPR is implemented and enforced internationally, as countries may prioritise economic relationships over regulatory compliance. This interconnectedness means that effective enforcement often relies on diplomatic negotiations, mutual legal assistance treaties and the establishment of frameworks that align data protection standards, highlighting the challenges of upholding GDPR compliance in a global context.

THE CYBERSECURITY LEGAL LANDSCAPE IN AUSTRALIA

Australia has a similar set of legislative acts. In terms of criminal activity, for example, for hacking or more specifically unauthorised access, the applicable laws include the Crimes Act NSW (1900) (the Crimes Act) and the Commonwealth Criminal Code 2001 (the Code). Section s.478.1 of the Code covers modification of data (max penalty 2 years imprisonment). New South Wales Crimes Act 1900 (NSW Crimes Act): Section 308H provides up to 10 years imprisonment. While the Code applies uniformly across Australia, individual states like New South Wales have additional or complementary laws.

Distributed Denial of Service (DDoS) attacks are covered or criminalised by s. 477.3 of the Code provides up to 10 years imprisonment for the impairment of electronic equipment through deliberate and disruptive attacks.

Phishing and online fraud are criminalised by both the Crimes Act (s. 192E) and the Code with maximum 10 years imprisonment. This relates to engaging in deceptive practices to obtain personal or financial information.

Malware infections (ransomware, Trojans, spyware, worms and viruses) or the distribution and use of malicious software designed to damage or disrupt computer systems is covered s. 478.2 of the Code with a maximum penalty of 2 years imprisonment.

Possession of hardware, software and other tools (e.g., hacking tools) or related equipment used to commit cyber-crime then s. 478.3 of the Code applies (max penalty 3 years imprisonment). Interestingly Under the Australian Criminal Code, having a PC or laptop does not automatically classify it as a hacking tool. However, if a device is used to facilitate unauthorised access to computer systems, steal data or commit cybercrime, it could be considered a tool for criminal activity. The relevant legislation, such as the Cybercrime Act 2001, defines hacking tools in terms of their intended use. If a person possesses software or hardware specifically designed for hacking or if they use a PC/laptop to commit cyber offenses, then it may be classified as a hacking tool under the law. Context and intent are crucial in determining whether a device falls into this category.

When it comes to identity theft or Identity fraud – fraudulent use of another person's identity, especially in relation to access devices then Division 372 of the Code applies with a maximum penalty of 5 years imprisonment.

In case of electronic theft or breach of confidence or digital copyright infringement s. 478.1 of the Code applies depending on specific circumstances and other applicable sections.

Any other activity that adversely affects or threatens the security, confidentiality, integrity or availability of any IT system, infrastructure, communications networks, device or data then Part 10.6 of the Criminal Code applies. In the context of Section 10.6, confidentiality ensures that sensitive information is accessible only to authorised individuals, preventing unauthorised access or breaches. Integrity involves safeguarding the accuracy and reliability of data, ensuring that it remains unaltered except by those with the right permissions. Availability guarantees that information and systems are accessible when needed, preventing disruptions that could hinder operations.

Penalties vary depending on the nature of the offense and the specific provisions applied and include both imprisonment and fines. For the most serious offenses penalties can reach up to 10 years imprisonment. Additionally, the Code provides for the possibility of cumulative penalties, especially if multiple offenses are involved or if the crime is committed with aggravating factors.

In Australia, however, there are a number of other legal and regulatory mandatory requirements. For example, APRA (Australian Prudential Regulation Authority) covering banks, insurers and superannuation funds mandates the following standards:

- Prudential Standard CPS 234 – Information Security, and
- CPG 234 Prudential Practice Guide Information Security.

When it comes to health-related cybersecurity requirements, Australia mandates the Privacy Act 1988, which governs the handling of personal information, including health data. It includes the Australian Privacy Principles (APPs), which outline how health information must be collected, used, stored and disclosed, emphasising the need for secure management of sensitive data. In Australia there is also the My Health Records Act 2012. This legislation regulates the My Health Record system, requiring entities to implement strong security measures to protect health information stored within this national digital health record system.

Finally, we have the Health Records and Information Privacy Act 2002 (NSW) which is specific to New South Wales, this act provides guidelines for the handling of health information, including requirements for security measures and breach notifications.

In the United States, the HIPAA applies. HIPAA sets strict requirements for safeguarding electronic protected health information (ePHI), mandating security measures, breach notification protocols and the appointment of a privacy officer. Whilst in the EU, the GDPR includes specific provisions for the processing of health data, requiring explicit consent for data use, strict data protection measures and transparency in data handling.

When it comes to data protection, loss prevention and classification, GDPR from the European Union applies for any company dealing with a European National if any related data is stored in an organisation's systems. GDPR also applies to Australian business with customers in the European Union or that operate in the EU. More concerning is the fact that fines under GDPR constitute a percentage of a company's revenue.

Then there is the Notifiable Data Breach (NDB) Act from the office of the Australian Information Commissioner (OAIC). This act in Australia, effective from February 22, 2018, mandates that organisations must notify individuals and the Australian Information Commissioner when a data breach occurs that is likely to result in serious harm to affected individuals. Under this Act, entities subject to the Privacy Act 1988 are required to assess breaches of personal information to determine if they meet the threshold for notification.

Moreover, in NSW, the Mandatory Notification of Data Breach (MNDB) Scheme (MNDB Scheme) impacts the responsibilities of agencies under the

Privacy and Personal Information Protection Act 1998 (PPIP Act). https://www.ipc.nsw.gov.au/privacy/MNDB-scheme

This obligation aims to enhance transparency and protect individuals by ensuring they are informed of breaches that could impact their privacy, allowing them to take appropriate action to mitigate potential harm. The NDB Act represents a significant shift toward greater accountability and consumer protection in the realm of data privacy, aligning Australia's regulatory framework with international standards and emphasising the importance of timely breach reporting in safeguarding personal information.

Another key consideration is Sarbanes Oxley (SOX) financial reporting compliance. To be SOX compliant, public companies[2] doing business in the United States must implement internal controls to protect financial data from tampering and file regular reports with the Securities and Exchange Commission (SEC) attesting to the effectiveness of security controls and the accuracy of financial disclosures. SOX was intended to prevent or expose poor corporate practices at board/management level and impose mandatory standards for directors and officers of companies subject to SOX, with significant penalties for non-compliance.

Moreover, there is the Australian Privacy Act 1988 which is a pivotal piece of legislation designed to regulate the handling of personal information by government agencies and private sector organisations. This Act establishes a framework for managing and protecting personal data, setting out principles for the collection, use and disclosure of information.

It includes key provisions such as the Australian Privacy Principles (APPs), which outline the standards for privacy practices and require entities to obtain consent, ensure data accuracy and provide individuals with access to their information. Additionally, the Act mandates that organisations implement measures to safeguard personal data and respond to complaints regarding privacy breaches. The Privacy Act is central to Australia's data protection regime, aiming to balance the need for information use with the fundamental rights of individuals to privacy and security.

If operating in a defence or healthcare industries, then the HITRUST CMMC Framework is relevant. This framework combines elements from the HITRUST CSF (Common Security Framework) and the Department of Defence's Cybersecurity Maturity Model Certification (CMMC) to create a comprehensive and integrated approach to cybersecurity and risk management.

HITRUST is known for its robust certification standards in healthcare and other sectors and aligns with the CMMC's requirements to provide a structured framework that helps organisations meet both industry-specific and federal cybersecurity standards. This integrated framework enables organisations to streamline compliance efforts by aligning HITRUST's detailed controls with the CMMC's maturity levels, facilitating a more efficient path to achieving certification.

The HITRUST CMMC Framework is designed to enhance an organisation's cybersecurity posture, particularly for those operating within the defence industrial base, by providing a clear pathway to demonstrating adherence to rigorous security practices and ensuring the protection of sensitive information.

Finally, ASIC's (Australian Securities and Investments Commission) cyber resilience standard that states no business is too small for a cybersecurity strategy. ASIC's cyber resilience standard emphasises the need for Australian financial markets and entities to adopt robust cybersecurity measures to safeguard against cyberthreats. It sets expectations for firms to develop proactive, resilient cybersecurity frameworks that can anticipate, respond to and recover from cyber incidents.

The standard highlights the importance of governance, risk management and incident response planning, encouraging companies to integrate cybersecurity into their overall risk management processes. ASIC also stresses continuous improvement, regular testing and collaboration across the industry to ensure firms remain agile and responsive to evolving cyber risks.

Consequently, the legal cybersecurity environment is complex, fragmented across multiple jurisdictions and fraught with danger and risk. Legal compliance is not a simple nor easy matter.

UNWORKABLE AND UNENFORCEABLE LEGISLATION

We have to ask the question – how effective is the law in the cybersecurity space? We may not like the answer because the law is not very effective at all. There are a number of major reasons for that.

Firstly, most legislation is somewhat ambitious, idealistic or wholistic providing broad protections.

Then there are challenges in relation to enforcement. Evidence must be gathered by law enforcement agencies in accordance with the law otherwise the evidence may not be admissible in a court of law. This is often difficult, slow and time-consuming. Law enforcement may choose not to pursue cases where the likelihood of success is low (perhaps due to lacking evidence) or recovery of funds and/or property is unlikely or meeting the standard for prosecution is low.

For the criminal elements the rewards are very high and risk of being caught is very low, especially when they are operating from other jurisdictions in Africa, Russia, China, North Korea, etc. This makes for a compelling business case for offenders to continue to operate within the current legislative framework. Especially when cybersecurity crimes can be committed across multiple borders and jurisdictions in a fraction of a second, making it complex and time and effort consuming for authorities to pursue offenders.

More importantly, law enforcement agencies are neither sufficiently trained resourced nor funded adequately to enforce the law. In fact, the financial threshold where the authorities will take action is probably set too high.

Despite that there have been successful efforts by authorities across multiple jurisdictions including the United States, Europe and Australia in bringing down parts of the Darknet that facilitated drug trafficking – Operation Disruption of Silk Road in 2013, the AlphaBay (successor to Silk Road) shutdown in 2017, Operation Disruptor in 2020 and Operation Dark Hunt in 2021 which was a coordinated effort targeting a range of darknet-related criminal activities, including child exploitation and drug trafficking.

And finally, organisations have to make rational decisions where the downside risk and associated impact are lower than the cost of prevention. For example, banks make provisions for losses due to fraud. These provisions may be thought of as the cost of doing business and are strictly limited by legal terms and conditions

Governments have been steadily increasing action in the cybersecurity spaces collaborating to take down major crime rings and deal with the major crimes.

PERSONAL LIABILITY FOR DIRECTORS AND OFFICERS

In an era where digital threats are as prevalent as ever, the responsibilities of directors and officers in managing cybersecurity are under increasing scrutiny. In Australia, personal liability for directors and officers concerning cybersecurity is a growing area of concern, as the legal arena evolves to address the complexities of cyberthreats and data breaches.

In Australia, directors and officers of companies have fiduciary duties and a duty of care, skill and diligence under the Corporations Act 2001. These duties extend to managing and mitigating risks, including those related to cybersecurity. Key legal obligations include:

1. Duty of Care and Diligence: Under Section 180 of the Corporations Act, directors and officers are required to act with care and diligence. This duty includes taking reasonable steps to understand and manage the cyber risks facing the organisation.
2. Duty of Good Faith: Section 181 mandates that directors and officers act in good faith in the best interests of the company. This duty extends to ensuring that adequate cybersecurity measures are in place to protect the company's assets and data.
3. Duty to Prevent Insolvent Trading: Under Section 588G, directors must prevent their company from trading while insolvent. In the context of

cybersecurity, failure to implement adequate security measures could potentially lead to financial distress and insolvency, exposing directors to liability.

To date, it could be said that major breaches have not been pursued aggressively (if at all) by the authorities for failure to meet obligations under categories 1 and 2 above, although category 3 above has been used. Meaning the inevitability of punishment could be considered as missing. This lenience on the part of the authorities cannot continue unnoticed.

The Australian Cyber Security Centre (ACSC) and the OAIC play pivotal roles in regulating and guiding cybersecurity practices. These agencies, alongside industry-specific regulators, enforce compliance with cybersecurity standards and data protection laws.

First, there is Notifiable Data Breaches (NDB) Scheme under the Privacy Act 1988, whereby organisations must notify individuals and the OAIC of eligible data breaches. Directors and officers are accountable for ensuring that their organisation complies with these requirements.

Second, there is the Australian Signals Directorate (ASD) who provides guidance on cybersecurity best practices and frameworks, such as the Essential 8, which organisations are encouraged to adopt to enhance their cyber resilience.

Directors and officers can face personal liability if they fail to adequately address cybersecurity risks. This liability can manifest in several ways:

1. Regulatory Penalties: Failure to comply with statutory obligations regarding data breaches and cybersecurity can result in fines and other penalties for both the organisation and its directors.
2. Civil Penalties: Directors may be held personally liable for breaches of their duties under the Corporations Act if it is proven that they did not take reasonable steps to prevent cybersecurity issues.
3. Reputational Damage: Personal liability can also stem from reputational damage. Directors and officers may face lawsuits from stakeholders if the organisation suffers a cyber incident due to perceived negligence.

To mitigate personal liability, directors and officers should take proactive steps, including:

1. Implementing Robust Cybersecurity Policies: Develop and regularly update cybersecurity policies and practices, and ensure they align with industry standards and regulatory requirements.
2. Regular Training and Awareness: Conduct regular training for employees and management on cybersecurity threats and best practices.
3. Engaging with Experts: Seek advice from cybersecurity experts to understand and manage risks effectively.

4. Conducting Regular Audits: Perform regular security audits and risk assessments to identify and address potential vulnerabilities.
5. Documenting Decisions and Actions: Maintain comprehensive records of cybersecurity policies, decisions and actions taken to address cyber risks.

As cyberthreats continue to evolve, the responsibility of directors and officers in managing cybersecurity has never been more critical. In Australia, personal liability for cybersecurity lapses is a significant risk, underscoring the importance of diligent oversight and proactive measures. By understanding their legal obligations and taking appropriate actions to safeguard their organisations, directors and officers can better protect themselves and their companies from the ever-present risks of the digital age.

NOTES

1 Although the legislation states that the 72 h starts from the moment an organization becomes aware of a breach. The clock does not start when the breach occurs, but rather when the organization has sufficient knowledge to report it. Consequently, it is very fine line on whether a suspected breach is actually a real breach.
2 This includes Australian companies or companies outside the United States.

Conclusion

So, you have just turned over the last page of the last chapter of this book. And possibly you feel disappointed – the book did not give you any explicit recipes for improving your organisation's cybersecurity posture. And in a sense, this is true, as the purpose of the book is not to provide any recipes (as a recipe that is great for one organisation may be useless for another), but to provide historical background and subsequently prompt thinking process.

One can throw their hands in the air and ask what can I do in this dire situation. I can't move away from von Neumann architecture.... Yes, this is true, but there are many things one can do to improve (and it takes time) their organisation.

The very first step is to arrive to a decision about what is most important: compliance/conformity (with ISO/IEC 27001 or Essential 8) that provides foundation for plausible deniability or actual cybersecurity posture. Having said this, it is not to deny positive impacts of conformance but to understand its place in cybersecurity. When it is rationalised and understood, one can shift focus on foundational hygiene cybersecurity aspects such as DNS, domain/subdomain and certificates' management, proper network segmentation, role-based identity and access management – these are major pragmatic steps that can and should be pursued.

Arresting the sprawl and shadow IT, applications stocktake and putting in place IT ecosystem simplification program is another major step that can and needs to be done. And yes, it takes time (and money) and will not happen overnight.... This step (or actually steps, as there are several steps here) require(s) significant education and stakeholder management activities.

When (and if) the most senior stakeholders are prepped, one can and should encourage them to look at digital transformations and agile approach through the lens of cybersecurity and, possibly, reconsider what, how and how fast organisation wants to achieve.

For Product Safety Concerns and Information please contact our EU
representative GPSR@taylorandfrancis.com
Taylor & Francis Verlag GmbH, Kaufingerstraße 24, 80331 München, Germany